T0302323

Refining Phylogenetic Analyses

Species and Systematics

The *Species and Systematics* series will investigate the theory and practice of systematics, phylogenetics, and taxonomy and explore their importance to biology in a series of comprehensive volumes aimed at students and researchers in biology and in the history and philosophy of biology. The book series will examine the role of biological diversity studies at all levels of organization and focus on the philosophical and theoretical underpinnings of research in biodiversity dynamics. The philosophical consequences of classification, integrative taxonomy, and future implications of rapidly expanding data and technologies will be among the themes explored by this series. Approaches to topics in *Species and Systematics* may include detailed studies of systematic methods, empirical studies of exemplar taxonomic groups, and historical treatises on central concepts in systematics.

For more information visit:
www.crcpress.com/Species-and-Systematics/book-series/CRCSPEANDSYS

Species and Systematics

For more information about this series, please visit:
www.crcpress.com/Species-and-Systematics/book-series/CRCSPEANDSYS

Refining Phylogenetic Analyses

Phylogenetic Analysis of Morphological Data: Volume 2

Pablo A. Goloboff

CRC Press
Taylor & Francis Group
Boca Raton   London

CRC Press is an imprint of the
Taylor & Francis Group, an **informa** business

First edition published 2022
by CRC Press
6000 Broken Sound Parkway NW, Suite 300, Boca Raton, FL 33487–2742

and by CRC Press
4 Park Square, Milton Park, Abingdon, Oxon, OX14 4RN

CRC Press is an imprint of Taylor & Francis Group, LLC

Library of Congress Cataloging-in-Publication Data
Names: Goloboff, Pablo A., author.
Title: Refining phylogenetic analyses. Volume 2, Phylogenetic analysis of morphological data / Pablo Goloboff.
Other titles: Phylogenetic analysis of morphological data
Description: First edition. | Boca Raton, FL : CRC Press, 2022. | Includes bibliographical references and index.
Identifiers: LCCN 2021061922 (print) | LCCN 2021061923 (ebook) | ISBN 9780367420277 (hardback) | ISBN 9781032274676 (paperback) | ISBN 9780367823412 (ebook)
Subjects: LCSH: Phylogeny. | Morphology.
Classification: LCC QH351 .G65 2022 (print) | LCC QH351 (ebook) | DDC 576.8/8—dc23/ eng/20220228
LC record available at https://lccn.loc.gov/2021061922
LC ebook record available at https://lccn.loc.gov/2021061923

ISBN: 978-0-36742-027-7 (hbk)
ISBN: 978-1-03227-467-6 (pbk)
ISBN: 978-0-36782-341-2 (ebk)

DOI: 10.1201/9780367823412

Typeset in Times
by Apex CoVantage, LLC

Contents

Chapter 10 Scripting: The next level of TNT mastery 229

    10.1   Basic description of TNT language 230
    10.2   The elements of TNT language in depth 233
          10.2.1   Getting help 233
          10.2.2   Expressions and operators 233
          10.2.3   Flow control 234
                10.2.3.1   Decisions 234
                10.2.3.2   Loops 234
          10.2.4   Arguments 235
          10.2.5   Internal variables 236
          10.2.6   User variables 240
                10.2.6.1   Declaration 240
                10.2.6.2   Assignment 241
                10.2.6.3   Access 245
          10.2.7   Efficiency and memory management 246
    10.3   Other facilities of the TNT language 247
          10.3.1   Goto 247
                10.3.1.1   Handling errors and interruptions 248
          10.3.2   Progress reports 248
          10.3.3   Handling input files 248
          10.3.4   Formatted output 250
                10.3.4.1   Handling strings 251
          10.3.5   Arrays into and from tables 252
          10.3.6   Automatic input redirection 253
          10.3.7   Dialogs 253
          10.3.8   Editing trees and branch labels 254
          10.3.9   Tree searching and traversals 254
          10.3.10 Most parsimonious reconstructions (*MPR*s) 256
          10.3.11 Random numbers and lists, combinations,
                permutations 256
    10.4   Graphics and correlation 258
          10.4.1   Plotting graphic trees 258
          10.4.2   Bar plots 259
                10.4.2.1   Heat maps 260
          10.4.3   Correlation 260
          10.4.4   Scatter plots 262
    10.5   Simulating and modifying data 263
    10.6   A digression: the C interpreter of TNT 264
    10.7   Some general advice on how to write scripts 266

References 269

Index 287

Preface

Why This Book?

The field of phylogenetic reconstruction has seen tremendous advances in the last few decades. The sources for those advances have come from different subdisciplines of biology (evolution, genetics, or molecular biology), and even outside biology (computer science, mathematics, philosophy). But a big part of the contributions to phylogenetic methodology have come from workers interested mostly in systematics and taxonomy. Systematists were instrumental in establishing phylogenetics as an important field in the late 1960s and early 1970s. The general outlook of researchers in systematics and taxonomy, and the kinds of problems they are interested in, tended (and still tend) to differ from those of people approaching phylogenetics from evolutionary biology or computer science. Systematists focus more on the relationships between species, are more interested in diversity than in model organisms, and tend to rely much more heavily on morphological characters—which are used as part of standard taxonomy. More evolutionarily oriented phylogeneticists, instead, tend to focus on the use of trees for testing specific hypotheses. Both are different kinds of problems, and legitimate fields of inquiry; personal preferences lead to focusing on one or the other type of problem. As someone with an upbringing rooted in systematics and taxonomy, I felt that no recent book summarized the main advances in this field from this perspective. This is the first reason to write this book.

Molecular phylogenetics exploded in recent decades and became the dominant approach in the field. There is only one "Systematics" and the goals and principles are ultimately the same for all of it; the principles for "morphological" and "molecular" systematics are exactly the same. However, it also true that different types of data pose different challenges and attract different groups of people. The enormous interest in molecular data is probably the combined result of molecular evolution being much more amenable than morphology to modeling and mathematical treatment, of the fascination of many scientists with new technological innovations, and of the intrinsic importance and interest of molecular biology and evolution. But despite that predominance, phylogenetic reconstruction using morphological data continued being developed in recent decades, and still presents many interesting challenges. Phylogenetic analysis of morphological data continues to be very important for researchers working in the systematics of groups where sequencing of most species is unfeasible (not just fossils, but also groups that are hard to collect, or rarely represented in collections by fresh material). New molecular results continue being tested against morphology—the main taxonomic groups, taken as reference in any analysis, have been established, and continue being recognized, on the basis of morphology. And morphology (in the wide sense of phenotypic characteristics, including behavior and their associated adaptations) will forever continue being the main reason behind the interest and fascination that animals and plants exert on us. It is thus logical to continue focusing on how to use morphological data for making inferences about phylogeny, for understanding morphology in a phylogenetic context, and for complementing the results of molecular data.

Despite the fact that morphology continues being an important source of data for phylogenetic analysis, however, no recent book on phylogenetics devotes much space to the peculiarities and problems intrinsic to the analysis of morphological datasets. No other book summarizes the analysis of morphological datasets, certainly not including a balance between concepts and more technical developments of the last 20 or 30 years. This is the second reason to write this book. Problems specific to molecular data are discussed in this book only insofar as they serve to illustrate differences with, or to improve understanding of, the problems of morphological data.

In addition to covering methods for morphological phylogenetics in more depth than usual, I also aspired to provide practical guidance for taxonomists in a way in which most books do not. This is the third reason to write this book. Some books help translate theoretical concepts into practice, by discussing real implementations in detail, but all these deal almost exclusively with molecular sequences. No book dealing with morphological phylogenetics and systematics includes an in-depth, authoritative treatment of computer implementations. Discussion of abstract concepts is important, but I also aim at facilitating application of those concepts to real-world problems. In recent decades, computational applications for phylogenetics have evolved toward a few major programs and a host of small, highly specific programs, useful for different parts of the analysis. While it is of course impossible for a single program to keep up with the ever-increasing repertoire of methods in phylogenetics, implementing most of the tasks required for a phylogenetic analysis in a single, general purpose program, has some definite advantages: the internal coherence of assumptions, not being limited by the possibilities of interconnection between programs, and so on. Given our belief that the detailed assumptions required for model-based (or "statistical") phylogenetics are not warranted in the case of morphology, I consider parsimony as the best justified method of analysis for morphological datasets. The practical parts of this book thus use TNT—which focuses on parsimony. It is not the only program for phylogenetic analysis, but (having developed it over the course of more than 20 years and having taught its use in numerous courses and seminars) is one that I know very well.

A few words on what this book is *not*. It is not a neutral summary of the literature on phylogenetics; other types of books, and primary literature, must be consulted for that—I take a position on many issues. The fact that this book takes positions is (hopefully) one of the aspects that make it interesting. As a consequence, this is not by any means a complete, comprehensive treatment of phylogenetic analysis; even within its focus, many books, papers, or approaches will go unmentioned. It is not intended, either, to provide a full historical account on the origin of all the different ideas presented. To some extent, many omissions of alternative methods and proposals are purposeful and (barring involuntary oversights) implicitly provide an overview of what I consider the higher and lesser contributions to the field. As in any communication process, what is not there *is* also part of the message. So, I take responsibility for omissions.

The book comprises two volumes, and the general development of themes follows from my experience teaching courses in cladistics for almost three decades. Teaching taught me (among many other things) how people who are starting to work in systematics and phylogenetics view the problems, as well as common misunderstandings and prejudice. Much of the flow in the different chapters of this book is the result of my experience teaching and explaining these problems to people who hear them for the first time. The sequence of topics loosely follows that of an analysis, from gathering of data to deciding an optimality criterion and finding appropriate trees (in the first volume), to summarizing results and more carefully considering the evidence (weighting) and non-standard types of data, such as continuous and morphometric characters (in the second volume). Hopefully, I have managed to achieve at least a fraction of the clarity that I aimed for, and this book helps upcoming systematists in their first phylogenetic steps. All the educational scripts accompanying this book can be downloaded from www.lillo.org.ar/phylogeny/eduscripts.

Finally, debts of gratitude. My approach to systematics has been shaped in interactions with many colleagues; I thank all of them both for their agreements and disagreements, but mostly for what I have learnt from them: Lone Aagesen, Victor Albert, Dalton Amorim, Salvador Arias, Ronald Brady (†), Andrew Brower, James Carpenter, Santiago Catalano, María M. Cigliano, Joel Cracraft, Jonathan Coddington, Jan De Laet, Julian Faivovich, Joseph Felsenstein, Gonzalo Giribet, Norberto Giannini, Charles Griswold, Peter Hovenkamp (†), Jaakko Hyvönen, Mari Källersjö, James Liebherr, Lucia Lohmann, Melissa Luckow, Camilo Mattoni, Rudolf Meier, Marcos Mirande, Martín Morales, Jyrki Muona, Gareth Nelson, Kevin Nixon, Norman Platnick (†), Diego Pol, Martín Ramírez, Robert Raven, Randall Schuh, Mark Siddall, Mark Simmons, Michael Steel, David Swofford, Claudia Szumik, John Wenzel, Christiane Weirauch, Ward Wheeler, Kipling Will, and Mark Wilkinson. I only imposed on a few of them the task of reading the manuscript; I both apologize for the burden and thank them for their useful comments: Andrew Brower, James Carpenter, Jan De Laet, Camilo Mattoni, Diego Pol, Christiane Weirauch, and Ward Wheeler. Santiago Catalano helped me with Chapter 9, providing literature, materials, and advice. I thank Chuck Crumly for his help and advice during the editorial process. Last but most important, I especially acknowledge María E. Galiano (†), world-renowned salticidologist who (in the late 1970s) introduced me to taxonomy, arachnology, and critical thinking, and James "Steve" Farris, who led phylogenetic systematics into the modern era and (in the late 1980s) helped me jump onto that train when it was already moving fast. Steve also originated the TNT project in the late 1990s (approaching Kevin Nixon and me with the proposal to join forces and write a comprehensive phylogeny program), and still continues contributing to it. Steve never wrote a book (he was always too busy developing new methods and ideas), but his influence all along this book is in fact so strong that—if my vanity can be excused—I'd like to think this book is not too different from the one he might have written.

Author Biography

Born in Buenos Aires, Pablo A. Goloboff became interested in spider biology and systematics in the late 1970s, in the Museo Argentino de Ciencias Naturales. His first papers (published during the 1980s) were on spider systematics, but he soon became more interested in systematic theory and phylogenetic methods. He graduated with a Licenciatura in Biology in 1989, from Universidad de Buenos Aires, and then pursued doctoral studies in Cornell University and the American Museum of Natural History, in New York, between 1989 and 1994. He published his first methodological papers in the early 1990s, gradually switching his research from spider systematics to systematic theory. During his stay at Cornell University, he became more involved with quantitative methods for parsimony analysis, and wrote his first computer programs. He moved to Tucumán in 1994, to work for the CONICET, and continued working on theory and methods for systematics and historical biogeography. He has published over a hundred scientific papers and about a dozen computer programs, the best known of which are Nona, Piwe, TNT (for phylogenetics), and VNDM (for biogeography). He is a Fellow Honoris Causa of the Willi Hennig Society, and served as President of the society from 2004 to 2006. Since 1995, he has been regularly teaching courses on phylogenetics in Argentina and about a dozen other countries.

Abbreviations

AIC	Akaike information criterion, a method to choose among statistical models based on minimizing loss of information
BS	Bremer supports
C	Consistency index
CBS	Combined Bremer supports
DC	Distortion coefficient, a measure of tree (dis)similarity
DMC	Discrete morphological characters
EA	Explicitly agree, a quartet-based measure of congruence proposed by Estabrook (1992)
FWR	Frequency-within-replicates, a method to summarize results from resampling
GC	In resampling, frequency of the group (**G**) minus the frequency of the most frequent contradictory group (**C**)
GLS	Procrustes generalized least squares superimposition (Gower 1975)
GM	Geometric morphometrics, the field for the mathematical study of shape
HTU	Hypothetical taxonomic unit, a reconstructed ancestor
Implik	Emulation of implied weighting under maximum likelihood
IW	Implied weighting
LSI	Leaf stability index, a measure of taxon stability
MAST	Maximum agreement subtree
MCMC	Monte Carlo Markov Chain
ML	Maximum likelihood
MP	Maximum parsimony
MP$_{lik}$	Equal weights maximum parsimony emulated under maximum likelihood
MPR	Most parsimonious reconstruction; a set of parsimony-optimal ancestral state assignments
MP-set	Set of states that occur in a **MPR**, at a given node
MPT	Most parsimonious tree
MRC	Matrix representation with cliques, a method for building supertrees
MRP	Matrix representation with parsimony
NCM	No common mechanism, a model of character change (Tuffley and Steel 1997)
NNI	Nearest-neighbor interchange
PBS	Partitioned Bremer supports
PC	Positional congruence, standard version
PCR	Positional congruence, reduced version
PP	"Profile Parsimony", a method for differential character weighting (Faith and Trueman 2001)
RAS	Random addition sequence
RBS	Relative Bremer supports

REP	Relative explanatory power (Grant and Kluge 2007), a measure of group support
RF	Robinson-Foulds distance
RFD	Relative fit difference
RFTRA	Resistant fit theta-rho superimposition (Siegel and Benson 1982)
RI	Retention index, a measure of relative homoplasy
SAW	Successive approximations weighting
sCF	Site concordance factors (Minh et al. 2020), a measure of group supports
SE	"Strongest evidence" (Salisbury 1999), a method for differential character weighting
SPR	Subtree pruning regrafting
TBR	Tree bisection reconnection

6 Summarizing and comparing phylogenetic trees

Large amounts of perfect data analyzed by infallible methods will always produce a single tree—the true phylogeny. It would be so nice if the world was so perfect! In reality, there is no perfect data, no infallible method, and the amounts of data available for phylogenetic inference are often insufficient to prefer a single tree over all others. When the data are insufficient to fully resolve a phylogeny, results are ambiguous, with several trees fitting the data equally well. A proper analytical method must be able to identify and handle this ambiguity, and produce multiple trees. In addition, different criteria for selecting trees used on the same dataset, often produce different results; the same is true for different datasets analyzed with the same criterion. It is then necessary to have tools to summarize and compare the different results. Note that most methods to summarize or compare trees simply work on the tree topologies, not evaluating (at least not directly) the fit of characters to trees. These methods are therefore fully independent of the criterion (parsimony, likelihood, or any other quantitative criterion) used for selecting phylogenetic trees.

There are two ways to deal with ambiguity, depending on the goal of the analysis. One of those is creating summary trees, that is, trees that synthesize the information common to several trees. The information to be summarized can be of different types—depending on which aspect of the trees is interesting, or relevant for the analysis in question. Trees summarizing information are generally called consensus (when all trees have the same sets of taxa) or supertrees (when the trees have different taxon sets). Depending on the type of common information to be used for creating the summary trees, and the goals of the analysis, different types of consensus or supertrees must be used. Perhaps the most important distinction is in whether the consensus is meant to summarize the information common to the multiple trees resulting from the analysis of a single dataset, or to summarize the trees resulting from analyzing different datasets. Evaluations of similarity among trees for different datasets can serve to answer questions about methods themselves, when we are uncertain about choice of method. A better method for phylogenetic analysis will produce, for different sources of data, trees that are more similar. This was done, for example, in the now classic studies of Farris (1971) and Mickevich (1978) demonstrating that cladistic methods produce more congruent classifications than phenetic methods (deeply infuriating pheneticists, see Sokal and Rohlf 1981; Rohlf and Sokal 1981; Rohlf et al. 1983).

The other way to evaluate ambiguity is by measuring the similarity among the trees—this uses some criterion of similarity (or distance) among trees. In the case

DOI: 10.1201/9780367823412-6

1

of trees, there are different aspects that can be compared, and this leads to different measures of similarity. There is a fundamental difference between consensus or supertree methods on the one hand, and measures of tree similarity on the other. Consensus and supertrees provide summaries of what is common to sets of trees, operating on several trees simultaneously and reducing them to a single summary tree. Measures of tree similarity, instead, are done only on a pairwise basis. The two aspects—summaries and comparisons—can be interrelated. For example, the degree of resolution of the consensus can be used as a proxy for the similarity among the trees consensed, and some consensus methods summarize sets of trees by finding a tree which satisfies a specific criterion of distance to each of the trees in the set to be summarized.

While seemingly simple, both summaries and comparisons can be complicated by many factors. Sometimes the trees in a set do not share much information in common, due to some taxa greatly varying their positions in the set of input trees (e.g. so-called "wildcards" or "rogue" taxa). The consensus summarizing the entire trees may have no groups in common, but displaying the information common to a subset of the taxa may be much more useful—if the relationships among the taxa in the subset are near-constant in the input trees. Some means are then necessary to identify the wildcards and extract the shared taxonomic information from the input trees. This chapter discusses the types of summaries and measures of similarity that are appropriate in different contexts, as well as methods to identify wildcard taxa, and different problems related to ambiguity, with particular attention to the implementation in TNT (Goloboff et al. 2003b; Goloboff et al. 2008b; Goloboff and Catalano 2016).

6.1 CONSENSUS METHODS

Depending on the purpose and context of the analysis, different types of consensus methods are appropriate. The series of trees to be summarized (*input* trees) is subjected to some procedure, and the *output* is a single tree—the *consensus*. Depending on the type of information that they are intended to summarize, consensus methods can be separated in those that reflect common information on clusters (i.e. groups, sets of taxa), and those that reflect common information on relative positions of taxa (e.g. whether most trees display two taxa as more closely related to each other than to a third taxon, even if those two taxa do not form a monophyletic group; this is often called information on *nestings* or *triplets*; see next).

In addition to such separation between methods based or not on clusters, it is possible to define methods for summarizing trees based on minimization of some measure of distance. In that way, the tree obtained represents a sort of "average" of the input trees. The exact sort of average, and whether the resulting "consensus" is useful for taxonomic work, depends on the measure of distance used. In general, only consensus methods based on groups (i.e. strict, combinable components, majority rule, and frequency difference consensus) have a direct taxonomic interpretation. Other methods serve as indirect tools to explore the similarities and differences between the input trees, more than as a means to summarize results.

As they can contain polytomies (or resolutions not found in any of the trees, in the case of majority or Adams trees), consensus trees can be longer (or less likely) than the individual trees used to construct them. For that reason, consensus trees should not be optimized (at least, not as "hard" polytomies, see Chapter 3). For clade diagnoses, or studies of character evolution, the individual trees should be optimized, with the synapomorphies or character-state sets plotted on the corresponding nodes of the consensus (see Chapter 3).

6.1.1   Cluster-Based Methods

6.1.1.1   Strict consensus trees

A "strict" consensus is a tree that displays the groups that are present in each and every one of the input trees. The first authors to use strict consensuses were Schuh and Polhemus (1980) and Schuh and Farris (1981), building on the work of Nelson (1979); the term "strict" was proposed by Sokal and Rohlf (1981). Nixon and Carpenter (1996b) provided a discussion on history and terminology.

In having a straightforward interpretation, the strict consensus is the natural choice for summarizing trees in a wide variety of contexts. An example is in Figure 6.1a; for example, the groups EF and CDEFGHI are present in some of the input trees, but not all. The strict consensus therefore does not include those groups. The strict consensus is the only consensus method that can legitimately be used for identifying supported groups in the analysis of a given dataset. When multiple equally optimal trees are found for a given dataset, the absence of a given group in one or more trees unequivocally indicates that, under the criterion used to choose trees, there is no support for the group. A group can be said to be supported by a dataset only if providing maximum fit requires that the group is included in the tree. Thus, the strict consensus is the only way to accurately indicate supported groups. In its basic application to summarize the equally optimal trees resulting from the analysis of a single dataset, the strict consensus tree is unobjectionable.

As with any consensus method, using the strict consensus tree instead of the input trees entails some loss of information. Consider the case of two input trees (A(BC)) and (B(AC)); their strict consensus is (ABC), which allows as a possible resolution (C(AB)), found in neither of the input trees. To the extent that the strict consensus is less resolved, the possibility increases that the consensus allows for some resolution that does not actually occur among the optimal trees (or among the input trees, more generally). Displaying the individual input trees may be the only way to faithfully represent the information common to all of them, but prohibitive for space reasons. Consensus trees, or pruned consensus trees (see Section 6.3), are usually the best tradeoff.

A commonly cited problem (e.g. Swofford 1991) with strict consensus trees is that they are poorly resolved—that's why they're called *strict*! When there are several otherwise identical trees, with a few taxa jumping from widely different positions, most of the groups will collapse in the consensus tree, perhaps even creating a full polytomy. In this case, the input trees are very similar—they are the same tree,

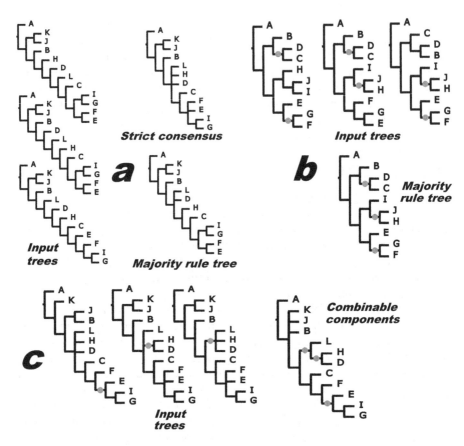

FIGURE 6.1 (a) The strict consensus produces a tree with the groups present in each and every one of the input trees (and only such groups); the majority rule produces a tree with every group present in more than half the trees (and only such groups). The resolution of the strict consensus is always the same as, or less than the resolution of the majority rule tree (in the case of two input trees, strict and majority are identical). (b) The majority rule can be completely resolved and yet different from every one of the input trees. (c) The combinable components consensus displays all the groups present in one or more trees, and contradicted by none.

differing only in the position of one or a few taxa (a minor difference). This shows that using the degree of resolution of the strict consensus is not always the best way to indicate congruence among source trees, not a problem with the strict consensus per se. In cases such as this, the best approach is to use pruned consensus trees. And, of course, if the intent is to measure degree of similarity among input trees, do not use the strict consensus, but instead appropriate methods for comparing the trees.

6.1.1.2 Majority rule consensus trees
In many cases, it may be desirable to output a tree that shows the groups that are present in the majority of the input trees. Consider the case where several independent datasets (with comparable amounts of evidence) have been analyzed separately.

If the majority of the analyses found a given group as monophyletic, and only one or two cases failed to recover the group as monophyletic, then we can conclude that the group is probably monophyletic. This is exactly what the majority rule consensus tree (Margush and McMorris 1981) does: it displays the groups that are present in more than half the input trees. For example, in Figure 6.1a, the group EF is present in the top and middle input tree, and absent from the bottom tree. With a frequency of 2/3, the group is included in the result. Likewise for group CEFGHI. It is common to include in the majority rule consensus, for each group, the frequency among the input trees, plotted on the corresponding branch.

The majority rule tree will display every group displayed by the strict consensus tree, and typically several additional groups. In the case of only two input trees, the majority rule tree is identical to the strict consensus. There are cases where the use of majority rule consensus is ill-advised. This is especially true in the case of multiple equally optimal trees resulting from analysis of a single dataset. Many researchers prefer to summarize their results by means of a majority rule consensus, simply because it produces a more resolved tree and thus the feeling of having succeeded in resolving the phylogenetic tree. That feeling is entirely illusory. The identical optimality scores indicate that every one of the trees is equally "probable", and thus concluding that a group that can be absent from some trees is more probably correct is self-contradictory. Consider the case of a well-structured matrix with T taxa that supports a single most parsimonious tree (MPT), and the addition of a single taxon X, scored entirely with missing entries. There are $2T - 3$ positions in the tree where X can be inserted with equal parsimony. The groups composed of only two taxa will be non-monophyletic only when X is placed inside the group—two out of the $2T - 3$ trees. Groups composed of three taxa will be non-monophyletic when X is placed in four possible positions—four out of $2T - 3$ trees. The example is similar to the ones shown in Figure 4.2a (but with missing entries in the long branch, instead of randomized character states) and Figure 8.5c. Smaller groups are then concluded to be, if using a majority rule consensus as indication of "confidence", much more strongly supported than larger groups. Taxon X results in a well-defined position in the consensus. The data, however, do not truly support such conclusion; as X is entirely unknown (i.e. it has only missing entries), it could perfectly well be included within a group of only two taxa; there is nothing in the data against such possibility. Only the strict consensus appropriately indicates (lack of) support in that case (see earlier), and that is entirely unresolved. This does not mean that the majority rule consensus is without use, only that it has to be used in the proper context. Note that when using the majority rule to summarize trees resulting from different datasets, the correct interpretation of a group in the majority rule consensus tree is now that the group is supported by most *datasets*.

The majority rule tree cannot display groups that are not present in any of the input trees, but it can display *combinations* of groups that are not present in any of the input trees (as in Figure 6.1b). Therefore, the majority rule tree can be entirely resolved, yet different from every one of the input trees (thus, less parsimonious, not just by virtue of displaying polytomies as the strict consensus, but by virtue of displaying combinations of resolutions that are contradicted by every single tree).

Although the majority rule consensus is often better resolved than the strict, when the input trees are too different, it will also be poorly resolved (with most groups occurring in very low frequencies). The same will happen when a few taxa can jump, in the input trees, between distant alternative positions. In that case, displaying the common skeleton of the input trees, by pruning the jumping taxa, may produce a better summary of the trees (see later).

6.1.1.3   Combinable components consensus

This type of consensus was first proposed by Bremer (1990), but some of the ideas in this method can be traced back to Nelson (1979). The same method was called a *semi-strict* consensus by Swofford (1990), and a *loose* consensus by Barthélemy et al. (1992). Bremer (1990) noted that consensus trees are often used as a tool to explore congruence among datasets, and the strict consensus is not best suited for such purpose. The exploration of congruence among datasets is best served when the comparison allows for a group to be supported by one of the datasets, and indifferent for the other. After all, different parts of the phylogeny may be resolved by different datasets. As long as there is no *conflict* between the groups supported by each dataset, the datasets can be said to be congruent. Bremer (1990) thus proposed to use a consensus of *combinable components* for exploring congruence. The combinable component consensus is a tree including all the groups (and only such groups) which are present in one or more of the input trees, and contradicted by none. Figure 6.1c shows a worked out example. Groups DH, DHL, and EIG are each included in a single input tree, and then would not be included in a majority rule tree, yet the combinable component consensus includes them—they are not contradicted by any of the trees. On the other side, group KJ is included in two of the trees, but contradicted by another. The majority rule consensus would include group KJ, but the combinable component consensus does not.

The combinable component consensus should not be used to summarize the multiple equally optimal trees for a single dataset (Nixon and Carpenter 1996b). Trees from the analysis of a single dataset can display polytomies when zero-length branches are eliminated. Even if some other tree displays a possible resolution of that polytomy with some character changes acting as synapomorphies, this does not detract from the fact that optimal explanation of the data can be achieved without postulating the group (i.e. in the polytomous tree). Thus, the combinable component consensus, when used to summarize the collapsed trees that result from a single dataset, may display groups that are unsupported by the data.

A more appropriate use of the combinable consensus is to evaluate congruence between trees for different datasets. However, in such case, it is necessary to be cautious about the trees chosen to represent each dataset (Nixon and Carpenter 1996b). Consider the case of two matrices, *1* and *2*:

```
    1      2
A   0000   00
B   1111   00
C   1100   11
D   0011   11
```

Matrix *1* produces two trees, (A(D(BC))) and (A(C(BD))), the consensus of which (strict, majority, or combinable components) is the full polytomy, (A(BCD)). Matrix *2* produces a single tree, (A(B(CD))). Using the consensus tree for each of the matrices as input for the combinable component consensus produces the tree for the second matrix, (A(B(CD)))—there is no conflict between the polytomy in the consensus for matrix *1* and the resolved consensus for matrix *2*. *Yet the individual trees are incompatible* (it is only the loss of information in the consensus for matrix *1* what makes it compatible with the consensus for matrix *2*). Any of the two trees produced by matrix *1* is in conflict with the tree produced by matrix *2*. Thus, the consensus of combinable components must use as input the *individual* optimal trees for each of the matrices, not their consensus. In the case of very large numbers of trees, practicality may mandate using consensus trees instead of the individual trees, but in that case it must be kept in mind that the results are only an approximation.

It is clear, therefore, that the input for the combinable component consensus must consist of the individual trees for each dataset. However, it is also necessary to eliminate branches unsupported by synapomorphies; those resolutions are arbitrary and would incorrectly indicate the existence of support for conflicting groups. Consider two additional matrices *3*, and *4*:

	3	4
A	000	000
B	000	110
C	000	101
D	000	011

Matrix *3*, containing no informative characters, produces three distinct binary trees: (A(B(CD))), (A(C(BD))), and (A(D(BC)); matrix *2* produces (as we have seen) a single tree (A(B(CD))). The combinable component consensus of the four trees is unresolved, suggesting that there is incongruence between matrices *2* and *3*, when in fact matrix *3* cannot be "incongruent" with anything—it is devoid of grouping information. If zero-length branches are eliminated, then matrix *3* produces a single tree (A(BCD)), which—when combinably consensed with the tree for matrix *2*—correctly indicates compatibility. But even if using individual collapsed trees for each matrix the proper interpretation of the combinable component consensus may not be entirely clear. Consider the case of matrix *4*, which even when zero-length branches are eliminated supports the same three trees as matrix *3* supports when zero-length branches are retained. The combinable component consensus of the trees for matrix *2* and matrix *4* is a full polytomy: some of the optimal trees for matrix *4* contradict the single tree supported by matrix *2*. Yet, in a sense, matrix *4* cannot really contradict anything; due to character conflict, any possible tree is an equally good solution for that matrix. The polytomy in the combinable component consensus for the trees of matrix *2* and matrix *4* results from conflict *within* matrix *4*, not from the conflict *between* matrices *2* and *4*. Perhaps a better evaluation would choose, in the case of matrices supporting multiple equally parsimonious trees, alternative combinations of trees from both matrices. The complexity of such process belies the utility of the combinable component consensus for evaluations of congruence between matrices.

6.1.1.4 Frequency difference consensus

In the context of summarizing results of bootstrap or jackknifing analysis (see Chapter 8), Goloboff et al. (2003a) proposed the frequency difference consensus, as a better alternative than the majority rule consensus for measuring strength of group support. The frequency difference consensus is formed by calculating, for every group, the difference between the frequency of the group, minus the frequency of the most frequent contradictory group; this is known as the *GC* value (for "Group present/Contradicted"). The frequency difference consensus contains all groups with positive *GC*. It is typically more resolved than the majority rule consensus tree. The properties of the frequency difference consensus (and other consensus methods) have been studied by Dong et al. (2010b).

6.1.2 METHODS NOT BASED ON CLUSTERS

Methods that are not based directly on clusters are rarely of use to taxonomists, although they can be useful to identify particular aspects of the trees.

6.1.2.1 Adams consensus

This method (Adams 1972) is based on creating a tree where the nestings found in a majority of the trees are retrieved. A *nesting* is the relative closeness between taxa; for example, both trees (A(B(C(DE)))) and (A(B(E(CD)))) share taxa E and D as closer to each other than they are to B, even if E+D do not form a monophyletic group in both trees. Taxa in widely distant positions in a common skeleton are placed as polytomies at the lowest common ancestor of their possible locations, and some authors find this property helpful for identifying wildcard taxa. As it is not based on groupings (=clusters), the Adams consensus can display groups that are not present in *any* of the input trees. An example is Wheeler (2012: 342, fig. 16.3); another is the trees (A(C(D(BE)))) and (A(E(B(CD)))), which produce (A(CE(BD))) as Adams consensus (both C and E are lowered to their common node, creating group BD, contradicted by each of the input trees). Adams consensus trees are not implemented in TNT. They can be calculated with several programs, such as PAUP* (Swofford 2001) or Euncon (Wheeler 2021b).

6.1.2.2 Rough recovery consensus

In their analysis of a large matrix for Eukaryota, Goloboff et al. (2009a) noted that many large groups were not recovered exactly in their analysis, but this was due only to incorrect inclusion or exclusion of just a few taxa. When the tree displays a group comprising 99 of the 100 taxa of the group, and only a single terminal has been placed somewhere else in the tree, the group is very close to have been recovered. The idea of measuring how strongly a non-monophyletic group deviates from monophyly had already been suggested by other authors before (e.g. Stinebrickner 1984; Sanderson 1989). This "degree" of recovery is evaluated on the basis of similarity in group composition, using both the number of taxa that belong to the reference group but are excluded from the group in question, and the number of taxa that do not belong to the reference group but are included in the group in question. If the sum of those two numbers, relative to group size, is below a certain threshold, the group can be considered to have been "roughly" recovered. TNT uses as degree of recovery the proportion $D = S / (S + E \times F_{out} + I \times F_{ins})$, where E is the number of taxa that belong to reference group but were excluded in the group checked, I is the number of taxa that do not belong to reference group but are included in the group checked, and S ("shared") is the number of taxa that belong both to reference and checked group. F_{out} and F_{ins} are factors to differentially weight exclusion of members or inclusion of non-members (unity by default). In Goloboff et al. (2009a), no F_{out} or F_{ins} factors were used (i.e. weights were always unity). An almost identical index (which they call the "transfer index") was used by Lemoine et al. (2018) in combination with boot-strapping. The implementation in TNT (unlike the one described in Goloboff et al. 2009a) inverts the partitions, if too few taxa are left outside for the groups (just like including one taxon in a 2-taxon group is a big distortion of the group, including one more taxon in a group that excludes only two of the taxa in the dataset is also a big distortion; the biggest value is used). For every input tree, the group with the highest value of D is identified (when recovery is perfect, $D = 1$). This *rough recovery consensus* is implemented in TNT only for the groups in a specified reference tree. Note that when groups mostly agree, with only a few wrong taxa being included or excluded for each group, the rough recovery of groups in the reference tree will be higher. This type of consensus, therefore, when displaying many groups with rough recovery close to unity, indicates cases where majority rule or frequency difference consensus trees can probably be improved by pruning just a few taxa.

6.1.2.3 Median trees

Another approach to forming a consensus consists of identifying the tree(s) that min-imize the sum of a specific distance to the input trees. The "consensus" is in that case a *median tree*, situated in the "middle" of the input trees, according to the distance measure used. The utility and interpretation of such trees depends on the specific measure of distance being minimized. An interesting equivalence, demonstrated by Barthélemy and McMorris (1986) is that the tree minimizing the sum of Robinson and Foulds (1981) symmetric distances (*RF*, see Section 6.7.1) to all the input trees is exactly the majority rule consensus tree. The Robinson-Foulds distance to one of the input trees increases both with failure to include a group present in the input tree (so that if group is present in more than half the input trees, including it in consensus will decrease the sum), and with inclusion of groups not present in the input tree (so that if group is absent from more than half the input trees, excluding it will decrease the sum). For even numbers of trees, there can be ties in the minimization of ΣRF, but in that case the strict consensus of all the trees minimizing that sum also minimizes the sum (Goloboff and Szumik 2015). The equivalence demonstrated by Barthélemy and McMorris (1986) has been useful in constructing an analog of the majority rule consensus in the supertree setting (Cotton and Wilkinson 2007; see Section 6.5.4). The majority rule tree is usually constructed by simply inspecting taxon clusters and their frequencies, but it could also be constructed by searching the tree with min-imum ΣRF, for example using heuristics. An important difference with heuristics used in standard searches for *MPT*s is that a partially resolved tree can have a better score (i.e. a smaller ΣRF) than a fully resolved tree (which can never happen when the score is calculated with parsimony or likelihood).

The distance to be minimized, of course, can also be one different from the *RF*. An example is the minimization of the number of steps in matrices representing each of the input trees, when mapped onto the candidate "consensus" tree. A matrix is said to *represent* a tree when it has as many "characters" as groups in the tree, with a character representing a group by assigning state 1 to each terminal included in the group, and state 0 to each terminal not included in the group. This had been used first by Farris (1973) as a means to compare tree shapes (see Section 6.7.4), but subsequently became known as *matrix representation with parsimony* (*MRP*), after Baum (1992), Ragan (1992), and Baum and Ragan (1993). Baum and Ragan originally proposed it as method for building supertrees, but it can also be seen as a consensus method, in which case its properties can be more directly compared to the property of other, well-characterized consensus methods. *MRP* (even in the consensus setting, where all trees have the same taxon sets) often creates groups that are contradicted by each of the input trees (Goloboff and Pol 2002; Goloboff 2005). Using parsimony to analyze the matrices representing the input trees can create groups "out-of-nothing" in the resulting tree because of taxa that did not belong to specific groups in the input trees (by mapping their "characters" as "reversals"), or because they were part of larger groups in the input trees (by mapping their "characters" as "parallelisms"); see Goloboff and Pol (2002: 515, fig. 1) for an example. Neither non-membership to a group, nor being part of a larger group, provide grounds for concluding monophyly. *MRP* trees are then difficult to justify, and cannot be considered as displaying groups present in the majority of the trees (Goloboff 2005; contra Bininda-Emonds 2004). Goloboff (2005) argued that any measure of tree distance which is asymmetric or measures failure to recover a group by degrees (as parsimony, where a group of an input tree could be separated in one or many parts in the result tree, thus requiring different number of steps) is bound to produce consensus trees that do not necessarily display the groups present in the majority of the input trees. Note that the Robinson-Foulds distance, which produces exactly the majority rule consensus, is symmetric and measures group recovery as all-or-none.

Bruen and Bryant (2008) also suggested that one could use the *SPR*-distances, seeking a consensus which minimizes the number of branch moves (as in subtree pruning regrafting, *SPR*) needed to convert it into each of the input trees. This was subsequently used in the supertree context by Whidden et al. (2014). As shown by Goloboff and Szumik (2015, Figs. 1–2), consensuses based on minimizing *SPR*-distances easily produce illogical results, possibly including groups contradicted by each of the input trees. Another problematic method is that of "minimum flipping", introduced by Chen et al. (2003), based on finding the trees which require the minimum number of changes of matrix entries for the matrix to represent the tree perfectly (every matrix cell is counted independently; see Eulenstein et al. 2004, for details). Minimum flipping can also produce "consensus" trees displaying groups contradicted by the majority of the trees (Goloboff 2005).

6.2   TAXONOMIC CONGRUENCE VS. TOTAL EVIDENCE

Both the use and interpretation of consensus methods are connected with the controversy of taxonomic congruence vs. total evidence. Pheneticists had expressed the expectation that using different subsets of data would produce similar classifications.

As a "theoretical" justification for this, they advanced the "hypothesis of nonspecifity" (Sokal and Sneath 1963), stating that "a genotype cannot be partitioned into disjointed classes of loci in such a way that any of those classes of loci affect exclusively a single class of phenotypic characters". There was no actual evidence for this "hypothesis", only the hope that it would be true, for pheneticists thought that only then could one expect different types of characters (e.g. muscle, bone, or dermal characters) to produce congruence among classifications. Farris (1971) first noted that there was another possible reason for expecting congruence among classifications based on different types of characters: phylogeny. If a method is effective at recovering phylogeny, then different subsets of data would produce congruent classifications. This made the hypothesis of nonspecifity superfluous. Not long after, Mickevich (1978) showed that, for several groups of organisms where different datasets were available, the classifications produced by cladistic analysis of the different datasets were more similar than the classifications produced by phenetic analysis. Mickevich (1978) measured congruence using degree of resolution of consensus trees, and her work was important in establishing the superiority of cladistic over phenetic methods (see Sokal and Rohlf 1981; Colless 1981, with immediate reply by Schuh and Farris 1981).

Moving forward in time, the same approach began being used to analyze datasets separately, and then using as the best hypothesis of relationships the consensus of the resulting trees (or their supertree, if matrices have different taxon sets; Bininda-Emonds 2004). This was often called the "congruence" approach. This is very different from exploring the congruence between classifications produced by different methods as a way to assess the validity of those methods. The rationale for the congruence approach is that the same conclusion reached independently on the basis of different lines of evidence is more likely to be correct. That is superficially appealing, but Kluge (1989) noted that scientific conclusions should be based on evaluations of all available evidence, and that the only way to achieve such a global evaluation is by including all known characters in a matrix and analyzing them without constraints. Kluge (1989) called this approach "total evidence"; Nixon and Carpenter (1996a) agreed with the idea, although they preferred the name "simultaneous analysis". More recently, a similar discussion involved the "supertree vs. supermatrix" controversy (combining the trees from separate matrices, proposed by Purvis 1995; Bininda-Emonds et al. 1999; Liu et al. 2001; Ranwez et al. 2010; vs. combining all data in a single matrix analyzed simultaneously). Terminology aside, Kluge's (1989) point seems quite obvious, and it is clear that, whenever possible, all available evidence should be analyzed simultaneously. Simply using the consensus of the trees resulting from analysis of each individual matrix debilitates the link between conclusions and actual evidence, much more so if results are summarized by means of methods akin to majority rule consensus or supertrees, rather than strict consensus methods (an obvious point many authors have made, e.g. Gatesy et al. 2002; Wheeler 2012; Janies et al. 2013). Consensus trees are created just from tree topologies, with no direct way to take into account the actual fit of characters to alternative trees; some characters but not others may be duplicated in the two matrices; and a small matrix and a large matrix are, each, represented by a single tree. This problem is so serious that having a group supported by two wholly independent datasets gives no guarantee that combining the two datasets (thus globally considering all the evidence simultaneously) will also support that same group (Barrett et al. 1991). Note that we do not even refer here to the group being historically correct (a much more difficult question!); we only refer to the parsimony scores of trees having or lacking the group. An example is shown in Figure 6.2a, where a group present in all the *MPT*s for two independent datasets is contradicted by the *MPT* for the combined dataset. This may seem surprising, but it can happen because of character interactions. The globally

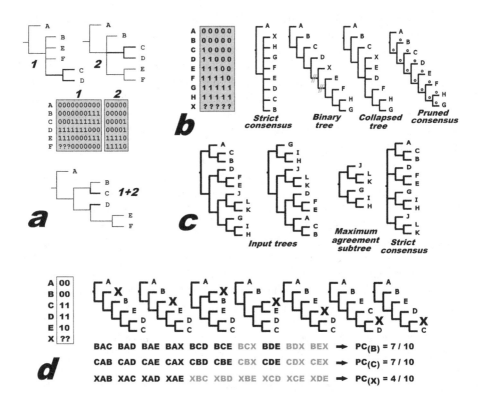

FIGURE 6.2 (a) A group separately supported by two datasets (CD, in the example), may well be absent from the trees supported by the combined dataset. Therefore, the groups present in the strict consensus for the trees produced by the separate datasets may well be contradicted by all of the evidence analyzed simultaneously. (b) Rogue taxon X can float among different positions in a common skeleton, and it strongly decreases the resolution of the strict consensus. The pruned consensus displays only the common skeleton, possibly indicating the alternative locations (circles) of the rogue taxon (or taxa). To produce a pruned consensus, if zero-length branches are to be eliminated, special precautions need to be taken. In a binary (most parsimonious) tree, the branch both below and above X will be supported only ambiguously, so that collapsing tree first and then removing taxon X does not improve the resolution of the consensus. (c) The maximum agreement subtree is the tree with the minimum possible number of taxa pruned for the input trees to become identical. This maximizes identity, not consensus resolution. (d) The triplet-based measure of stability, Explicit Agreement (*EA*), applied to taxa B, C and X. X is the least stable taxon, and has the lowest value of *EA*.

most parsimonious tree—the one that best allows explaining observed similarities by reference to common ancestry—depends on all the characters taken together.

Note that despite these solid arguments against establishing taxonomic conclusions based on the congruence of different datasets, the problem of whether different sources of data lead to similar phylogenetic conclusions continues being an interesting problem, regardless of the fact that the best conclusion is always the one achieved by combining all the evidence. And although the ideal approach is to combine all available evidence and analyze it with the best available methods, there can be room for some exceptions, as it is often necessary to make practical compromises. Goloboff et al. (2008a), for example, noted that implied weighting seems to produce better results for morphological datasets, and that this may pose a dilemma for combined analysis of sequence data and morphology. Sequence data are (at least in the view of many authors; see Wheeler et al. 2006; Wheeler 2012) best analyzed using programs that do not require a prior alignment. Such an approach is computationally demanding, and few programs implement it. One of those programs is POY (Wheeler et al. 2006; Varón et al. 2010), which uses a problematic approach for calculating implied weights for sequences (it uses the total homoplasy of each sequence, Wheeler et al. 2006: 258, so that sequence length comes into play more than actual homoplasy; see discussion in Goloboff 2014a: 262). One is then left with the choice of analyzing a combined dataset (with morphology plus sequences) either using a method of analysis that is more appropriate for sequences (in POY), or using a method of analysis that is more appropriate for morphology (in TNT). The choice of analyzing both datasets separately and exploring the degree to which they support similar conclusions may then be legitimate, at least when faced with lack of better alternatives. This does not detract from Kluge's (1989) general advice that all the evidence must be considered whenever possible; in an example like the one just discussed, the best way to consider the morphological evidence is by means of one program, and the best way to consider sequences is by means of another. While the ideal world would have programs and methods capable of doing everything simultaneously, we live in a less than ideal world, and careful judgment must be used instead of ironclad rules.

6.3 PRUNED (=REDUCED) CONSENSUS AND IDENTIFICATION OF UNSTABLE TAXA

When the input trees share most of the structure, but a few taxa (or groups) can move between distant positions in a common "skeleton", the consensus will be poorly resolved, even if the underlying trees are in fact quite similar. This is one of the most serious arguments against using the degree of resolution of consensus trees as a proxy for tree similarity. The problem is especially acute in (but not exclusive to) the strict consensus. Obviously, a strict consensus that is completely resolved (or almost so) indicates with certainty that all the input trees are identical (or very similar). The converse, however, is not true: a poorly resolved strict consensus does *not* indicate that the trees are very different. The problem affects not only the use of consensus as proxies for tree similarity (a question which is best approached by using appropriate measures of tree similarity), but also the problem of summarizing the input trees in such a way that useful information is displayed.

Early work on reduced consensus was by Wilkinson (1994a, 1995, 1996), Thorley and Wilkinson (1999), and Thorley and Page (2000). Goloboff and Szumik (2015) provided a recent general discussion of problems associated with pruned consensus and summarized previous work. Unstable taxa that de-resolve the consensus by jumping positions in optimal trees are also called *wildcard* or *rogues*, particularly when they can occupy a large number of alternative positions. This happens often when the matrix has many missing entries (as in paleontological datasets; Kearney and Clark 2003), but it can also happen in a variety of other circumstances. A single terminal moving among just a few distant positions may be enough to substantially decrease resolution of the strict consensus. The same problem occurs when tree scores are minimally decreased by taxa moving to widely different positions, which affects measures of support (Chapter 8). A common course of action when consensus trees are poorly resolved because of unstable taxa is to eliminate those taxa from the trees, displaying a *pruned* consensus. An example is in Figure 6.2b, where the relationships between all taxa are well defined by characters without conflict, but taxon X is scored as a missing entry for each of the characters (i.e. a wildcard, strictly speaking). X can be placed with equal parsimony at any of the remaining 13 branches of the tree. The strict consensus for all the taxa is completely unresolved, but pruning X from the trees and calculating the pruned consensus shows the relationships between the other taxa. TNT can also indicate the positions at which the pruned taxon (or taxa) can be introduced (all available positions in this case); when more than a terminal is pruned, TNT uses a single-letter code to identify the branches of the tree where each pruned taxon can be located (using upper and lowercase, this can show up to 52 simultaneous prunings). The single pruned tree shown in Figure 2.6b, with indication of the alternative positions for the pruned taxon, allows reconstructing completely each of the original 13 trees (this will sometimes not be the case, e.g. when the pruned consensus still has some polytomies, but it is still much more informative than an almost unresolved consensus).

Indication of possible locations of removed taxa aside, pruned consensuses correctly specify the relationships of *some* of the taxa in the matrix, and say nothing about the positions of the excluded taxa. Pruned consensuses thus display only some of the information present in the input trees, but any information they actually display is entirely correct (i.e. indeed present in the input trees). The choice of what part of the information in the input trees to display may depend on the nature of the problem and the interest of the researcher. Alternative sets of pruned taxa may produce trees that are more useful, or that better display the relationships that the researcher is interested in resolving; in this sense, no set of prunings is "correct" or "incorrect". Thus, designing a general criterion of optimality for choosing more useful sets of prunings is difficult. A simple maximization of the number of nodes in the consensus (Deepak et al. 2012; they showed that this is an NP-hard problem), or sums of frequencies in remaining groups (as in Aberer et al. 2013), will not always result in prunings that are most useful from the perspective of the researcher (Goloboff and Szumik 2015).

An important distinction is that between removing the unstable taxa from the matrix, or only from the trees. Only removal from the trees can be justified. Removing taxa from the matrix and reanalyzing may change the relationships between the remaining taxa. If an unstable taxon has a character combination that affects the relationships of the other taxa, removing it from the matrix amounts to ignoring part of the evidence (of course, if the unstable taxon does not affect the relationships of the other taxa, the question of removing it is irrelevant). The only possible reason for not including a terminal in a matrix is uncertainty in the scorings (e.g. due to poor preservation, to data taken from old literature, etc.), or doubts in the identification of the taxon itself; and those problems would point toward excluding the taxon from the dataset, regardless of whether it behaves as a wildcard.

The main problem with unstable taxa is that, when the input trees are numerous and/or comprise many taxa, discovering the taxa which are decreasing consensus resolution may be difficult, and a number of methods have been proposed that can assist with that task.

6.3.1 Maximum Agreement Subtrees (*MAST*)

Gordon (1980) and Finden and Gordon (1985) first posed the problem of finding the largest set of taxa such that all the input trees become identical. There may be different sets of prunings that make input trees identical, and so there can be multiple *MAST*'s for a set of input trees. This problem has been well studied by mathematicians (e.g. Steel and Warnow 1993; Hein et al. 1996; Semple and Steel 2003), and is well known as an NP-complete problem (due to its strongly combinatorial nature). The number of taxa in the *MAST* is often used to indicate the similarity of the input trees (as in Eulenstein et al. 2004; Kuhner and Yamato 2015). However (and similarly to the case of the strict consensus), a large number of taxa in the *MAST* indicates a high similarity, but a low number of taxa need not indicate that the trees are very dissimilar. The *MAST* is designed to prune taxa until input trees become *identical*; the goal is not to produce similar trees or more resolved consensuses. Thus, if large groups can switch position in trees with otherwise only minor differences in the rest of the topology, it may be necessary to remove large groups of taxa just to solve those few differences (Goloboff et al. 2006). An example is shown in Figure 6.2c. The strict consensus, with 8 out of 10 possible nodes, is a better indicator of the high similarity among input trees than the *MAST*, which has only half the taxa and would suggest large differences between the two trees. Unlike the methods discussed next, therefore, the *MAST* rarely produces a better resolved consensus.

6.3.2 Brute-Force Methods

The naive approach to identify unstable taxa is by brute force, checking the effect of alternative prunings on the consensus. In the case of the strict or combinable component consensus, the polytomies can be resolved one at a time, by removing the terminals or groups directly connected to the polytomous node. For example, in Figure 6.1a, the polytomy involving L, H, D, and CEFGI in the strict consensus, could never be improved by pruning taxa C, G, or J—those taxa do not directly connect to the polytomy. Thus, the process can be speeded up by creating reduced subtrees (e.g. in the case of Figure 6.1a, the subtrees created for trying to improve the node just discussed would represent group CEFGI as a single "terminal" unit). In many cases, several nodes may have to be pruned together for improving the consensus, if several wildcards move independently. The maximum number of nodes that can be usefully pruned in a polytomy of degree n is $n - 3$; pruning a node to improve resolution of a trichotomy is pointless. For n nodes connected to a polytomy and a maximum of k nodes allowed for simultaneous cut, this may require checking up to $\sum^k_{i=1} n! / ((n - i)! \times i!)$ sets of prunes, which can be a large number, especially for large n (e.g. for up to 15 nodes cut together in a polytomy of 25 nodes, about 7.12×10^6 reduced consensuses need to be calculated, close to prohibitive). The brute-force approach can therefore be time-consuming. The implementation in TNT has been available since the late 1990s (see Section 6.8.4); it first tries prunings of fewer nodes, increasing number of simultaneously pruned nodes only if lower numbers do not improve the consensus. All the prunings that add nodes to the consensus are reported (or stored for subsequent use). Obviously, if pruning a taxon or a group q increases the resolution in x nodes, then subsequent prunings of q combined with other taxa are reported if and only if they increase resolution in more than x nodes.

6.3.3   TRIPLET-BASED METHODS

In one of the earliest attempts to quantify identification of unstable taxa, Estabrook (1992) proposed the use of quartet methods. Quartet methods are useful for unrooted trees, but those methods must examine all combinations of four taxa, which is time-consuming. When trees are rooted (as is often the case in phylogenetic analysis; this is the way in which TNT treats its trees), it is possible to work with triplets. Thorley and Wilkinson (1999) modified Estabrook's methods to work on triplets (requiring rooted trees, but much less numerous). Estabrook's (1992) measure EA ("Explicit Agreement") calculated the proportions of triplets (quartets, in his case) in which a terminal participates which are resolved identically in all the trees under consideration. Consider the example of Figure 6.2d, with a simple matrix, a wildcard taxon X, and 7 trees (i.e. all the locations of X in the common skeleton of the remaining taxa). Taxon A, used to root the trees, is always the outermost taxon in any three-taxon combination (i.e. $EA_{(A)} = 1.0$). Taxon B participates in 10 triplets (BAC, BAD, BAE, BAX, BCD, BCE, BCX, BDE, BDX, and BEX), of which only those involving X (but not A) are resolved differently in some of the trees (the triplets marked in gray in Figure 6.2d). For example, in all seven trees the triplet BCD is resolved as B(CD), and BDE as B(DE). Therefore $EA_{(B)} = 0.70$. As X moves around in the trees, a higher proportion of the triplets in which it participates are resolved in different ways in different trees; thus, $EA_{(X)} = 0.40$, the lowest of all values, correctly identifies X as the most unstable taxon. Thorley and Wilkinson (1999) extended Estabrook's EA to use the average difference in frequencies of the different resolutions of each triplet (calculating for each triplet the difference between most frequent resolution and second most frequent resolution); they called their measure LSI ("Leaf stability index"). The EA is potentially useful for helping resolve strict consensus trees, and the LSI for helping resolve majority rule trees.

   In principle, EA and LSI can be used to identify candidates for pruning manually, but this can be laborious. Pol and Escapa (2009) also discussed some shortcomings of these measures, and proposed solutions. Pol and Escapa (2009) discussed improving the strict consensus, and then referred to EA rather than LSI. They noted that *EA* (which they called *PC*, "positional congruence") is affected by taxa that move together, in groups. For improving specific polytomies of a strict consensus, it is more useful to examine only the nodes directly descended from the polytomous node, as units, and to calculate *PC* values for that reduced subproblem; they called this *PCR* (the *R* stands for "reduced"). They also noted that once the node (either a terminal, or an entire group) is removed, the values of *PCR* for the rest of the nodes can change, and then to choose the next taxon to prune, it is best to recalculate the *PCR* values for the new problem (i.e. with already pruned taxa removed). They proposed to iteratively remove nodes and recalculate *PCR* values, and they called this process *IterPCR*. For every taxon removal, one can check whether the consensus is actually improved (i.e. whether it has additional nodes). The difference with the brute-force process is that the selection of the taxa (or groups) to remove is guided by the *PCR* values, instead of an exhaustive enumeration, thus saving time in the case of big polytomies. Note that removing a taxon or a node does *not* change the resolution of the triplets among the remaining taxa or nodes; it only changes the total number of possible triplets, eliminating those triplets involving eliminated taxa or nodes. Therefore, saving lists of identically resolved triplets allows a faster recalculation of *IterPCR* values for each cycle. Pol and Escapa (2009) implemented the method in a primitive, proof-of-concept TNT script—this showed that the method produces good results, but it can be used to solve only relatively small problems. Goloboff and Szumik (2015) described an efficient implementation in TNT.

6.3.4 IMPROVING MAJORITY RULE OR FREQUENCY DIFFERENCE CONSENSUS

Improvement of this type of consensus trees is much more difficult than for strict or combinable component consensus. In strict or combinable component consensus, the nodes in the consensus tree act as solid delimiters of subproblems. In the case of consensus methods including groups that can be entirely absent or contradicted in some (a minority) of input trees, the situation is different. Consider the simpler case of the majority rule consensus, in the example of Figure 6.3a, where there are 6 *MPTs*. One of those trees is as shown in Figure 6.3a, displaying group EF as monophyletic. The other five trees result from moving B, D, or J to be sister of E, or moving H or L to be sister of F. The frequency of group EF is then 1/6. Note that improving the frequency of group EF requires removing taxa separated from EF by branches with frequency up to 5/6 (0.833). The low frequency of EF results from many different taxa possibly entering group EF, but every one of those taxa does so in only one of the input trees. Thus, the improvement of the frequency of a group may require pruning any terminal in any other part of the entire tree. In general, for maximum branch frequency f_{max} separating the candidate for pruning from the group, the maximum possible gain in frequency for the group is $1 - f_{max}$. Thus, pruning each of B, D, H, J, or L separately in the example of Figure 6.3a can increase the frequency of EF in at most 1/6 = 0.166. Raising the frequency of EF above a certain threshold will require pruning several terminals together; whether or not this is desirable, will depend on the goals of the analysis and whether EF is a critical group. In general, the multiple prunings to improve EF will not be selected by criteria that measure optimality as the total sum of frequencies in the resulting consensus, because several groups of a relatively high frequency (CD, GH, IJ, KL, and the group of all taxa but AB) then have

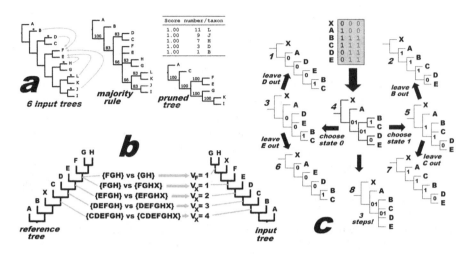

FIGURE 6.3 (a) Example of the difficulties in finding taxa to prune for improving majority or frequency difference consensus, with six alternative trees (each moving one of the taxa indicated as sister of E or F). Improving the frequency of a group with low frequency (EF) requires simultaneously pruning taxa that belong in groups with a high support. (b) The principle used by the `prunmajor` command of TNT to identify taxa reducing group frequencies. Group-to-group comparison between the two trees identifies the closest group, and the minimum number of taxa that have to be removed for the groups to become identical, increasing a counter of the instability value *V*. This correctly identifies X as the most unstable taxon. (c) A simple case illustrating problems with elimination of zero-length branches. See text for discussion.

to be removed from the consensus, decreasing overall sums of frequencies. Group EF is small, but this can happen for groups of any size; Goloboff and Szumik (2015, their fig. 8) provided another example, with an actual matrix, where the group whose frequency can be improved is much larger (in their case, the low frequency results from potential members separately abandoning the group in different input trees).

Pattengale et al. (2011), Aberer and Stamatakis (2011), and Aberer et al. (2013) proposed several algorithms for improving bootstrap frequencies. Those are based on a simple optimality criterion for maximizing sums of frequencies, implemented in the program RogueNarok (Aberer et al. 2013). Their algorithm is a heuristic (recall that finding sets of prunes to maximize sums of frequencies is an NP-hard decision problem; Deepak et al. 2012), designed for fast analysis of large datasets, based on quick recalculation of group frequencies for each pruning tried. Goloboff and Szumik (2015), and Wilkinson and Crotti (2017) have showed simple examples where the criterion in RogueNarok may produce undesirable results.

TNT implements a different approach (in the `prunmajor` command), based on comparing groups in a reference tree, against the groups in each of the input trees. Unlike RogueNarok, the `prunmajor` command seeks prunings to improve the frequency of the groups in a reference tree (a list of groups of interest in the tree can also be specified). The method keeps a counter of prune value, *V*, for each terminal taxon. All prune values are first initialized to 0. Then, for each group in the reference tree, the method compares each of the input trees one at a time. For every input tree, the group in the input tree closest in composition to the group of the reference tree is identified (using the same criterion described for the rough recovery of groups, see Section 6.1.2.2). Identical groups are skipped. Otherwise, any taxon present in group of reference but not in group of input, or group of input but not in group of reference, could be pruned to make the groups in reference and input identical (recall that the criterion of rough recovery minimizes the number of extraneous members included, or members excluded; these two alternatives can be given different weights). Every time a group comparison identifies some taxa that would make the groups identical if pruned, the prune value V of those taxa is increased. After comparing every group in reference tree with all the input trees, the taxa with the highest prune values are displayed (or saved to a taxon group). The method is shown in Figure 6.3b, when using one of the trees produced by the matrix in Figure 6.2b (with taxon X scored only with missing entries) as reference, and another tree as input tree. Group {FGH} in reference is equally close to {GH} or {FGHX} in input. F is included in group of reference but not in input tree in the first case, and then V_F is increased ($V_F = 1$). X is included in group of input tree but not in reference in the second, and then V_X is increased ($V_X = 1$). The next group of reference tree, {EFGH}, is closest to {EFGHX}; as X is in group of input tree but not in reference, V_X is increased again ($V_X = 2$). Subsequent comparisons also increase the value of V_X only. Final pruning value for X is then $V_X = 4$, and for F is $V_F = 1$. Values are rescaled on output so that largest prune value equals 1.

In the case of Figure 6.3a, the heuristic of prunmajor correctly identifies B, D, H, J, and L, if using group E+F as target for improvement. The algorithm of RogueNarok would report no improvement in that case—the sum of frequencies in the complete consensus $(100+83+83+66+83+66+83+83=647)$ is less than the sum of frequencies when pruning B, D, H, J, and L $(100+100+100+100+100=500)$. Even if the frequency of the target group EF is improved with every one of the individual prunings of B, D, H, J, or L, the overall sum of frequencies (due to the disappearance of groups of high frequency, e.g. IJ or KL) is decreased.

6.3.5 SWAP AND RECORD MOVES

All the methods described so far to identify unstable taxa rely on the tree topologies only, with no regard for the underlying matrix. Particularly for discovering unstable taxa that reduce group supports, branch-swapping with tree bisection reconnection (*TBR*) allows quickly exploring the fit of alternative topologies to the data. If only rearrangements producing equal score are accepted, taxa or groups that can be cut and moved to more positions, or to more distant positions, are good candidates for helping improve the consensus. For a near-optimal tree for Källersjö et al.'s (1999) classic rbcL dataset (2,594 taxa), TNT identifies 29 taxa that can move to distances of 6 or more nodes from original location with *TBR* rearrangements of equal score, in 1.19 sec.[1] When slightly suboptimal rearrangements are accepted (with user-defined limits for absolute and relative fit differences; see Chapters 5 and 8), then pruning the

[1] This timing, as well as all timings reported, corresponds to an Intel i7–3770 processor (3.40–3.90 GHz) under Windows.

taxa that move to distant or numerous positions is likely to improve group supports for the relationships between the remaining taxa.

For the example of Figure 6.2b, only taxon X can be moved to more than three positions. In any one of the 13 distinct binary *MPT*s for that dataset, the taxa next to X can also move to positions around X; for example in the tree used as reference for Figure 6.3b, B can move to be sister of X, or to be sister of CDEFGH, with equal cost. But the only taxon that can move to 3 or more positions with equal score is X, and thus it is quickly identified as an unstable taxon by recording number and distance of moves during *TBR*.

6.3.6 IMPROVING PRUNE SETS WITH AN OPTIMALITY CRITERION

All the preceding methods (including brute-force methods) use a criterion of local improvement to accept a taxon as candidate for pruning. Those are simply heuristic initial estimates, and they can include as candidates for pruning taxa which (if pruned simultaneously) do not improve the tree. Recall that the potential pruning combinations can be very large, particularly for improving majority or frequency difference consensus when measuring group supports. By narrowing the set of pruning candidates to a lower number, those methods provide the raw material for a method based on a more global criterion. By refining the prune sets (contained, e.g. in a taxon group), different combinations of those prunes can be tried, placing the prune sets that improve the result in another taxon group. The result of a candidate prune set is evaluated with $E = (\sum S + P) / (T - 2)$, where $\sum S$ is the sum of support values across all branches of pruned tree, P is a penalty for pruning taxa, and T is the number of taxa with least possible prunings. The penalty P is defined as $100 \times R \times (1 - F^2)$, where R is the number of taxa removed, and F is the factor to penalize taxon removal (with larger values imposing a stronger penalty; default is 0.5).

With the initial set of candidate prunings and the penalties defined, the method can work in an "up" fashion (starting from the tree for the complete taxon set, and pruning sets of candidates from the tree), or a "down" fashion (starting from the tree with all candidates pruned, reinserting sets of candidates into the tree). For every set tried (whether up or down), the method recalculates E, and finally saves the set of candidates that most improve E. To reduce the number of sets (i.e. simultaneous reinsertions or prunings) to try, a maximum set size of three is used by default (this can be changed by the user). The optimal set of candidates (for reinsertion or pruning) defines a new set to continue improving, in successive rounds. The taxa to prune or reinsert simultaneously are chosen avoiding combinations of terminals separated by branches with high frequency (or high frequency differences, if improving that measure); those branches act as *separators* (default separator value is 75).

This method is useful to refine candidate sets for pruning obtained from the other, more heuristic methods. It was incorporated into TNT as Goloboff was collaborating in the analysis of the dataset of Pei et al. (2020), to automate as much as possible the final choice of pruned taxa as the dataset continued being developed and new versions analyzed. Pol and Goloboff (2020) used it to improve the results of candidate prunings obtained by assessing ambiguity in each individual jackknife replicate, and it led to the best results.

6.4 ZERO-LENGTH BRANCHES AND AMBIGUITY

A special consideration must be given to ambiguity in results from pure lack of information—absence of characters distinguishing some groups of species. In the simplest case, this can result from all species in a group being identical. In that case, each of the resolutions of the group will have the same fit to the data. For example, for a matrix with 5 identical taxa, each of the 105 possible trees will be equally parsimonious, and the strict consensus of those 105 trees will be a bush (i.e. a fully polytomous tree). Presenting the bush as result indicates the 105 trees as possible resolutions, and is thus the most economical way of presenting the result.

That is not the same thing as saying that the polytomous tree is "the best phylogeny" for the group in question: each of the resolutions of the polytomy in the consensus is an equally good explanation of the data—the characters can be mapped with the same number of steps on a binary tree, and thus the explanatory power of both trees is exactly the same. That is, any binary resolution of the polytomous tree provides as good an explanation of the data as the polytomy itself. The polytomy, in other words, does not forbid any of the possible binary resolutions, just being less defined. Although there is no difference in terms of requirements of homoplasy for the polytomous and the binary tree, the polytomous tree is a much more economical way of presenting the information common to the 105 possible binary resolutions. Rather than forcing the user to look at 105 possible trees, therefore, many parsimony programs (TNT included) eliminate branches which are not supported by synapomorphies. For each of the 105 possible binary resolutions in this example, collapsing (or *condensing*) the branches unsupported by synapomorphies produces the same polytomy. Put differently, all 105 trees become identical when zero-length branches are removed, and thus a single polytomous tree (containing as potential resolutions each of the binary trees) is reported by the program.

It is important to keep in mind that, during tree searches, parsimony programs always operate on trees that are binary (=dichotomous). Tree-collapsing is done as a subsequent step, before the tree is compared to previously saved trees and (if different) stored in the memory buffer. This is because optimization and search algorithms can be more easily defined for binary trees. Although the basic notion of tree-condensing is a simple one, and involves no special theoretical considerations, there are some practical considerations.

6.4.1 IDENTIFICATION OF ZERO-LENGTH BRANCHES AND COLLAPSING RULES

The simplest way to identify zero-length branches is by comparing the most parsimonious state sets (*MP*-sets) of ancestor and descendant nodes. For a given character, if the sets of ancestor and descendant share one or more states, then it follows that there is at least one reconstruction for which there is no change along the branch. In that case, the branch can be collapsed, assigning the shared state to the node that results from merging ancestor and descendant nodes into one, and the number of changes along tree branches for the character will remain the same. Thus, whenever the *MP*-sets for ancestor and descendant for every character along a given branch share some state, the branch can be collapsed without increasing overall tree length.

There is, however, one important complication when collapsing *several* zero-length branches simultaneously: when two or more branches are collapsed at the same time, the state that has to be assigned to the equivalent node may not be the same. Mutually exclusive assignments for equivalent nodes lead to increases in tree length, even if each of the branches can be collapsed *individually* without increasing length. An example (taken from Coddington and Scharff 1994) is shown in Figure 6.3c. Inspection of the matrix shows that the group BCDE is supported by two characters (in gray). The character creating the ambiguity is the first (black) character, and all numbers of steps discussed next (and in the figure) refer to that character (all the trees in Figure 6.3c have the same numbers of steps for the gray characters). Among possible resolutions of the group BCDE the central tree, number 4, is the one with an ambiguous optimization—given that tree, the first character has 2 steps, and there is no unambiguous support for either group BC or DE. Given that (in tree 4) a state is shared between the state sets for the DE node and its ancestor, collapsing that node (creating a trichotomy, tree 3) and selecting that shared state (0) for the trichotomous node, will result in a tree with 2 steps. Likewise for group BC in tree 4—state 1 is shared, so that collapsing the branch leading to BC and creating the trichotomous tree 5, also results in 2 steps. Thus, either group BC or group DE can be collapsed without increasing tree length. But collapsing both BC and DE together at the same time, produces tree 8—with 3 steps! That is because group DE can be collapsed without increasing tree length when state 0 is selected for the ambiguous BCDE node; but group BC can be collapsed when state 1—instead of 0—is assigned to that same node. The central tree, 4, can produce *two* alternative trichotomous trees of minimum length, 3 and 5. The trichotomies in trees 3 and 5 also imply that each alternative resolution of the corresponding polytomy is of the same length. Thus, tree 3 can be resolved as trees 1 and 6, and tree 5 can be resolved as trees 2 and 7, all four of the same length (tree 4 is, of course, also a possible resolution of either tree, 3 or 5). The five distinct binary trees of minimum length for the dataset of Figure 6.3c are then 1, 2, 4, 6, and 7.

Figure 6.3c shows one of the simplest cases where some branches of the tree can have, in different most parsimonious reconstructions, different possible lengths, with a minimum of zero. How should the most parsimonious trees for these data be presented? There are several options available, and these have been called collapsing "rules". Most of the problems associated with collapsing zero-length branches were discussed by Swofford in the PAUP Manual (Swofford 1993).

First option is to collapse all the branches together, and consider the tree thus produced as a consensus, not as a "tree". The tree obtained by collapsing all branches simultaneously is the same as the strict consensus of all the trees produced by the individual collapsings that do not increase length. More generally, it is the strict consensus of the three binary parsimonious resolutions implied by collapsing one at a time each of the unsupported branches. This criterion is easy to implement during a search: a branch is kept only if there is at least one character for which state sets of ancestor and descendant nodes do *not* share any states. Put differently: if, for a given character, among all possible reconstructions, there is at least one reconstruction for which there is no change along that branch, you then recognize that the character cannot provide support for that branch. As this option collapses the branch when there is at least one reconstruction where the branch has length of zero, it is called "Min Branch Length" in PAUP*. In TNT, it is simply called *"Rule 1"* (following Swofford 1993), or collapse level 3, and is the default rule for collapsing. Since the resulting polytomous tree is a sort of consensus, then it is not to be optimized—it can be longer. The advantage of using this option is that it generally decreases the number of trees considered as distinct, over the number of trees produced by "Rule 3" (see Section 6.4.3), thus allowing easier diagnoses and comparisons of results. For the case of Figure 6.3c, this rule produces three distinct trees, trees 3, 5, and 8: of the five binary trees, trees 1 and 6 collapse to produce tree 3, trees 2 and 7 collapse to produce tree 5, and tree 4 collapses to produce tree 8.

The second option is to consider that, when *some* reconstruction of *some* character has change along the branch, then the branch is to be retained. This requires considering possible reconstructions, and the idea that branches are *potentially* supported by some character-state reconstruction. This potential support can be checked during the up-pass of the optimization: if the set of final states for the ancestor is a subset of the set of preliminary states for the descendant, then no reconstruction will imply change along that branch. This is usually called *"Rule 3"* (Swofford 1993). The tree thus produced cannot be less resolved than in the case of Rule 1: some branches kept under the previous rule will be kept under this one, and any branch eliminated by this one will also be eliminated by the previous one. The trees having the same or more resolution than under Rule 1 means that the same or fewer trees become identical after collapsing some branches. Therefore, Rule 3 often produces more distinct trees than Rule 1. When the tree collapsed under this criterion is optimized, it still has minimum length—but some branches have no unambiguous synapomorphies. For the case of Figure 6.3c, this criterion for collapsing produces 3 distinct trees, 3, 4, and 5: binary trees 1 and 6 collapse to produce tree 3, trees 2 and 7 collapse to produce tree 5, and tree 4 remains as binary. Tree 4 is an example of a tree retained by this rule which has change along its branches only in *some* reconstructions.

The third option is to retain all the trees which have both (a) minimum length and (b) all their branches with at least one unambiguous synapomorphy. This is known as "Rule 4", and was proposed by Coddington and Scharff (1994). A practical problem with Rule 4 is that a given tree (like tree 4 in Figure 6.3c) can be collapsed to produce several alternative trees (3 and 5, in this case), for the several possible reconstructions. Thus, this criterion is difficult to implement during a search: every time a tree is found, the alternative combinations of collapsings followed by reoptimization (or collapsings implied by alternative reconstructions) would have to be tried, and every one of the trees resulting from collapsing the binary tree just found would have to be compared to each of the preexisting trees, in search of duplicates. No current program implements this criterion during the search. In TNT, it is implemented only as a subsequent step. For the case of Figure 6.3c this criterion produces only two trees, 3 and 5.

TNT (and Nona/Pee-Wee before; Goloboff 1993b, c) implements also an additional criterion for removing zero-length branches (called "`ambig=`"), which retains a branch if the ancestor-descendant *MP*-sets for one or more characters are different. This causes condensed trees to be of minimum length, but collapses a little more strictly than Rule 3 (an example of the two criteria producing different results is a dataset where every terminal has a different state of a nonadditive character: state sets of internal nodes are then all the same).

The sets of trees considered as distinct will be progressively smaller as the criterion becomes stricter; the numbers of trees will be binary ≥ Rule 1 ≥ identical state sets ≥ Rule 3 ≥ Rule 4. For studies intended to assess character evolution by mapping characters on the resulting trees, all these options are not relevant: there is no reason to discard any binary tree as inferior, as long as it is of maximum parsimony. Studies of character evolution should simply consider alternative binary resolutions (or a sample thereof).

Different defaults for collapsing trees are an important source of differences between different parsimony programs. The default collapsing is Rule 1 in TNT, but Rule 3 in PAUP*. This is a reason why many studies report PAUP* as finding more equally parsimonious trees than TNT—the trees were being collapsed with the default option in PAUP*, less strict than the default in TNT and then producing larger number of trees. Hennig86 (Farris 1988) collapsed trees with Rule 3 as only option (see Platnick et al. 1991), and also produced unnecessarily large numbers of trees for many datasets.

6.4.2 CONSENSUS UNDER DIFFERENT COLLAPSING RULES

An extremely important point is that, regardless of the collapsing criterion used, the strict consensus of all the trees considered as distinct will always be identical. This is often the source of confusion, and it is not uncommon to see papers needlessly reporting that the strict consensus trees for different collapsing criteria were the same. Any other result would imply a deficient search, or a bug in the program used. For the example in Figure 6.3c, Rule 4 will consider two trees as distinct, 3 and 5; Rule 1 will consider three trees, 3, 5 and 8; Rule 3 (as well as the criterion of identical state sets) will consider three trees, 3, 4 and 5; and there are five distinct binary trees of minimum length, 1, 2, 4, 6 and 7. Note that the strict consensus of the trees found by each of the rules is always the same, identical to tree 8. Therefore, if all trees distinct under the criterion in question are saved, taxonomic conclusions are not affected by the collapsing rule used. But note the conditional: "if all distinct trees are saved". Practical matters are different (see next section).

6.4.3 NUMBERS OF TREES, SEARCH EFFORT

For practical purposes, it is highly desirable to collapse as strictly as possible. Using stricter collapsing criteria can enormously reduce the number of trees considered as distinct. As example, consider the numbers of trees for Goloboff's (1995a) matrix on nemesiid spiders. Under Rule 1, *TBR* (starting from a single optimal tree) finds and swaps 72 trees (in a tenth of a second). Under Rule 3, it finds 23,328 trees (in some minutes). Saving all distinct binary trees, TNT found one million trees after swapping on the first 190,000 trees, at which point the search was interrupted, after about 135 hs. (the rate at which new trees were accumulating during swapping suggests that at least 3 million trees exist, although finding them would be very time-consuming). The strict consensus of the 72 trees collapsed under Rule 1 is, as pointed out in the previous section, exactly the same as the strict consensus of the millions of binary trees.

Since the trees distinct under Rule 1 tend to be more different (i.e. Rule 1 is stricter than Rule 3), then saving low numbers of trees per replication, during random addition sequences, tends to better identify the groups which are unsupported than if the trees are compared according to Rule 3. For the example of Figure 6.3c, saving only two trees considered as distinct under the stronger Rule 1 is guaranteed to produce the same consensus as saving all the trees: the strict consensus of 3+5, 3+8, and 5+8 is identical to tree 8. But saving only two trees distinct under the weaker Rule 3 will lead (in a third of the cases, if all trees are equiprobably found) to erroneously conclude that some unsupported group is supported: finding trees 3 and 5 produces the correct consensus (tree 8), but finding trees 3 and 4, or 4 and 5, leads to conclude that either BC, or DE (respectively), are supported; the only way to *guarantee* that the correct consensus is found, if trees are collapsed under Rule 3, is finding *all three* possible trees. Of course, things are even worse if trees are saved as binary: only saving *four* or more trees guarantees that the correct consensus is found in that case. These differences (as well as the probability of finding the correct consensus for alternative collapsing criteria in the case of Figure 6.3c) are summarized in Table 6.1. For more complicated cases, with more taxa and characters, the differences in numbers of trees needed to correctly identify all unsupported groups may be enormous.

The differences just discussed must be kept in mind when only a sample of *MPTs* can be saved during a search. The best approach when there are too many trees to save them all, is simply calculating the consensus from multiple independent hits to minimum length trees (as discussed in Chapter 5), but without extensive branch-swapping to save multiple trees. In that case, at the time of calculating a strict consensus, it is better to collapse the trees as strictly as possible. This reduces the chances that a group that is unsupported by the data (i.e. absent from some *MPTs*) is present in the consensus. To make collapsing even stricter, TNT also implements the possibility of collapsing with *SPR* or *TBR*. This method swaps the tree under *SPR* or *TBR*, and

TABLE 6.1
Probability of Calculating the Correct Consensus by Saving Different Numbers of Trees

	$P_{(right,1)}$	$P_{(right,2)}$	$P_{(right,3)}$	$P_{(right,4)}$
Binary trees	0	0.40	0.80	1.00
Rule 3	0	0.33	1.00	—
Rule 1	0.33	1.00	1.00	—
Rule 4	0	1.00	—	—

Note: With Alternative Criteria of Collapsing Zero-Length Branches. Values are for the dataset of Figure 6.3c (0 indicates impossibility, 1.00 certainty). $P_{(right,n)}$ is the probability of calculating the correct strict consensus tree by saving n trees collapsed according to the corresponding criterion (dashes indicate that fewer than n trees exist under the criterion). Calculations assume that the search algorithm can find any of the 5 binary trees with equal probability.

every time a move produces a tree of equivalent score, marks all the nodes between source and destination node to be collapsed—this is equivalent to calculating the strict consensus of the trees before and after the move, but does not require saving the actual trees to memory (thus saving RAM and time; see Goloboff 1999). If not all trees can be saved to RAM, then this option of *SPR* or *TBR* collapsing is especially useful. The trees produced by searches from independent starting points are likely to be more different than the trees produced by swapping from the same initial tree (which differ in just a few *TBR* rearrangements at the most), but collapsing the resulting trees more strictly decreases even more the probability that some unsupported group is included in the conclusions. This can be used to save time in searches (see Chapter 5), by doing several independent hits to minimum length, and collapsing them strictly, instead of doing a few hits and saving large numbers of equally parsimonious trees via branch-swapping.

Similar considerations can be made for the influence of collapsing criteria on the effectiveness of tree searches based on saving and swapping multiple equally parsimonious trees, particularly when saving few trees (e.g. in multiple random addition sequences, *RAS*, followed by *TBR*; see Chapter 5, Figure 5.1h). The searches become more effective—in terms of finding shorter trees—as the collapsing criterion used is stricter. A more relaxed criterion for collapsing will lead to quickly filling the memory buffer with trees that are almost identical, while a stricter collapsing criterion will save only trees differing in more "radical" rearrangements. When the trees saved are more different, it is more likely that swapping one of them will lead to a shorter tree (if such exists); therefore, collapsing more strictly tends to produce shorter trees when low numbers of trees are saved per replication (Goloboff 1996, 1999; Davis et al. 2005).

6.4.4 TEMPORARY COLLAPSING

In the case of TNT the default collapsing rule is Rule 1; since the trees collapsed with this rule may be longer than the original binary resolution, TNT saves only those trees that would be different if zero-length branches were to be collapsed, but does *not* physically collapse the branches. The original binary resolution is retained instead. This means that, for consensing, the trees must be collapsed. To avoid losing track of the original binary (shortest) resolution, the trees are condensed temporarily, only for the sake of calculating the consensus, and then returned to their original binary status. Permanently collapsing the trees may be desirable in some situations, but the user must ask the program to do so explicitly. In that case, the trees must be used for consensing without collapsing them—eliminating zero-length branches again may lead to errors.

6.5 SUPERTREES

While summarizing the common information in trees with the same taxon sets is relatively straightforward, things become a lot more complicated when the trees have different taxon sets. All the caveats about using consensus trees to establish phylogenetic conclusions by summarizing results from different datasets apply to the case of supertrees, but even more forcefully. Yet, in some cases, it may be legitimate to inquire the common implications of trees produced by datasets with only partial taxonomic overlap. Thus, such implications should be established with the appropriate logic.

The basic principle in supertrees is that trees with partial taxonomic overlap can often be combined, with unambiguous implications. In this case, the trees are said to be *compatible*; more properly, two input trees are compatible when there is a tree including each and every one of the taxa in the two input trees, but for which each input tree is a possible subtree. Consider the case of two trees, such as trees 1 and 2 in Figure 6.4a. If what both trees say is true, then certain things must also be true for their combination 1 + 2. Tree 1 states that D is more closely related to C than it is to B; tree 2 states that D is more closely related to A than it is to C. If both things are true at the same time, then that means that the tree shown in Figure 6.4a as combining 1 + 2 must also be true. No other tree can include the information contained in both trees 1 and 2, other than the one shown. This is not a question of a specific method to produce supertrees, it is a matter of inescapable logic.

The first methods for producing supertrees for two compatible input trees were proposed by Gordon (1986) and Steel (1992). Even in the absence of conflict between input trees, some strange implications of supertrees arise from the fact that supertrees, instead of actually evaluating conflict as in the case of consensus, assume that the relationships that are not explicit in the input trees (because of incomplete taxon coverage) must have been resolved alike in the whole trees that the input trees were carved from. Much of what is done in systematics is based on generally expecting congruence between results obtained independently; in the case of consensus, the lack of congruence is actually observable, but for supertrees, the lack of congruence is much more elusive. This leads to properties of supertrees entirely unlike those seen in consensus trees. For example, several consensus methods deal with unrooted trees, but unrooted supertrees are simply not possible: any unrooted supertree method will have an irreducible arbitrariness (such as dependence of input order; Steel et al. 2000).

Back to Figure 6.4a, assume that a third tree (e.g. from a third dataset) is included in the input, as tree 0. While combining trees 1+2 produces as only possible conclusion the tree (B(C(AD))), combining tree 1 with 0 produces a different conclusion, (A(B(CD))). The third possible combination, among trees 0 and 2, produces yet a third inescapable conclusion, ((AD)(BC)). Every one of the three possible combinations of input trees produces an unambiguous and conflict-free conclusion. Yet all three conclusions are different. This shows that any two of the input trees in Figure 6.4a can be true at the same time, but not all three of them; and the only way to discover conflict among the partially overlapping input trees is by collectively examining them all. Pairwise evaluations of conflict fail miserably in this regard. The strict consensus is a commutative method, but supertrees are clearly not commutative.

In addition to the restrictions demonstrated by Steel et al. (2000), Goloboff and Pol (2002) showed other problems that may arise with supertrees. Consider the case of the two trees with identical taxon sets shown in Figure 6.4b; they have some conflict, and their consensus displays E as closest to F. By pruning different taxon sets from each of the trees (C and E, from tree 0, and F, from tree 1), the resulting reduced trees are no longer in conflict. They can now both be combined in a unique, fully resolved supertree. Again, this is without conflict, and does not depend on a specific

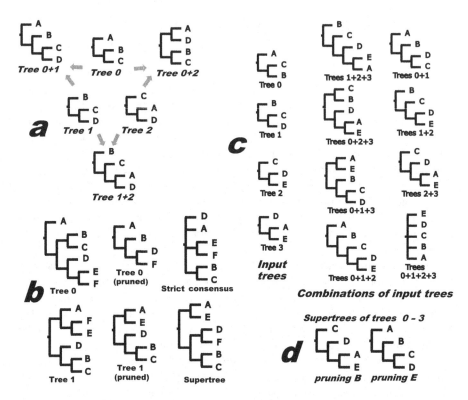

FIGURE 6.4 (a) Each pair of trees 0–2 can be combined, with no conflict. However, the three possible combinations (0–1, 0–2, and 1–2) imply different (and incompatible) trees; combinability of a set of trees can only be checked globally, not in a pairwise fashion. (b) Two trees containing incompatibilities can be pruned to become compatible; in that case, a group (EF) shared by the two complete trees can become contradicted by the supertree of the pruned trees. (c) Problems determining number of trees supporting a given group can arise when trees are incomplete. The group ADE can be either supported or contradicted, without conflict, by each of the input trees, depending on the other trees with which it is combined. This makes it difficult to define an analog of the majority rule in the supertree setting. (d) The conflict in the global combination of trees 0–3 in (c) can be eliminated by pruning some taxa, but different (and contradicting) results are obtained for different taxon prunings.

method for creating the supertree; the supertree is just the only logical conclusion possible if both reduced input trees are correct. That supertree displays E as closest to A instead of closest to F (as in the consensus); D is now closest to F. Of course, the original complete trees did contain conflict; that conflict was removed by pruning, from each original tree, different taxon sets. The supertree displays the conclusion one would be forced to draw if the original trees had lacked conflict, but they did not lack conflict! Supertrees, therefore, always have to be interpreted with caution, even in cases of apparent compatibility. Note that taxa are commonly pruned to improve the resolution of consensus trees, but in the consensus setting the prunings and the remaining taxon sets are always identical.

Goloboff and Pol (2002) also discussed the problem of conflict resolution. Combinations of trees with different taxonomic coverage have some unexpected properties in the absence of conflict between the trees, but things get much worse when trees (or combinations of trees) conflict. Methods that are capable of resolving conflict are then desirable. Conflict resolution can be achieved in a manner similar to the combinable component consensus (i.e. collapsing groups for any single case of contradiction). Goloboff and Pol (2002) proposed a method for creating a semi-strict supertree, with properties matching those of the combinable component consensus. In many cases, however, it may be desirable to resolve conflict more liberally, as in a majority rule consensus tree, so that groups supported by many trees (and contradicted by few) appear in the result. One of the surprising findings of Goloboff and Pol (2002) is that such a supertree apparently cannot be defined: a given input tree can equally well "support" or "contradict" a group, depending on the trees with which it is combined. An example (taken from Goloboff and Pol 2002, fig. 10) is reproduced here, in Figure 6.4c, with four input trees. Each of the input trees can *support* group ADE if combined with some trees (i.e. both 1+2+3 and 0+2+3 combine without conflict to produce ADE). But then, each of the input trees *contradicts* ADE, if combined with some *other* trees (i.e. 0+1+3 and 0+1+2 combine without conflict to contradict ADE). As each of the 3-tree combinations consists of compatible trees, this example does not rely on any specific method to combine trees; it just uses the logical implications of possible combinations (the same is true of 2-tree combinations; the supertree for the four trees shown in Figure 6.4c is the semi-strict supertree, which resolves any conflict by collapsing). Given that one cannot decide whether a given tree supports or contradicts a group—it could do either, depending on the trees it is combined with—it is not possible to count number of trees supporting each group. Cotton and Wilkinson (2007), in a reply to Goloboff and Pol's (2002) work, proposed an alternative way to define what they call a "majority rule supertree", which, although interesting, is not entirely without problems (see Section 6.5.4).

Another strange property of supertrees involving conflict is that pruning taxa from the input trees produces a very different effect from the one produced in consensus trees. The four input trees of Figure 6.4c (when collectively considered) have conflict; one could be tempted to resolve that conflict by pruning some of the taxa causing it. Pruning B, the supertree of the four input trees implies—without conflict—the supertree (C(D(AE))), and pruning E, the supertree (A(B(CD))). When pruning B, D is closer to A than to C, but when pruning E, D is closer to C than to A. That is, the resolution of the relationships between the *remaining* taxa depends—unlike the consensus case—on which taxa are pruned.

These difficulties seem unsurmountable, and suggest that supertrees always need to be interpreted very cautiously. The following subsections describe the main methods for building supertrees, as well as additional problems of each own.

6.5.1 Semi-Strict Supertrees

Goloboff and Pol (2002) noted that conflict between trees must be examined collectively, and then proposed algorithms to produce supertrees which would contain all the groups that are supported by some input tree or combination of input trees, and exclude all groups that are contradicted by some input tree or combination of input trees. For a matrix representing the groups in the input trees (as usual, taxa excluded from a tree are scored with a missing entry), this would be the largest set of mutually compatible characters not contradicted by any character or combination of characters, a concept akin to the maximum clique (e.g. as in Le Quesne 1969; Estabrook et al. 1977). Goloboff and Pol (2002) called this uncontradicted set of characters the *ultra clique*, and proposed heuristics to find it. Day et al. (2008) discuss some possible shortcomings of the semi-strict supertree. However, the semi-strict supertree is the only supertree method with a direct conceptual analog—the combinable component consensus—among consensus methods. It has the obvious shortcoming that a group present in 99 trees but contradicted by a single tree will not be present in the result. Just like the combinable component or strict consensus, the semi-strict supertree is useful when well resolved, but of difficult interpretation when resolved poorly.

6.5.2 MATRIX REPRESENTATION WITH PARSIMONY (*MRP*)

The idea of representing each input tree by a matrix, and then analyzing the matrix with standard parsimony algorithms, has already been discussed, in the context of consensus trees. Beyond automation of the recoding of trees into matrices, which is a simple task, this method did not require the development of any special algorithms—it could simply coopt existing algorithms for parsimony tree searches. Therefore, it has been the most widely used method for building supertrees.

The most serious problem of *MRP* is that it can create spurious groups, groups that are contradicted by every one of the input trees. Even those who defend the method as a valid alternative (Bininda-Emonds and Bryant 1998; Bininda-Emonds et al. 1999) acknowledge that the method can create this kind of group (which they euphemistically call "novel"). Goloboff and Pol (2002, their fig. 1) showed a simple example, where two trees, (A(B(CD(EF)))) and (C(D(E(A(BF))))), produce a supertree ((AB)CDEF) when using *MRP*. Group AB is contradicted by both input trees. Examples of such pathological behavior abound in the literature criticizing *MRP* (Pisani and Wilkinson 2002; Gatesy et al. 2002). Bininda-Emonds (2004) claimed (despite Goloboff and Pol's 2002 argument on the impossibility of meaningfully defining a majority rule supertree method) that *MRP* is expected to resolve the supertree in favor of the groups more frequently supported by the input trees, thus being analogous to the majority rule consensus tree. However, Goloboff (2005, his fig. 2) showed an example (with identical taxon sets) where several of the groups present in a single tree, and contradicted by a tree with nine copies, are nonetheless included in the *MRP* supertree. That results from the asymmetry in mapping number of steps—given two trees A and B, the number of steps required to map the matrix for tree A onto tree B is different from the number of steps required to map the matrix for tree B onto tree A. In the case of only two tree topologies, this asymmetry deviates the supertree toward the topology that requires the fewest steps, and if the asymmetry is important, this effect can prevail even when there are many more copies of one of the two topologies. The other problem with *MRP* (Goloboff 2005) is that it does not measure the recovery of a group in the input trees as all-or-none—rather, it counts the steps for the character representing a group, steps which can then need to be minimized by secondarily joining taxa that were not members of some groups in the input trees (i.e. "reversals"), or were members of larger groups broken into parts in the supertree (i.e. "parallelisms"). Useful as those concepts are when applied to characters representing actual morphological observations, they are devoid of meaning when referring to groups in trees.

6.5.3 Other Methods Based on Matrix Representation

Instead of analyzing the matrix representing the input trees by means of parsimony, there have been proposals to analyze the matrix by means of cliques (MRC; Ross and Rodrigo 2004). In the case of fully resolved trees with identical taxon sets, the number of characters with homoplasy for the reciprocally mapped matrices is symmetrical, and thus the MRC is much closer to majority rule consensus trees (and necessarily identical when only two resolved trees are compared; Goloboff 2005). It can, however, also produce trees displaying groups present in the minority of the input trees when there are more than two input trees (see examples in Goloboff 2005).

Chen et al. (2003) proposed the method of "minimum flipping", based on minimizing the number of matrix entries that have to be changed to make the matrices representing two trees identical. The method has problems similar to those of MRP or MRC, but even more acute (including group-size effects: larger groups require changes in more matrix cells). Goloboff (2005) also gave examples where minimum flipping produces a supertree displaying the groups present in a minority of the input trees.

6.5.4 Majority Rule Supertrees

In one of the most interesting papers on supertrees, Cotton and Wilkinson (2007) proposed a method they call "majority rule supertrees". As the majority rule consensus tree minimizes the sum ΣRF of symmetric Robinson and Foulds (1981) distances to the input trees (Barthélemy and McMorris 1986), Cotton and Wilkinson (2007) proposed modifications of the Robinson and Foulds (1981) distance for the case of different taxon sets, to use it in creating supertrees. They proposed (p. 447) to use as supertree the strict consensus of all the tree(s) minimizing this sum. Cotton and Wilkinson (2007) discussed two modifications of the Robinson-Foulds distance, the (—) and (+) variants. The (—) consists of pruning from the candidate tree the taxa not present in an input tree; the (+) consists of adding the missing taxa to the input trees, at every possible position, and choosing those positions that minimize ΣRF to candidate supertree (see also Dong and Fernández-Baca 2009). This is a rather clean extension of the majority rule consensus method, and there is little doubt that (among existing supertree methods) it will generally produce the results closest to a majority rule tree. However, the problem pointed out by Goloboff and Pol (2002) still exists: majority rule consensus trees are not useful because they produce minimum ΣRF, they are useful because they display clades in a majority of the input trees, and that cannot be guaranteed (or even evaluated) in the case of different taxon sets. In addition, Goloboff and Szumik (2015) showed a fundamental difference between seeking trees of minimum ΣRF when taxon sets are identical, or when they are different. For identical taxon sets, there can be several trees minimizing ΣRF; in that case the strict consensus of those trees is identical to the majority rule tree and *also* minimizes ΣRF. When taxon sets are different, there can be several trees minimizing ΣRF, but their strict consensus can have a non-minimum ΣRF (see the example in Goloboff and Szumik 2016, fig. 4). The "majority rule supertree" of Cotton and Wilkinson (2007), defined as the strict consensus of the trees minimizing ΣRF, is not in itself a tree of minimum ΣRF. If what defines a "majority" is the minimization of ΣRF, then such minimization can sometimes only be achieved by alternative, mutually contradictory trees, thus being defined only ambiguously—as claimed by Goloboff and Pol (2002).

A practical problem with majority rule supertrees is that the tree(s) minimizing ΣRF are difficult to compute; they can only be found by heuristic searches, or exhaustive enumeration. Bansal et al. (2010) and Dong et al. (2010a) have described some potential approaches, but none of those methods has made it yet into any major program for phylogenetics. Finally, it must be noted that, under certain conditions (i.e. if the probability of inferring a tree with a given RF distance decreases monotonically with its distance to true tree, and uniformly for different taxon subsets; see Steel and Rodrigo 2008:244), the majority rule supertree can also be seen as a "maximum likelihood" supertree. In Steel and Rodrigo's (2008) approach, if the probability of recovering a tree varies with a distance different from the RF (e.g. number of SPR moves), then a method other than majority rule supertrees will produce maximum likelihood supertrees.

6.6   ANTICONSENSUS

Consensus trees allow visualizing the relationships shared by two trees. In some contexts, it can be useful to visualize the groups that are *not* shared—this could be called an *anticonsensus*. Obviously the groups present in a tree A but not in tree B cannot be displayed, on a single tree, together with the groups present in tree B but not in tree A. Every one of those two types of groups, however, can be displayed as a tree. Consider as example the two input trees, 1 and 2, shown in Figure 6.5a. Their strict consensus resolves 5 groups (not counting the node that separates the outgroup). Tree 0 displays three groups not present in tree 1, and tree 1 displays three groups not present in tree 0; those groups are shown in Figure 6.5b. When the two trees being compared are fully resolved, then the number of nodes in each of the two anticonsensuses is the same, but it can be different if the trees compared have a different degree of resolution. When there are polytomies, it may be useful to output only the groups that are not shared by one of the trees, and are contradicted by the other tree, or only those which are absent but are resolved compatibly (i.e. if a possible resolution of a polytomy in the other tree).

6.7   TREE DISTANCES

A careful comparison between trees must rely on measures of the differences among the trees. This is an important aspect of any serious comparative study, and must be appropriately quantified. Trees being complex structures, there are different

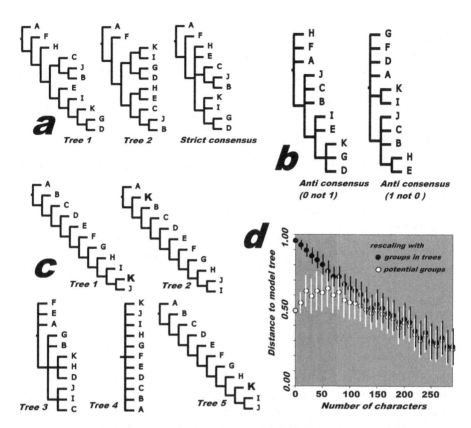

FIGURE 6.5 (a) Two trees and their strict consensus. (b) The "anticonsensuses" of trees in (a), showing the groups of tree 0 not present in tree 1, and the groups of tree 1 not present in tree 0. (c) Five trees, used to exemplify measures of tree similarity in Table 6.3. (d) Distance to model tree, for simulated datasets with increasing number of characters, as measured by a Robinson-Foulds distance rescaled by the number of potential groups, or by the number of groups actually present in the input trees. The measure rescaled with number of potential groups favors unresolved trees (so that maximum distance to a bush is 0.5, not 1.0), and has the absurd implication that datasets with fewer characters produce better estimates of the model tree.

components to their similarity. Depending on the contexts, one or other type of comparison may be desirable; they all reflect different aspects of the "similarity" among trees. One could consider that trees are similar when they differ only in the positions of few taxa; alternatively, one could consider that similar trees are those that share many groups, or nestings. Different measures of tree distance that have been proposed better reflect one or another aspect of the similarity, or establish a tradeoff such that one dominates over the others. Measures of tree distance have rarely been discussed or compared; most papers just discuss one or a few. St John (2017) discussed several measures of tree distance in the context of phylogenetic searches, and the analysis of Smith (2020) is very useful. Another exception is Kuhner and

TABLE 6.2

Comparison of Different Measures of Tree Distance

	RF	RF'	GrpSim	DC	DC'	SPRdist	SPR scaled	NNIdist	Triplets
Detects whether absence of a group is due to one or many taxa moving	No	No	Yes	Yes	Yes	Yes	Yes	Yes	Yes
In single taxon moves, reflects distance of the move	Too much	Too much	OK	OK	OK	Too little	OK	OK	OK
Distance to almost unresolved trees	≈0.50	Large	Large	Large	Large	Min(max)	Min(max)	Min(max)	Large
Calculation	Very fast	Very fast	Fast	Very fast	Very fast	Slow (NP-complete)	Slow (NP-complete)	Slow (NP-complete)	Slow
Symmetric	Yes	Yes	Yes	No	Yes	Yes	Yes	Yes	Only if both trees binary
Obeys triangle inequality	Yes	Yes	No	No	No	Yes	No	Yes	Yes

Yamato's (2015) paper, but their analysis focused mostly on whether distances accurately track expected deviations from a model tree, by generating data under different conditions. The discussion that follows focuses more on intrinsic properties of the most common measures. A general comparison of the properties of different measures of tree distance is given in Table 6.2.

6.7.1 ROBINSON-FOULDS DISTANCES (RF), AND DERIVATIVES

This is one of the most widely used measures for comparing trees, proposed by Robinson and Foulds (1981). Two advantages of the RF are its simplicity and ease of interpretation, and that it is very easy to calculate (in time linear with number of taxa; it can be approximated even faster, Pattengale et al. 2007). For two trees A and B, the absolute RF, or RF_{abs}, is based on simply summing the number of groups present in tree A but not tree B, plus the number of groups present in tree B but not tree A. Note that this is the number of groups in both anticonsensuses. The sum of groups in both directions makes the value necessarily symmetric (the measure is also known as "symmetric distance"). The method counts simply the number of groups that are not exactly recovered, without considering whether the group is absent due to a single misplaced taxon or to more substantive differences. It then shares the behavior of the strict consensus, possibly producing large distances for trees differing only in the position of a single taxon, if the taxon jumps between two distant locations (as in trees 1 and 2 in Figure 6.5c; see summary of rescaled RF distances in Table 6.3). When both trees are fully resolved, RF_{abs} is a function of the number N of nodes in their consensus, $RF_{abs} = 2 \times (T - 2) - N$ (where T is the number of taxa).

TABLE 6.3

Comparison Between Trees in Figures 6.5a and 6.5c, with Different Measures of Similarity

Comparison	RF	RF′	GrpDist	DC$_{ab}$	DC$_{ba}$	DC′	SPRmov	SPRdist′	Trip$_{ab}$	Trip$_{ba}$
Fig. 6.5a	0.375	0.375	0.289	0.158	0.188	0.172	3	0.375	0.275	0.275
Fig. 6.5c = 1/2	1.000	1.000	0.535	0.320	0.320	0.320	1	0.125	0.300	0.300
Fig. 6.5c = 1/3	0.687	1.000	1.000	0.920	0.857	0.906	4(4)	0.500(0.500)	0.617	0.558
Fig. 6.5c = 1/4	0.500	1.000	1.000	1.000	—	1.000	0(8)	0.000(1.000)	1.000	—
Fig. 6.5c = 1/5	0.125	0.125	0.111	0.004	0.004	0.004	1	0.125	0.008	0.008

Note: In the case of *SPR*-based measures, the values given in parentheses are those counting moves to resolve polytomies. The triplet distance is the proportion of triplets in one tree that are not resolved in the same way in the other (not counting triplets involving outgroup, which is fixed). *RF* is rescaled with $2T - 6$ (maximum possible), and *RF′* is rescaled with number of observed groups (depending on the trees, not on number of taxa).

As RF_{abs} distance is unbounded, it is common to rescale it. The maximum number of groups that can differ among two binary trees is $2T - 4$ (or $2T - 6$ if both trees rooted on the same taxon and root node is not counted), and thus the rescaled distance is $RF = RF_{abs} / (2T - 4)$. In Figure 6.5a, $RF_{1,2} = 6 / 16 = 0.375$. Note that for trees 1 and 2 of Figure 6.5c, $RF_{1,2} = 16 / 16 = 1.000$. Those trees are in fact rather similar, because they differ just in the position of taxon K, but the *RF* distance (scaled or not) fails to detect that similarity.

Another peculiarity of the *RF* distance scaled by maximum number of possible groups is that it produces intermediate distances to trees with many polytomies. Consider the comparison between trees 1 and 3 of Figure 6.5c—tree 3 has few groups that could be missed by tree 1, but the rescaling is still done relative to the maximum number of possible groups. Trees 1 and 3 are then considered as closer to each other (at distance 0.687, see Table 6.3) than trees 1 and 2 (at distance 1.000), but tree 2 is in fact more similar to tree 1 (differing in a single rearrangement) than is tree 3 (differing in many). In the extreme case of a fully unresolved tree, the rescaled *RF* distance has a minimum value of 0.500 (as in the comparison of trees 1 and 4 of Figure 6.5c; see Table 6.3). Depending on the context, this may or may not be desirable. For example, it is common to compare methods of phylogenetic analysis by simulating data from a given model tree, inferring trees from those data with different methods, and then checking which method produces the trees closest to the model tree. A common measure to use in such analyses is the *RF* distance, as done by Puttick et al. (2018). Puttick et al. (2018), however, generated datasets with much homoplasy (as much homoplasy as in random data), and therefore the use of the rescaled *RF* distance leads to preferring Bayesian analysis simply because it is (at least with the options used by Puttick et al. 2018 for parsimony analyses) the most conservative method—always produces a full polytomy, thus never exceeding distances of 0.500 to model tree. Smith (2019) noted this, and proposed to rescale the *RF* distance dividing by the number of available groups in the trees (instead of the number of groups the trees would have had if fully resolved). When this is done, the distances to model trees in Puttick et al.'s (2018) simulations are similar in both Bayesian and parsimony analysis (see Chapter 4). The *RF* as rescaled by Puttick et al. (2018), in penalizing polytomies too mildly, has peculiar implications when used to measure distance of poorly resolved conclusions to model trees. Consider the example of Figure 6.5d, where random trees (with all branch lengths such that $P_{(stasis)}$ along any branch equals 0.80) of 50 taxa were used to generate datasets with different numbers of characters; for each number of characters, 500 matrices were generated, each analyzed with 10 *RAS+TBR* (saving up to 5 trees/replication and calculating the consensus with *TBR* collapsing). The standard rescaling of the *RF* distance (with maximum number of possible groups, white circles in Figure 6.5d) implies that as one begins adding characters to a dataset, the quality of the inference *decreases*—the *RF* distance increases, from about 0.50, to about 0.60–0.65. According to the standard rescaling of the *RF*, we can have more confidence on the results for a matrix with 10 characters, than on the results for a matrix with 60–70 characters, and only when adding characters beyond the first 70 or 80 the inference begins to improve. Rescaling instead with number of groups actually present in both trees as proposed by Smith (2019), producing distance *RF'*, gives a more meaningful evaluation: as characters are added to the dataset, the average quality of the inference always improves (black circles in Figure 6.5d).

Other modifications of the *RF* distance can be useful in specific circumstances. For example, a distance that increases only with groups that are contradicted (instead of just not recovered) can be useful to measure conflict between consensus trees derived from different very conservative genes. Such genes are likely to produce poorly resolved consensus trees, so failure to recover a group is of no special meaning; finding contradictory groups is. An *RF* measure for contradicted groups only can be easily implemented with TNT scripts, taking advantage of the fact that the tcomp command allows saving the anticonsensus of contradicted groups only, and commands or scripting expressions can count numbers of nodes (see Sections 6.6 and 6.8.3, and Chapter 10).

6.7.2 GROUP SIMILARITY (ROUGH RECOVERY)

A useful modification of the *RF* distance is based on the degree of similarity of groups, as discussed under "Rough recovery consensus". Rather than considering groups as a 0/1 variable (i.e. absent or present), this method finds the closest group in each tree, and calculates the similarity among the groups (between 0–1). A related (but not identical) approach was proposed by Böcker et al. (2013). The main difference with the traditional *RF* is that just one or a few taxa moving to distant positions increases the perceived distance between the trees only mildly. For example, trees 1 and 2 of Figure 6.5c were more different than trees 1 and 3 according to *RF*, but according to this measure of group recovery, they are more similar. Just one taxon being subtracted from (or added to) a group changes the group only slightly; tree 1 displays a group {FGHIJK}; the closest group in tree 2 is {FGHIJ}, which is not identical, but is not too different in composition. By calculating the recovery of groups in both directions (tree A onto B, and B onto A) the measure is made symmetric, in the same way as the *RF* distance. In TNT, this measure is normalized always by the number of groups present in the trees (thus, analogously to *RF'*).

6.7.3 Rearrangement Distances

Another way to quantify the differences between trees is by using the (minimum) number of branch moves that need to be done to convert one tree into the other. Depending on the type of rearrangements to be done, this defines *NNI-*, *SPR-*, and *TBR*-distances. Calculation of the minimum number of moves between two trees (for any of the three kinds of rearrangements) is an NP-complete problem (DasGupta 1997; Bordewich and Semple 2005). Early proposals to use number of *NNI* rearrangements as a measure of similarity were made by Robinson (1971) and Waterman and Smith (1978).

The number of *NNI* moves between two binary trees is related to the *RF* distance: it cannot be lower than their distance RF_{abs} / 2. No widely available phylogeny program, to our knowledge, currently implements calculation of *NNI*-distances (some R packages do, like Smith et al.'s 2021 "TreeDist"). Another widely used measure is the number of *SPR* moves. This has been used especially in the context of evaluation of horizontal gene transfers (e.g. Maddison 1997; Nakhleh et al. 2005; Baroni et al. 2006; Beiko and Hamilton 2006). Goloboff (2007) proposed heuristics for calculating *SPR*-distances that outperformed then existing implementations. Goloboff's (2007) algorithms have been outperformed by faster methods subsequently developed by Whidden and Zeh (2009) and Whidden et al. (2010), but their being implemented in TNT facilitates their use. The method used in TNT explores different paths, and increasing the number of paths (i.e. replications) increases its precision; it produces reliable results when the distances between trees are relatively small, although it fails more often (overestimating distance) when the distances are very large. The *SPR*-distance does not suffer from the effect of single taxon long-distance moves as the *RF* distance, but then it has the opposite behavior—a single taxon move to a distant position is just one move, exactly like a single taxon move to a close position. For example, according to the *SPR*-distance, trees 2 and 5 are equally distant from tree 1 in Figure 6.5c, but tree 5 differs in a shorter move. As a remedy, Goloboff (2007) proposed to "weight" the moves by distance (although calculation is in that case more error-prone, requiring more intensive heuristics).

Bruen and Bryant (2008) showed that, for nonadditive characters, the *MPT* is the tree that minimizes the sum of *SPR*-distances to the trees that represent each of the characters without homoplasy, thus being a sort of "consensus". They do note that constructing consensuses (or supertrees) by minimizing sums of *SPR*-distances to the input trees is a different problem, computationally too difficult at that time. With faster algorithms than available in 2008, Whidden et al. (2014) proposed to construct supertrees median on the *SPR*-distance. Goloboff and Szumik (2016) showed that those *SPR*-supertrees can display undesirable groupings (e.g. groups in a minority of the trees, or contradicted by each of the input trees). This reinforces the idea that a measure being meaningful to indicate degree of similarity between trees does not automatically mean that trees median on that measure are also meaningful consensuses.

6.7.4 Distortion Coefficient (*DC*)

In one of the earliest proposals to measure the similarity among trees, Farris (1973) proposed to represent one of the trees as a matrix of binary variables, counting the number of extra steps h_i ("homoplasy") needed to map the "character" i onto the other tree. For each one of the n variables, the maximum possible number of extra steps on any tree is $hmax_i$ (see Chapter 3, Section 3.10.3 for how to calculate minima and maxima). In Farris' (1973) original proposal, the *distortion coefficient* was defined as $DC = (\sum h_i / hmax_i) / n$—the average of the ratios between observed and maximum possible homoplasy, for all variables. As with much of Farris' early work, this proposal has been given much less recognition than it should, by subsequent workers. The distortion coefficient has some advantages over other methods. It is bounded between 0 and 1, and it can be calculated very quickly (with just two passes over the trees). A move of a single terminal to a different place in the tree will add a step to the variable representing each of the groups to which the terminal belonged, but a single step—still far from *hmax*. Thus, single taxon moves increase the perceived difference more when they are more distant (behaving better than *SPR*-distance in this regard), but they can never make the distance between trees maximum (thus behaving better than the *RF* in this regard).

The implementation of the distortion coefficient in TNT uses the modification proposed by Farris (1989) with the *retention index*. Define $H = \sum hi$ and $Hmax = \sum hmax$. Then, the retention index is $RI = 1 - H / Hmax$ (i.e. ratio of sums instead of average ratio). Table 6.3 reports as *DC* values the complement of the *RI*. The most serious problem with the retention or distortion coefficients, at least as originally proposed, is that they are not symmetric measures. Mapping tree A onto the matrix representing tree B may have a different number of steps than mapping tree B onto the matrix representing tree A, even when the trees are fully resolved. In addition to the possibly different numbers of steps, the values of *Hmax* can be different, changing the normalizing factor used in the two directions. For some kinds of comparisons, this may not be a problem. For example, the question asked can be how much of the information in one tree the other tree fails to show. For that question, the asymmetry may be acceptable and the use of a metric is then not mandatory, as noted by Smith (2020: 5009): an unresolved tree displays none of the information of a binary tree, but the unresolved tree has no information that the binary tree could fail to display. However, if the measure is to be used to indicate the similarity between the trees, it must be symmetric. The measure can be modified (Goloboff et al. 2017) so that the reciprocal numbers of H and $Hmax$ are used. For that, define $H_{A,B}$ as the homoplasy produced by mapping the matrix representing tree A onto tree B, and $Hmax_A$ as the maximum homoplasy any tree could require for the matrix representing tree A. The symmetric distortion coefficient, DC', between trees A and B is then $DC' = (H_{A,B} + H_{B,A}) / (Hmax_A + Hmax_B)$. Although it fulfills the condition of symmetry, this modified DC' differs from most measures discussed in this chapter in not obeying the triangle inequality (because of the influence of *Hmax*).

6.7.5 TRIPLETS AND QUARTETS

Counting the proportion of taxon quartets (in unrooted trees) or triplets (in rooted trees) that are identically resolved in both trees provides a measure of similarity among trees that generally agrees with intuitive expectation. Estabrook et al. (1985) made the first proposal to use quartets in this manner. The number of possible quartets is larger than the number of possible triplets (i.e. for T taxa, there are $(T - 3) / 4$ times more quartets than triplets), and thus triplets are preferable in the case of rooted trees. The proportion of triplets in a tree recovered by the other responds well to moves of a single taxon that make many groups non-monophyletic (decreasing perceived difference only slightly), and to the distance of the moves (with longer moves preventing recovery of more triplets, thus costing more than shorter moves). When both trees are binary, the measure is symmetric (number of triplets of tree A recovered by tree B is always the same as number of triplets of tree B recovered by tree A), but this changes when one or both trees are not fully resolved. There are two potential problems with proportion of triplets recovered as a measure of tree similarity. The first is that the size of groups moved may influence the measure of similarity, even for otherwise identical moves (as they affect different proportions of triplets). The second is that the counts of triplets are not wholly independent, as the resolution of some of the triplets necessarily implies specific resolutions for others. Both of these problems had been pointed out for the use of triplets in analyzing character data (e.g. by De Laet and Smets 1998), and they also occur when using triplets to compare tree shapes.

6.8 IMPLEMENTATION IN TNT

As it should be obvious by now, even for the analysis of a single dataset, a phylogenetic study may well require much more than just assembling the matrix and finding the *MPT*s. Summarizing results in the face of ambiguity can also be a significant challenge and require a careful choice of options. Most of the methods described in this chapter are implemented in TNT; a summary of the relevant commands is in Table 6.4.

6.8.1 CONSENSUS TREES

TNT can calculate strict consensus trees, with the `nelsen` command (the name of the command is taken from Hennig86, a play on words with "Nelson" and "consensus"). The command can be followed by the list of trees to consense, and the taxa to exclude, separated by a slash. Default is to use all trees, excluding no taxa (but see `unshared` command, Section 6.8.2). The taxa to exclude are specified by the user (e.g. previously identified with commands for detection of unstable taxa). The input trees themselves remain unmodified; the positions of the excluded taxa are ignored only temporarily, during consensus calculation. If the list of taxa to exclude is followed by another slash, the possible locations (among different input trees) of the taxa listed after the second slash are shown on the strict consensus of the remaining taxa, with letter codes. This is very useful, because it allows approximating the topologies of all the input trees from a single tree diagram. The consensus can be displayed (by default), or saved as last memory tree (`nelsen*`). In the latter case a tree-group named `strict_consensus` is created automatically. Saving the consensus to the tree buffer allows subsequent counting of number of nodes, diagnosis, saving to a file, and so on.

The same syntax is used for the combinable component (`comcomp`), majority rule (`majority`), and frequency differences consensus (`freqdifs`), saving the tree

TABLE 6.4
List of Commands for Consensus and Tree Comparisons

Command	Minimum truncation	Action(s)
chkmoves	chk	Effect branch-swapping (with current settings of suboptimality) and report moves
collapse	col	Set rules for collapsing zero-length branches
comcomp	com	Combinable component consensus (=Bremer, semi-strict, or loose consensus)
condense	con	Eliminate zero-length branches from trees in memory
freqdifs	fr	Calculate consensus of frequency differences for trees in memory
majority	m	Calculate majority rule consensus for trees in memory
matchtax	mat	Establish possible taxon matchings in trees with different taxon sets
mixtrees	mix	Calculate semi-strict supertree from trees in memory
mrp	mr	Create *MRP* for trees currently in memory (replacing current matrix)
nelsen	ne	Calculate strict consensus from trees in memory
pcrprune	pc	Apply *IterPCR* (Pol and Escapa 2009) to trees in memory
pruncom	pru	Improve resolution of combinable component consensus, by brute-force methods
prunmajor	prunm	Find taxa to prune for improving resolution of majority or frequency difference consensus
prunnelsen	prunn	Improve resolution of strict consensus tree, by brute-force methods
pruntax	prunt	Prune taxa from trees
prupdn	prup	Based in a list of potential taxon prunings, combinatorially find the prunings that maximize number of nodes or group frequencies
resols	res	Show effectively different resolutions of each polytomy in the strict consensus
rfreqs	rf	Frequencies of rough recovery (similar to majority rule consensus, but unaffected by few taxa moving among distant locations)
shpcomp	shp	Compares shapes of unlabeled trees
sprdiff	sp	Calculate *SPR*-distances between trees, using a heuristic
taxonomy	taxo	Define and test recovery of taxonomic groups
tcomp	tco	Compare trees, using several possible measures of distance; create anticonsensus of two trees
tequal	te	Compare trees, report identical ones

produced to the tree buffer with *, or displaying it by default, and specifying the trees and taxa to exclude (separated by a slash). The option to show the location of pruned taxa on the tree is not available for any type of consensus other than strict. The majority and freqdifs commands can also be used to show the frequencies (or frequency differences) for the groups in a reference tree, even if 0.50 or less in the case of majority, or 0 or less in the case of freqdifs. To do this, specify the number of the reference tree in square brackets. For example,

```
majority [0] 1. / mouse lion;
```

will calculate the frequencies of the groups in tree 0, on the remaining trees, without considering the position of taxa mouse and lion.

All these types of consensus and options are also available in Windows, from *Trees > Consensus*. The frequency for the groups in a reference tree is calculated with *Trees > Comparisons > Do group frequencies*.

As discussed in the chapter on optimization (Chapter 3), consensus trees should not be optimized as such—they can be longer than the trees used to produce them. For diagnosis of groups, or studies of character evolution, it is preferable to reconstruct characters on the individual trees, then map the synapomorphies (or character-state super-sets) on the corresponding nodes of the consensus. This can be done as options of the `apo` or `map` commands, or with the "Common" options for *Optimize > Synapomorphies and Optimize > Characters*.

6.8.2 TEMPORARY COLLAPSING OF ZERO-LENGTH BRANCHES, UNSHARED TAXA

By default, the trees found by search algorithms of TNT are retained as binary. This is so because collapsing with "Rule 1" (i.e. retain a branch only if every possible reconstruction has changes along the branch) may increase the length of the tree. Having tree 4 of Figure 6.3c tells us that groups BC or DE can both be absent from trees with a total length of 4; tree 8 (with total length 5) does not provide such information. As the trees produced contain arbitrarily resolved branches, those branches need to be eliminated temporarily for consensus calculation. The same rules in effect for the search are applied during consensus calculations. As in the case of taxon exclusion, the input trees themselves are unmodified during consensus calculation; only their zero-length branches are ignored. The calculation of zero-length branches uses the data (character states, activities, and additivities or transformation costs) at the time of the consensus calculation. This can be changed with the `collapse` command: `collapse temp` (the default) uses the temporary collapsing. With `collapse notemp`, every tree in the tree buffer is consensed exactly as is (polytomous or not). The latter is necessary for consensing trees from two different datasets. The trees should in that case be collapsed before reading into memory, either by consensing in every case (with temporary collapsing from its own dataset) or by physically eliminating zero-length branches according to the criterion in effect. The trees are physically collapsed (i.e. losing track of the original resolution) with the `condense` command (with no arguments, it condenses all trees; otherwise, it condenses the trees specified in the list). Keep in mind that if the input trees have been collapsed, then they *must not* be collapsed temporarily again for further consensing and comparing.

All the operations of temporary or permanent collapsing are done under the rules in effect. The rules for deciding retention or elimination are set with the `collapse` command. In order of increasing strength:

```
collapse none ;
collapse rule 3 ;
collapse amb ;
collapse rule 1 ;
```

```
collapse rule 4 ;
collapse spr ;
collapse tbr ;
```

The numbering of the "rules" follows Swofford (1993). Rule 4 or stronger cannot be applied during tree searches; if those rules are in effect, then the trees retained are those distinct under Rule 1 (the next strongest available option). In Windows, these rules can be changed from *Settings > Collapsing rules*, and the temporary collapsing is changed from *Settings > Consense options*.

Note that consensing with temporary collapsing and elimination of some taxa is *not* equivalent to first condensing the trees and then consensing with the same taxa removed, under collapsing Rule 1 or stronger. This is obvious in *SPR* or *TBR* collapsing, but less so with Rule 1. The behavior with Rule 1 can be exemplified by the dataset of Figure 6.2b. The reader can verify the differences by running some examples. The dataset in Figure 6.2b produces 13 binary *MPT*s; 11 of those are distinct under collapsing with Rule 1. Consider the binary tree shown in Figure 6.2b (one of the *MPT*s); it has two branches of minimum length zero (marked with gray), below and above X. This is because X has only missing entries; any synapomorphy of EFGH could as well be a synapomorphy of XEFGH on that tree. Thus, collapsing with Rule 1 eliminates the branches below and above X, producing the collapsed tree shown in Figure 6.2b. For each of the 13 binary *MPT*s collapsed in this way, X becomes part of a larger polytomy, and subsequent pruning of X from the collapsed trees does not improve the resolution. To prevent that, in the temporary collapsing for consensus calculation, the collapsing is effected as if separately identifying every individual zero-length branch and converting it into a trichotomy, then pulling the taxa to prune from that tree. For the case of Figure 6.2b, only this produces the correct pruned consensus in conjunction with temporary collapsing. If the trees need to be condensed before (e.g. to calculate the consensus without temporary collapsing), the list of taxa to be pruned must be passed to the condense command so that the resulting trees can be used directly (i.e. without temporary collapsing, without further removal of taxa) for the consensus calculation. Thus, for a case like that in Figure 6.2b, the following options produce the appropriate resolution for the pruned consensus:

```
mult ; bbreak ;          mult ; bbreak ;
collapse temp ;          condense / X ; collapse notemp ;
nelsen / X ;             nelsen ;
/* this is the           /* this is OK; X is eliminated when
default */               condensing */
```

These combinations (could) produce incorrect results:

```
mult ; bbreak ;          mult ; bbreak ;
condense ;               condense ;
collapse notemp ;        collapse temp ;
nelsen / X ;             nelsen / X ;
/* this is wrong, as     /* in this case, trees
```

```
post-pruning of X does      that had been collapsed are
not resolve polytomies      collapsed again—this
created by collapsing */     can lead to errors! */
```

Of course, the same thing could be done with less precautions if Rule 1 is not being applied:

```
collapse none ;
mult ; bbreak ;
nelsen / X ;
/* this is OK, as there is no removal of zero-length branches,
temporary or otherwise */
```

The difficulty in the last case is that the number of equally parsimonious trees can be very large and make the tree buffer unwieldy—but the results are, other than that, correct.

Another command that changes settings for consensus calculations is unshared. By default, the taxa that are not shared by all trees are excluded from the consensus. To include them, use unshared-. Note that including them is not equivalent to creating supertrees (for which, see Section 6.8.5). If a taxon included in one tree violates the monophyly of some group(s) in other tree which excludes the taxon, the group is polytomized in the result. As example, consider the trees (A(B((CD)(EF))) and (A(B((CD)(E(FX))))). With the default, unshared=, the strict consensus is (A(B((CD)(EF))). With the unshared- option it is instead (and given that all consensus calculations in TNT are rooted) the partially unresolved (XABEF(CD)).

6.8.3 Tree Comparisons and Manipulations

The anticonsensus can be calculated with the tcomp command. Tcomp 0 1 displays the groups of tree 0 that are not present in tree 1, and tcomp 1 0 shows the opposite. This is accessed, in Windows, from *Trees > Comparisons > Compare Groups*. By default, all groups are displayed; to display only groups that are contradictory, use tcomp]; to display only groups that are compatible, use tcomp[. As in the case of the consensus, using * as argument saves a tree with all those groups to tree buffer, and a list of taxa whose positions are to be ignored (many otherwise different groups can be made identical by ignoring the positions of some taxa) can be given preceded by a slash, so that the command sequence:

```
tcomp] * 0 1 / mouse lion;
```

saves to tree buffer the groups of tree 0 that contradict groups of tree 1, ignoring the position of mouse and lion.

The tcomp command can also be used to map node numbers between trees (node numbering in different trees is arbitrary, so that the same node number in two different trees can correspond to a different group, and vice versa). To establish the equivalence, use tcomp& M N, which plots tree N, showing the node number of the same group in tree M (if none=X). A list of taxa whose positions must be ignored to establish group equivalences can follow the second tree (preceded, as always, by a slash).

A useful feature offered in TNT is displaying the resolutions effectively contained in the tree buffer (which can be much less than the possible resolutions) for a polytomy in the consensus. Together with pruned strict consensuses where possible location of pruned taxa is indicated, this option allows giving more detailed information of the possible resolutions and input trees, using a compact format. From the menus, this option is accessed with *Trees > Comparisons > Show Resolutions*. The command is `resols`, and like the commands for consensus trees, this option can be given a list of taxa whose positions are to be ignored (which often decreases the number of possible resolutions). Temporary collapsing cannot be used for this command (if in effect, it is disconnected; beware that it is not connected again after execution of `resols`). Resols calculates and displays the strict consensus and reports the resolution for each polytomy separately; to specify the display of the resolutions of a single polytomy, give the number of polytomy within square brackets, as first argument. If finding the resolutions of a single polytomy in the consensus (N), the groups of trees having each resolution can be saved to (or removed from) groups of trees, for subsequent manipulation, with

```
resols [N]/>G;
```

where, the first resolution is saved to tree-group G and subsequent resolutions to successive groups. Using < instead of >, it removes the trees from groups. For example, the largest polytomy (a 7-chotomy) in the consensus of all distinct trees for the dataset example.tnt (the dataset of Goloboff 1995a, distributed with the TNT package, with 72 *MPTs*, if using the default, Rule 1) is node 104, with 6 possible resolutions. This, and the individual trees displaying every resolution, can be found out with the following sequence of commands:

```
mult; bbreak;                    /* search trees */
condense ; collapse notemp ;     /* collapse them and set tempo-
                                    rary collapsing off */
tgroup -. ;                      /* undefine all tree-groups */
resols [104] />0 ;               /* show resolutions, and add to
                                    tree- groups the trees dis-
                                    playing each resolution */
tgroup ;                         /* show the trees with each reso-
                                    lution */
```

This will create 6 groups, of 12 trees each (each group with a distinct resolution of the polytomy of node 104, named accordingly). Note that 6 resolutions is only a small fraction of the 10,395 resolutions possible for a polytomy of degree 7.

TNT also allows several forms of manipulation of the tree buffer, which can facilitate comparisons and preparation of tree files. In Windows, several options are under *Trees > Tree Buffer*. The command `tequal` (*Trees > Tree Buffer > Compare Trees*) shows duplicate tree topologies. The trees in the tree buffer can be examined (discarding duplicates) and compared with the commands `best` and `unique` (*Trees > Tree Buffer > Filter*). With no arguments, `best` eliminates any suboptimal trees; followed by a score value S, best eliminates the trees with a score difference beyond S from the best tree. With best] or best[, it discards the trees fulfilling or failing the currently defined constraints (defined with the force command). Unique simply checks duplicate topologies (with no further filters). Adding * as an argument, both best and unique will collapse the trees (according to presently enforced rules) before checking for duplicates. The condense command (described previously) collapses the tree; the same option is available under *Trees > Tree Buffer > Condense trees*. The number of nodes of the trees in memory is calculated with tnodes (or *Trees > Describe > Number of nodes*); for example to find out the exact resolution of a strict consensus you can use nelsen*;tnodes {strict} (i.e. taking advantage of the fact that nelsen* creates a tree-group with the tree it saves). The pruntax command (*Trees > Tree Buffer > Prune taxa*) removes from the trees listed the taxa specified (after a slash; use ! to just remove all inactive taxa). If the tree-list is replaced by *, the command removes the terminals specified from tag-tree (stored with ttag and tree-displaying commands; see Chapter 1), or collapses nodes with a specific label (using =label instead of a taxon list). The elimination of nodes from the tag-tree can also be done with the condense command (see help condense for details).

6.8.4   IDENTIFYING UNSTABLE TAXA

The identification of the taxa decreasing resolution of the consensus, or decreasing group supports, can be difficult in large datasets, or when there are numerous trees or unstable taxa. Most of the methods described in the section on theory are implemented in TNT. The unstable taxa can be displayed graphically—so that the user can manually specify the taxa to exclude from subsequent consensus calculations—or saved to taxon groups—so that the specification of taxa to exclude can be more easily automated. For saving lists of prunings to taxon groups, use >G after the list of taxa to exclude (or after the slash, if no list given for taxon exclusion); this adds the list of prunings to group G (use <G to remove taxa from group). All commands to identify wildcards take lists of trees to use, as well as lists of taxa to exclude from the outset (preceded by a forward slash, /). Most commands can also take a list of "untouchable" taxa (preceded by a backslash, \), that is, those taxa that *cannot* be removed (e.g. the taxon whose location in the tree the study is trying to determine!).

The prunnelsen command implements a brute-force method, which works well for simple cases. Prunnelsen calculates the strict consensus, displays it showing node numbers, and then shows all the prunings that improve each polytomy, and its effect on the polytomy. To improve a specific polytomy, specify the number in square brackets (keep in mind that node numbering refers to a strict consensus calculated with the same trees and no temporary collapsing). The minimum gain in nodes for a pruning to be accepted can be given as first argument (with >N), and the maximum number of taxa to cut simultaneously is given with =C:

```
prunnelsen [20]=4 >5 / lion \ mouse >0;
```

the example will find the taxa to prune (up to 4 simultaneously) to improve 5 or more nodes in the polytomy of node 20, for the consensus of trees excluding lion, but making sure that mouse is never excluded, and save the list to taxon group 0 (the second > is recognized because it is after the lists of taxa). Even if no more than 4 cuts are tried simultaneously, taxon group 0 can of course contain more than 4 taxa—for example if different subsets of taxa improve a polytomy beyond the 5 nodes minimum. Additionally, every taxon (or taxon set) which improves the consensus beyond 5 nodes is saved to the group, but pruning all simultaneously may not improve the tree (e.g. if the prunings are mutually exclusive). As in the case of other heuristics to approximate the initial pruning sets, the taxon groups saved can be subsequently refined with the `prupdn` command (see Section 6.8.4). This can be easily applied to Goloboff's (1995a) spider dataset (example.tnt, included in the TNT package) analyzed under equal weights, showing the possible locations of pruned taxa on a strict consensus:

```
p example.tnt; mult; bb; agr-.; prunn/>0; nels//{0};
```

producing results similar to those reported by Goloboff (1995a)—not identical, because that paper reported results under implied weighting with an earlier program. Note that before calling `prunnelsen`, the instructions given make sure that all taxon groups are undefined (`agr-.`). In Windows, most of the functionality of the `prunnelsen` command can be accessed from *Trees > Comparisons > PrunedTrees*. The command `pruncom` does the same as `prunnelsen`, except that it attempts to improve the resolution of the combinable component instead of the strict consensus.

The command options `tcomp!` and `tcomp!!` calculate, respectively, the *PC* and *LSI* indices (see Section 6.7.5). They can take a list of trees and (separated by a forward slash) a list of taxa whose position should be ignored. The output is a table where the *PC* or *LSI* value of every terminal is displayed. These values can be used to select taxa which, if pruned, will probably improve the tree. These calculations require checking all possible triplets, and therefore can be time-consuming. For *PC* values, when there are numerous trees and they are rather similar, using `tcomp!&` speeds up calculations by deducing the results for some triplets from previous ones (e.g. if all trees resolve both A(BC) and B(CD), then it follows that all trees resolve A(CD) and A(BD) as well).

The command `pcrprune` integrates the use of *PC* values in *IterPCR*. Based on *PCR* values, the program attempts prunings, and reports those that effectively improve the tree. The criterion for improvement is similar to that of `prunnelsen`. Specification of an individual node to improve, trees, taxa, untouchables, taxon group for saving results, minimum number of nodes to gain for a pruning to be acceptable, and maximum number of nodes to cut simultaneously, are specified as in `prunnelsen`. As in the `tcomp!` option, specifying & will infer triplets in advance, possibly speeding up calculations. For example, running

```
p example.tnt; mult; bb; agr-.; pcrp/>0; nels//{0};
```

should produce results similar to those for the `prunnelsen` command. When there are big polytomies in the consensus `pcrprune` can be faster than `prunnelsen`.

Another difference is that `pcrprune` allows improving majority rule consensus trees, using *LSI* values instead of *PC*. For this, use `pcrprune (majority F)`, where F is the frequency of resolution of a triplet above which it is considered as "compatible". In this case, the criterion for improvement is whether the sum of frequencies of the resulting majority rule tree is larger than in the entire tree, assigning a "weight" (with `pcrpprune:W`) to taxon remotions, so as to make the method more or less prone to accept prunings. Most of the options described for the `pcrprune` command can be accessed also from *Trees > Comparisons > IterPCR*.

The command `prunmajor` offers another heuristic to improve frequencies or frequency differences in majority rule trees. In this case, the program calculates the expected influence on the consensus, providing a list of taxa expected to influence resolution the most. Instead of displaying just best taxa, you can display the best N values, with `prunmajor =N`. The values are shown with an arbitrary rescaling, giving taxa expected to influence resolution the most a score of 1.00. The method used is based on identifying closest groups (as in rough recovery methods, see Section 6.3.4). The maximum proportion of intruders/eliminations for a group to still be considered as recovered is set with `prunnmajor &P`. The weight given to incorrect taxon inclusion (`Wi`) or removal (`Wr`) is specified with `prunmajor [Wi Wr`. The scores of expected improvement can be shown separately for each node of the reference tree, with `prunmajor +`. To try to improve frequency differences instead of just frequencies, use `*` before the number of reference tree. This command requires specification of a reference tree, the groups of which are the target for improvement. Unlike other commands for improving consensus, the `prunmajor` command can use a list of nodes of the reference tree to be improved (if no list given, it checks improvement of all nodes in reference tree). The list of nodes is given as a list of numbers, following the number of reference tree. The list of input trees follows the reference tree (and optionally, node numbers), preceded by a slash (`/`), and can be followed by a list of taxa to exclude from the outset (with another `/`), as well as a list of untouchables (with `\`). The taxa in the list displayed (i.e. either the best, or the best N values) can be added to (or removed from) group G, indicating (after lists of trees, taxa, and untouchables) the option >G or <G. The `prunmajor` command is accessed with *Trees > Comparisons > Quick Pruning Heuristic*.

Finally, a different type of criterion is implemented in the `chkmoves` command, based on branch-swapping and recording acceptable moves (i.e. moves producing trees of equal score, or within a pre-specified value of suboptimality, with the `subopt` command). `Chkmoves` identifies potentially unstable taxa or groups on the basis of character data, thanks to the speed with which moves can be evaluated during *TBR* (see Chapter 5). The method records, for each terminal and group in a series of trees, the number of acceptable moves, their maximum distance, and the maximum distance to a changed rooting ("rerooting depth"). Acceptance of a move depends both on the absolute (*A*) and relative (*R*) value of suboptimality (set with `subopt AxR`). The results are output in the form of table, sorted by number of alternative locations where each taxon/group can move (default), by distance (`chkmoves[]`), or by rerooting depth (`chkmoves&`). The list of trees to process follows the specification of sorting criterion, and this can be followed by the minimum (M) number of moves/distance/rerooting depth to report, as `/M`. Using `\N` instead, the largest N values of the sorting criterion chosen are output. This can be optionally followed by specification of a taxon group G where to add the taxa shown in the list, as >G (or to remove from list, with <G). If constraints have been defined and enforced, they are taken into account during swapping. In Windows, this is done with *Trees > Swap and Report moves*. This option can be applied to the example. tnt dataset of Goloboff (1995a), producing results quite similar to those of prunnelsen or pcrprune:

```
p example.tnt; mult; bb; agr -.; chkmov[0\2>0; nel//{0};
```

All the commands discussed so far are alike in using a criterion of local optimality for identification of unstable taxa. With the prupdn command, the unstable taxa found by those methods can be saved to taxon groups, and a more global criterion used to refine the selection of taxa, as discussed in Section 6.3.6. The syntax for prupdn first takes the number of taxon group with the set of candidate prunings (input group), then the number of taxon group where to save the refined set (output group). This can be followed by several options:

- whether to use an "up" (begin with no prunings, and select taxon combinations from input group, adding best combinations to output group) or "down" refinement (begin with all taxa in input group pruned, and un-prune combinations of taxa, saving the best combinations to output group). The "down" option is the default, or indicated explicitly with prupdn >; the "up" option is indicated with prupdn <
- the tree whose groups must be improved (default is majority rule or frequency difference consensus), indicated as [N]
- improving frequency difference (*GC*) values, instead of majority rule consensus, indicated with *
- a factor F for taxon removal penalty, with : F (default is 0.5)
- maximum number of taxa N to simultaneously prune or un-prune (depending on whether the "up" or "down" options selected), with =N
- cutoff support value C (use 100 to improve strict consensus), with &C
- maximum number R of successive replications to do, with ! R. This is usually unnecessary because the prunings stabilize with the first few replications
- use branches of support S as separators, with | S

These options are followed by the list of trees, and the list of taxa whose positions in the tree are to be ignored (with list of trees and taxa separated by a slash, as usual). In the prupdn command there is no option to define untouchable taxa, for the same effect is achieved by eliminating those taxa from the taxon group with candidate prunings. In the Windows version, this command can be accessed from *Trees > Comparisons > Optimize Prunesets*; prior to running this option, you need to have filled the input group, either manually, or by running one of the previous heuristic methods, then ticking on *save prunes to group*. As an example, the chkmoves command can be used to generate a larger set of pruning candidates (saving 7 taxa to taxon group 0), then refining it with `prupdn` and saving into taxon group 1:

```
p example.tnt;mult;bb;     /* read data set and search trees */
agr-.; chkmov[0\3 >0;      /* save initial candidates to group 0 */
prupd 0 1 :1.5 > &100;     /* refine group 0, save into group 1 */
nelsen//{1};               /* display consensus */
```

The *MAST* can also be calculated in TNT. `Prunnelsen+` calculates the exact *MAST* (which can be time-consuming), and `prunnelsen!`, calculates a quick heuristic. The heuristic can be used only when trees are binary, and it may give solutions with too few taxa left in the trees. As in commands for consensus trees, using a * saves the *MAST*s to the tree buffer (note that the ! option produces a single pruning). The *MAST* prunes taxa until the input trees become identical, not until a more resolved consensus is produced, but it can nonetheless help find unstable taxa or do other types of analyses. The number of taxa in the *MAST*s is sometimes used as a measure of tree similarity; for this, save the *MAST*s to tree buffer and count the number of taxa, with `prunnelsen*+; tsize`. In Windows, *MAST*s can be calculated with *Trees > Comparisons > Agreement Subtrees*.

6.8.5 SUPERTREES

The only native method for supertrees implemented in TNT is the semi-strict supertrees of Goloboff and Pol (2002), with the `mixtrees` command. The list of trees to process can be given first, followed by a slash and the list of taxa whose positions are to be ignored (remember that reduced supertrees behave very differently from reduced consensus trees in this regard). Using * as first argument, the supertree is saved to tree buffer. In Windows, semi-strict supertrees can be calculated with *Trees > Supertrees*.

TNT also includes a command to create the *MRP* from the trees in memory, with the `mrp` command (*Trees > Tree Buffer > Create MRP*). This replaces the dataset held in memory with the matrix representing the trees; every character is named for a tree and group. The matrix can be analyzed with parsimony (to calculate the *MRP* supertree; recall this is the strict consensus of *MPT*s for the *MRP*), or with cliques (*MRC* supertree). For cliques, you need to first set implied weights on before creating the *MRP* matrix, and (after creating the *MRP*) define a user-weighting function (see Chapter 7) where any step of homoplasy beyond the first has zero cost (i.e. a clique analysis):

```
piwe=; mrp; piwe[1 0; mult; bb; nel;
```

Only when all the input trees have the same taxon sets can you analyze the matrix representing them with the `kleex` command. The `kleex` command uses J.S. Farris' (2003) implementation of Bron and Kerbosch's (1973) algorithm for clique enumeration, which is faster and more effective at finding cliques than search heuristics under implied weights with a 0/1 weighting function. This option can be used, for example, for the cases shown by Goloboff (2005) to illustrate differences between *MRC* and majority or frequency difference consensus. Assuming the trees are saved to a file input.tre:

```
keep 0; p input.tre; mrp; kleex; coll none; nel;
```

Note that (as done in the earlier example) either zero-length branch-collapsing, or temporary collapsing, must be turned off for consensing clique trees.

6.8.6 MEASURES OF TREE DISTANCE

Three different measures (with additional variants) for tree comparisons are implemented in TNT. *SPR*-distances are calculated with the `sprdiff` command; the command takes as arguments the two trees to compare, followed by the number of replicates x stratifications (a stratification first accepts only moves that can improve the score a given number of steps, then switching onto accepting moves saving fewer steps; see Goloboff 2007 for details). The same is done with *Trees > Comparisons > SPR-distances*. The default setting uses only 10 replicates and no stratification; this works well only for very similar trees. Increasing the number of replicates to at least 100, and stratifications to 2 or 3, usually produces paths that are optimal or very close. Using * as first argument (i.e. preceding trees to be compared), the moves in the shortest path found are shown on tree diagrams. By default, in the case of polytomies, the number of moves needed to travel between the two closest resolutions of the polytomies is used (i.e. the distance between a binary tree and a bush is 0). With `sprdiff:poly`, the moves needed to resolve each polytomy are counted as well (i.e. the distance between a binary tree of T taxa and a bush is $T - 3$, the maximum possible).

The other measures for tree comparisons implemented in TNT are variants of Robinson and Foulds' (1981) *RF* distance, and the (complement of the) distortion coefficient *DC* of Farris' (1973). All of them are calculated with the `tcomp` command. Tcomp can take two lists of trees, separated by a slash, and reports the distance between each tree in first set to each tree in second set (and the smallest overall distance). If no list of trees is given, `tcomp` compares every tree in the buffer against each other. The second list of trees can be followed by a slash and a list of taxa whose position is to be ignored for distance calculation.

The standard *RF* distance, rescaled by maximum possible groups and unrooted, is calculated with options `tcomp<`. The rooted version (i.e. rescaling with one more possible group) is `tcomp>`. For rescaling with number of groups actually present in input trees, use `tcomp<<` (unrooted) or `tcomp>>` (rooted). For *RF* with rough recovery of groups (where a few unstable taxa do not strongly increase distance; see Section 6.7.1), use `tcomp:` (this is always rooted, and rescaled with number of groups actually present in input trees; factors to calculate degree of recovery are set from the `prunmajor` command, see above). The standard *RF* distances can be invoked from the menus, with *Trees > Comparisons > RF distances* (the standard rooted, rescaled by maximum possible groups).

The distortion coefficient (with the modification described in Section 6.7.4, H / $Hmax$, where $H = \Sigma\ hi$ and $Hmax = \Sigma\ hmax$) is calculated with tcomp=. This is reported as the complement of H / $Hmax$, i.e. a measure of similarity. The same is done with *Trees > Comparisons > Dist Coefficient*. This is possibly asymmetric. The symmetric version, using reciprocal calculations $[(H_{A,B} + H_{B,A}) / (Hmax_A + Hmax_B)]$, is invoked with tcomp== (available only from commands).

The degree of recovery of groups in a reference taxonomy can also be calculated, with the taxonomy command (using the same criterion of rough recovery for group matching), or with *Trees > Taxonomy*. Although the identification of groups themselves is the same as with other commands or options, the comparisons with taxonomy have the advantage of using the actual taxonomic names to refer to groups.

7 Character weighting

This chapter shows that considerations about character weights are fundamental in phylogenetic analysis, particularly in the case of morphological datasets. Character weighting is one of the most important topics in phylogenetic analysis, but also very controversial and poorly understood by many authors. Whenever two alternative groupings are each supported by their own set of characters, the relative weights of the characters supporting each group become relevant. Depending on which characters are considered to be more influential, the conflict may be resolved in favor of one or the other grouping. Note that such conflict occurs even when there are no exact ties in the numbers of conflicting characters. If a single very reliable character contradicts two extremely unreliable ones, following the dictate of the most reliable character may be justified. This chapter argues that differential weighting is more important in morphological than in molecular datasets, and at the same time limited to fewer possible options (because it is harder to make generalizations across characters).

In the early days of phylogenetics, differential character weighting was considered an anathema, because defining appropriate criteria for reliability seemed impossible:

> to recognize a character is to recognize a group. Weighting of characters therefore cannot proceed independently of a hypothesis of grouping and hence of relationship, or, to put it in other words: characters weight themselves, with those forming useful groups (in congruence with other characters) carrying more information content than the others.

(Rieppel 1988: 61, paraphrasing Patterson 1982)

Part of Patterson's (1982) argument was that correctly recognized homologues would define correct groups ("On weighting, it is argued that the systematist has no role to play: homologies weight themselves"; Patterson 1982: 21). That of course begs the question of what criteria could be used to recognize true homologies when characters are in conflict, which (as discussed in Chapter 2) cannot be done unless the characters have been assigned the weights they truly deserve. So ingrained was the idea that all characters must be weighted equally, that authors who leaned toward evolutionary taxonomy and had tried to come up with general criteria for *a priori* weighting (Hecht and Edwards 1976, 1977; Szalay 1977; Bock 1977) were among the most reviled by cladists:

> Recently, however, the opinion has been expressed that some hypotheses of synapomorphy are less useful and should have less weight... (Hecht 1976; Hecht and Edwards 1976, 1977). This viewpoint has also gained some approval from evolutionary systematists (Szalay 1977). .... We merely reiterate the fundamental observation that all characters are evolutionary novelties (synapomorphies) at some level. The problem is to find that level for each character, and not to assume that "conservative" features are

more or less valuable than "variable" ones, or that "non-adaptive" features are better or worse than obviously "adaptive" features".

(Eldredge and Cracraft 1980: 66)

Hecht and Edwards (1977: 15–16) have designed criteria of character weighting which attribute more informational content to "those states which are part of a highly integrated functional complex" than to a "simplification or reduction" of characters. . . . The problem, however, is that there is no unequivocal measure of complexity available at the present time, nor can the assumptions implicit in these criteria of weighting be justified by the present state of development of evolutionary theory.

(Rieppel 1988: 61)

The logical jump from the impossibility of well-justified *a priori* weighting, to the impossibility of *any* kind of weighting, seemed natural to these and many authors, and it seems to have force even today. I have been regularly teaching courses on cladistics for more than 25 years, and it is still the case that the vast majority of students feel very strongly against the idea of differential weighting—even if having only a superficial knowledge of phylogenetic methodology and without being able to point a finger at any actual reasons for doing so.

Ironically, it was (and is) widely recognized that not all characters provide equally strong evidence of relationships. This was explicit not only by the authors whose critiques to differential weighting were just cited, but even by some authors of a phenetic persuasion (e.g. Cain and Harrison 1960; Sokal and Sneath 1963; Kendrick 1965; McNeill 1978), who were among the most prominent defenders of the equal weights approach. The refusal to differentially weight characters stemmed more from the supposed impossibility to evaluate their relative reliabilities, than from a belief that all characters truly deserve the same weight. The usual recommendation of equal weights was then an admission of failure or ignorance (e.g. as in Sokal and Sneath 1963: 267) more than anything else. The expectation was then that by increasing the number of characters used, the "good" ones would outnumber the "bad", and the correct solution would be identified. While that may be true in the long run, morphological datasets usually consist of too few characters to have reached the point where one could reasonably expect such automatic filtering of good and bad characters. Weighting is therefore more important when the dataset consists of few characters. But not just any weighting—what is needed is weighting by *appropriate* criteria.

All of the criteria proposed for deciding character weights *a priori* (i.e. without considering the entire dataset) seem indeed flawed, or at least very difficult to justify in general. However, that does not necessarily mean that all characters need to be considered equally influential when analyzed by proper methods. While it truly seems impossible to achieve a justifiable decision on which characters are more reliable in the abstract (i.e. without reference to a specific dataset), appropriate methods for weighting characters can take into account evidence from the characters themselves, by considering how they interact with other characters and with phylogenetic hypotheses. In that regard, the conclusion that some characters are more reliable than others is a conclusion of the analysis, not a premise. Before describing such methods in detail, however, it is necessary to clarify some terms and to discuss some general properties of character weights.

7.1 GENERALITIES

One of the common misconceptions about weighting is that it implies additional assumptions over equally weighted analyses (Kluge 1997; Grant and Kluge 2003, 2005; Congreve and Lamsdell 2016). That idea is debatable, if not plainly wrong. Defining the weights as all equal still *is* a statement about the relative importance of the characters—and a questionable one. In other words, there is no such thing as an "unweighted" analysis; the correct expression is actually "equally weighted" analyses. Equal weights can be defended only as long as an argument can be made for all the characters to be considered equally reliable—which is a very strong statement. Appropriate methods for differential weighting (discussed later) allow for the weights to be different if the data so suggest, or all equal otherwise (e.g. for a set of congruent characters). Appropriate methods thus do *not* impose anything in terms of character weights; only equal weighting does so. In a sense, equal weights forces all characters to conform to a much stronger assumption.

In the early days of phylogenetics, Mayr (1969), an avowed anti-cladist, asserted that Hennig's (1966) proposal of grouping by synapomorphies constitutes an extreme form of weighting, effectively giving zero-weight to "plesiomorphic characters". As we have seen in Chapter 1, the notion of grouping by synapomorphies is actually an expression of parsimony. The notion that grouping by synapomorphies constitutes *any* form of weighting is incorrect; all character states are used to define groups at the proper level of the hierarchy. Paired appendages characterize vertebrates, paired appendages transformed into legs define tetrapods, legs transformed into wings define birds. In the absence of homoplasy, every character state is used just once in the hierarchy, at the level at which it first originates; from then on, it is just a plesiomorphy.

Grouping by parsimony does not, of course, render character weighting unnecessary, as there can be conflicting synapomorphies, supporting incompatible groups. Note that the criterion of parsimony is often misunderstood as the notion that all characters should be weighted equally and that the conflict between two sets of characters should be resolved always in favor of the grouping supported by the set with more characters (e.g. Turner and Zandee 1995; see reply by Goloboff 1995b). Parsimony is nothing of the sort. Parsimony tells us that, to the extent that a tree requires more homoplasy for a character, the worse it fits that character. The basis for parsimony is associating explanatory power with (a low number of) independent originations of similar features. This provides a measure of how well different trees allow explaining the distribution of a single character by genealogy. This does not in itself prescribe how conflict between characters must be resolved; conflict can be resolved only when the relative weights of the characters are taken into consideration. Saving one step in one character at the expense of postulating another step for a different character will produce a tree of the same explanatory power *only* as long as the characters can be considered equivalent. Farris' (1983) classic justification of parsimony has often been cited to justify the use of equal weighting (e.g. by Kluge 1997;

Grant and Kluge 2003; Miller and Hormiga 2004; Congreve and Lamsdell 2016), but it is clear that Farris (1983) himself did not view parsimony as precluding weighting:

> No one supposes, however, that characters in general all deserve the same weight—that they all yield equally strong evidence. Drawing conclusions despite conflicting evidence requires that some evidence be dismissed as homoplasy. It is surely preferable to dismiss weaker evidence in deference to stronger. A decision reached by weighting characters, at any rate, can hardly rest on a basis different from parsimony.
>
> **(Farris 1983: 11)**

Put differently, when characters conflict this means that at least one *ad hoc* hypothesis of homoplasy must be postulated for one of the characters. Clearly, not all *ad hoc* hypotheses need to be equally disturbing: some *ad hoc* hypotheses can be highly implausible, others may be more reasonable (Goloboff et al. 2008a). Thus, explanatory power is not necessarily decreased when weighting. On the contrary: because not all of the evidence demands to be explained with the same force, taking into account the relative weight of observations makes the measure of collective explanatory power more adequate. There is thus no philosophical requirement that weights be all equal (contra Kluge 1997). Revealingly, rather than dropping the idea of equal weighting, Kluge and Grant (2006, both of whom had before incorrectly cited Farris 1983 in support for the need of equal weighting) eventually chose to drop Farris' (1983) justification of parsimony altogether, and embraced a different one ("anti-superfluity", Baker 2003), one which in their view precludes differential weighting[1] (for a discussion of the many problems with Kluge and Grant's 2006 position, see Farris 2008; Giribet and Wheeler 2007).

Another common misconception is that weighting characters amounts to disregarding evidence. A downweighted character is not eliminated from the matrix—it is simply considered less influential. As long as no character is given zero-weight, there is no *disregarding* of evidence. What matters in a numerical analysis is the relative weights of the characters. Giving all characters in a matrix a weight of 2, except the last character a weight of 1, is exactly equivalent to giving them weights of 10 and 5, 200 and 100, or 1 and 0.5. As a consequence, weighting can have an effect on the results only when there is conflict between characters. If all characters are congruent, no weighting scheme (zero-weights aside) can produce a result different from equal weighting. Weighting is only a means to push resolution of character conflict in one or another direction. Most commonly, the term "weighting" is applied to entire characters, and then weighting the character is equivalent to multiplying the columns of the character, that is, giving a character a weight of 5 is like making 5 copies of the character. Optimization (i.e. assignments of ancestral states) does not change with character weights. The costs of transformation between states within a character (be it additive, or step-matrix) are of course a form of "weighting"; in this book at least, the term "cost" is used for cost of transformation between states. Those costs do change character-state optimization. In fact, when homologizing one state precludes homologizing another, changing ancestral assignments is precisely what those costs are *intended* to do! Just as in the case of characters, where weighting does not affect the results for a set of congruent characters, changing transformation costs has an effect on how the states are homologized only when it is impossible to homologize all states simultaneously (i.e. when a choice must be made for the states to be homologized).

One of the least cogent criticisms of current weighting methods is that they assume that the "weight" of a character is constant all over the tree; in other words, "weighting presupposes a character that behaves badly in one part of the cladogram . . . must do so as well elsewhere in the cladogram" (Kluge 1997: 335; Padial et al. 2014; Congreve and Lamsdell 2016). I made a similar argument myself (Goloboff 1991b) when I opposed differential weighting, before I realized (in Goloboff 1993a) that neglecting the implications about homoplasy and reliability of the inferred trees is illogical. It is true that most weighting methods assume constancy throughout the tree, but that can hardly be an argument for using equal weights—which *also* assume constancy throughout the tree, in addition to assuming constancy across characters. As discussed next, (a) it seems possible to do away with weights constant over all the tree (Goloboff et al. 2009b), but then the results cannot be called "parsimony", and (b) standard likelihood methods provide a way to consider changes more or less important, depending on the length of the branch where they occur, but they require that the reliabilities (rates) of characters in each partition be all uniform (otherwise risking overparameterization). Parsimony itself (i.e. the idea that explanatory power varies directly with the degree to which similarities can be attributed to common ancestry) seems to be what requires constant weights throughout the tree, and so considering that such constancy makes weighted parsimony methods inferior to equally weighted parsimony is illogical.

Some authors (e.g. Wheeler 1986; Turner and Zandee 1995) think that a weighting criterion is needed only when an analysis under equal weights produces ambiguous results (i.e. multiple *MPTs*). Such an idea is in some sense self-contradictory (Goloboff 1995b). If a weighting criterion can be properly used to select from among trees that happen to be of minimal length under equal weights, why is it not going to be appropriate to select from among trees that do *not* have the same length? Either the weighting method is appropriate to make a choice between *any* two trees, or is not appropriate at all.

Note that, in character weighting, the weight acts as a multiplier for every step occurring in any part of the tree. If equal influence is desired, what needs to be assigned a constant cost is all steps, and that is achieved by uniform character weights. In this regard, the cost of an individual step is not affected by the number of possibly distinct states in a character. A transformation between states in a character of unit weight and two states has the same cost as a transformation between states in a character of unit weight but six states. Additive characters are no exception to this, as they can be seen as representing hypotheses of homology at multiple levels (thus being equivalent to several binary characters, each with unit weight; see Chapter 3 for recoding techniques, and Chapter 9 for the special case of continuous characters). An ill-advised practice (which seems to have originated with Colless 1980, but was adopted by some phylogeneticists, e.g. Thiele and Ladiges 1988) is to lower the weight of multistate characters, so that the total number of steps (in case of no-homoplasy) is equivalent to the single step in a binary character. This is accomplished by giving the multistate character a weight $1/m$ (where m is the minimum possible number of steps

on any tree; see Chapter 3 for methods to determine minima). As the costs of transformations between states are no longer the same in different characters, the effect
of this so-called "rescaling to unit range" is in fact an *unequal* weighting (Farris
1990). That unequal weighting, however, is based on factors that have nothing to do
with character reliability: a character with more states does not necessarily exert a
stronger influence on the results (this is especially obvious in the case of nonadditive
characters, where a character with as many states as distinct taxa is uninformative).
Therefore, the very argument to use unit range rescaling for equalizing character
contributions is unfounded.

7.2   GENERAL ARGUMENTS FOR WEIGHTING

Two types of arguments can be advanced for considering some characters more influential than others; both are logically independent and not mutually exclusive. One
considers the extent to which the embodied observations can be trusted, and the second refers to properties intrinsic of the character. The former is an argument about
the observer, an admission that not all of the observations that go into a matrix are
made on an equal footing. Some cells are filled from personal observations, others
are filled from the literature. If filled from the literature, they could be from papers
that used better or worse observational techniques. If filled from personal observations, the material used—quality of preservation, reliability of taxonomic identification, number and sex of specimens—can also be highly variable. This could
provide a rational basis for differentially weighting some characters *a priori* (Neff
1986). Some of the arguments for differential weighting on the basis of complexity
are framed by reference to the possibility of the researcher making a mistake in
homology assessment—the simpler the structure, the fewer the parts or components
that can be compared to increase the certainty of having "the same" state—instead
of framing them by reference to the intrinsic properties of the character. This kind
of argument can be used only to establish *prior* weights; the results of a phylogenetic
analysis (no matter how they are arrived at) cannot logically affect the degree of
certainty of the observations.

The problem with this kind of argument is that, no matter how rational the downweighting of more doubtful observations may seem, it is very difficult to quantify
the uncertainty in observations. If quantifying what one knows is hard, a formal
framework for quantifying one's own uncertainty seems much harder. As a consequence, this kind of argument is very rarely used, and should be considered with
much caution, making the reasons for ignoring or downweighting some observations
very explicit. Most likely, discussion of those reasons on a case-by-case basis will be
needed.

The alternative is using weighting as a means to refer to intrinsic properties of
the characters. In the pre-cladistic days, many authors (starting with Darwin 1859;
Mayr 1969; Bock 1977; Hecht and Edwards 1977) argued that characters representing adaptations should not be used to indicate phylogenetic affinity. This makes
sense: if a character can be easily switched on and off as the same environment is
independently occupied or abandoned by different taxa, then the character seems a
poor guide to taxonomic relatedness. But the opposite argument can, and sometimes

has (e.g. Gutmann 1977; Manton 1977) been made: only characters representing adaptations or having a known function can be used to demonstrate relationships. This *also* makes sense: if a given character is fundamental for survival, it cannot be easily changed (e.g. mammary glands are clearly adaptive and functional, and they are indeed perfectly correlated with Mammalia). Thus, a character being affected by natural selection, or not, does not tell us much about its reliability. Sober and Steel (2015) reexamined the question, and concluded from formal arguments that the reliability of a character depends (at least in part) on whether it is under directional or stabilizing selection. However interesting Sober and Steel's (2015) results are from the theoretical point of view, no practical study has any way to know selective regimes of the past, and thus arguments about adaptive features (either for or against using them) seem moot. Furthermore, as several phylogeneticists noted (e.g. Platnick 1978; Schuh 1978; Eldredge and Cracraft 1980) it is difficult to see how the long term adaptiveness—or lack thereof—could be hypothesized in the absence of a phylogenetic tree. A phylogenetic tree can provide useful information to study adaptation only as long as the tree was constructed without assuming anything about the theory to be tested. In other words, a phylogenetic tree should ideally help demonstrate that some characters are better correlated with phylogeny, rather than assuming it. This is the train of thought that provided the basis for the flourishing field of comparative methods (starting from classic papers such as Felsenstein 1985a; Coddington 1988; Baum and Larson 1991, to cite just a few; much of the literature is reviewed in Grandcolas 2015).

Many authors have proposed to use complexity as an argument for differential character weighting (e.g. Mayr 1969; Hecht and Edwards 1976; Wägele 1995). Complexity, as mentioned, could be seen in principle as influencing the certainty with which decisions on homology can be made. However, it is quite difficult to quantify complexity (but see Ramírez and Michalik 2014). Another argument made to upweight more complex characters is that more complex characters are less likely to evolve in parallel. Again, as in the case of adaptive characters, the ideal situation is using phylogenetic results to test the idea that complex characters are less likely to evolve in parallel, rather than assuming it ahead of time. In addition, simpler morphologies are apparently very conservative in many groups, and reductive characters common. Thus, the notion that the degree to which characters are well correlated with groups depends on their complexity seems far from justified.

Both adaptiveness and complexity were proposed as a basis for *a priori* weighting of characters, and *a priori* weighting on this basis has been (correctly) rejected since the earliest days of phylogenetics (as cited at the beginning of the chapter). Many papers and textbooks (e.g. Wheeler 1986; Kitching et al. 1998; Schuh and Brower 2009) contrast *a priori* weighting (typically characterized as unavoidably subjective) with *a posteriori* methods. I confess that (when considering formal quantitative methods) it is unclear to me what the purpose would be for weighting characters, if truly done *a posteriori*. If the analysis has been made already, and groups have been established, then character weights do not matter anymore. Why weight characters if all groups have been formed? The very name of "*a posteriori*" is then somewhat misleading (other authors have noted this; e.g. Simpson 1961; Sharkey 1989). Ideally, a weighting method should allow for the character weights to be derived from the data themselves, and this would be done *during* the analysis, not after. Perhaps a more felicitous expression would be "empirical" weighting.[2]

7.2.1 HOMOPLASY AND RELIABILITY

Regardless of the reason for some characters being more reliable than others, it is widely agreed that the telltale symptom of unreliability is the existence of homoplasy. A binary character that has no homoplasy on the correct tree allows identifying one group with perfect confidence. To the extent that the character has more homoplasy, then the group that would be created by following the dictates of the character is more in conflict with the groupings in the correct tree. Put differently, when there is more homoplasy, a tree having the group defined by the character requires more branch moves to become identical with the correct tree (see Chapter 6). This is so, regardless of whether the homoplasy is a consequence of the character being adaptive, non-adaptive, fast-evolving, or so simple that it has a high probability of arising in parallel several times (Goloboff et al. 2008a: 760). Therefore, weighting based on homoplasy is the most agnostic approach. If homoplasy is considered as tied to (lack of) reliability, then it is important to note that in most empirical datasets, even when analyzed under the assumption of equal weights, some characters have much larger amounts of homoplasy than others (Figure 7.1a). Thus, even the results obtained when assuming equal weights reject the notion that all characters are equally reliable, indicating the need for appropriate weighting methods.

One possibility is using expected homoplasy for downweighting characters *a priori*, but of course this begs the question: on which tree should one calculate those amounts of homoplasy? On the correct tree? If we knew the tree, then we would not need weights or phylogenetic analysis anymore! Instead of an *a priori* approach, a more empirical method for weighting is needed, one which determines the weights from the data, not from prior assumptions about the tree. The first method to empirically determine weights on the basis of homoplasy was Farris' (1969) successive weighting (*SAW*).

7.3 SUCCESSIVE APPROXIMATIONS WEIGHTING (*SAW*)

Farris's (1969) idea was to assess the homoplasy from the trees produced in an analysis with some initial set of weights, use those amounts of homoplasy to reassign weights to the characters, and run a new analysis with the modified set of weights (Figure 7.1b). The process ends when the new set of weights is identical to those in the previous round. By doing this, Farris (1969) expected that the weights would successively approximate the "correct" weights. The method, way ahead of its time, saw little use until the late 1980s. Carpenter (1988) provided a general discussion of the method (using the mainframe program Physys; Farris and Mickevich 1982), and Farris (1988) included in Hennig86 the first implementation for microcomputers.

In the case of multiple trees, the implementation in Hennig86 reassigned weights using the lowest amount of homoplasy in any of the trees produced in the current round. The implementation in PAUP (Swofford 1993) also allowed the use of the highest or average amounts. One of these three choices is mandatory—when the

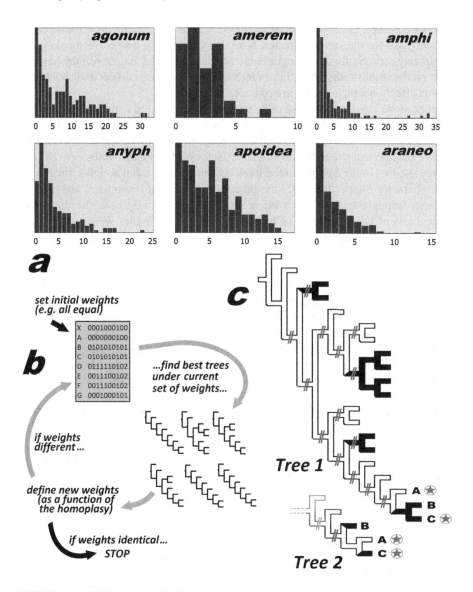

FIGURE 7.1 (a) Frequency of different numbers of extra steps, in trees optimal under equal weights, for empirical datasets taken from Goloboff et al. (2017). **(b)** Schematic of successive weighting. **(c)** Two hypothetical trees, differing in the resolution of relationships between A–C (the rest of the groups in the tree are well supported by other characters not shown). Group B+C is supported by the white/black character, and group A+C is supported by the star character (absent in the remaining taxa). According to these two trees themselves, the star character (with 0 or 1 steps of homoplasy) should be more reliable than the white/black character (with 3 or 4 steps of homoplasy), so that tree 2 is preferable.

current weights produce multiple *MPT*s, different trees may require different amounts of homoplasy for the same character. Note that if using the average homoplasy for reweighting, criteria for zero-length branch-collapsing (see Chapter 6) may have an effect on the results (other than this, branch-collapsing criteria do not and should not have an effect on which groups are considered supported).

As example, consider the case of Figure 7.1c, where analysis under equal weights produces trees 1 and 2. Assume that all groups in the tree (marked as //) are well supported by other characters not shown, except for the resolution of group ABC. Further assume that the only two characters relevant for the resolution of ABC are the ones shown in the figure. On the trees obtained under equal weights, the black/white character (supporting B+C) has three or four steps of homoplasy, and the star character (supporting A+C) has none or one. Reassigning weights to the characters on the basis of those amounts of homoplasy will give a higher weight to the star character, so that a second round of *SAW* will lead to prefer tree 2—thus saving steps in the character with less homoplasy, at the expense of adding steps to the character with more.

There have been some suggestions (e.g. Neff 1986; Swofford and Olsen 1990[3]) that the method amounts to circular reasoning. The accusation of circularity could stem from the expectation that the weights determined from a tree make that tree more strongly supported (or less strongly rejected). This may happen sometimes, but certainly not always: *SAW* from a set of trees often produces a different final answer, and the weights produced by reweighting from a single tree often lead to prefer *other* tree(s). This makes it clear that the method is not circular (Sharkey 1989; Carpenter 1994).

Although Farris (1969) did not discuss this explicitly, *SAW* is based on a criterion for self-consistency (Goloboff 1993a). A given tree will allocate different amounts of homoplasy to the different characters. In the case of conflict between two characters, the tree may be resolving the conflict in favor of the character(s) which (on the tree itself) have more homoplasy. That is the implication of reweighting the characters based on some tree(s), and finding that a *different* tree is shorter under such weights. Given conflict between two characters *A* and *B*, therefore, the tree is resolving it in favor of character *A*, but the tree also tells us that character *A* is less reliable than character *B*. The tree is in that sense self-contradictory. *SAW* provides a way to evaluate whether character conflict is resolved in a way that is coherent with the amounts of homoplasy observed on the tree(s).

7.3.1 WEIGHTING AND FUNCTIONS OF HOMOPLASY

In his original proposal, Farris (1969) used *c*, the *character consistency index* (Kluge and Farris 1969), to reassign weights in every round. The character consistency index is defined as m / s, the ratio between the minimum possible steps the character can have on any tree, m, and the observed number of steps, s. For informative characters this, of course, varies with different trees, as s varies. The *ensemble consistency index*, *C*, for the full matrix is M / S, where $M = \Sigma m_i$ and $S = \Sigma s_i$, which also varies with different trees and is maximum on the *MPT* (as this tree minimizes S). If character weights are different, the sums M and S take this into account. Note that the homoplasy h is the number of steps beyond the possible minimum, so that $h = s - m$, and therefore the complement of c measures the fraction of total changes in the tree that are homoplastic:

$$1 - c = 1 - m / s = 1 - m / (m + h) = h / s$$

When a character is binary $m = 1$, so that c reduces to $1 / (1 + h)$. As shown in Figure 7.2a, c is a concave decreasing function of the homoplasy.

In the present context, the interest of the consistency index is only in its single-character form. Note that the ensemble consistency index C used to be reported routinely for parsimony analyses, and it was often considered as giving some indication of confidence in the results. Interpreting the consistency index as indicating confidence is certainly wrong; the consistency index measures homoplasy, and tree discrimination is minimal when every tree has the same length, not necessarily when the homoplasy is maximal (Goloboff 1991a). A number of papers confusing ability to discriminate among trees and homoplasy have criticized the consistency index, mostly on the grounds that it could not achieve minimum values of zero on any dataset (e.g. Archie 1989; Sanderson and Donoghue 1989; Klassen et al. 1990; Meier et al. 1991).

Only a measure of relative homoplasy, not absolute homoplasy, can achieve ensemble values of zero on *MPT*s, when the observed homoplasy equals the one used as reference. Some of the papers already cited used different kinds of data randomizations to provide reference values of ensemble homoplasy, but this is bound to create problems (like minimum negative values for some datasets; negative values cannot possibly measure confidence or ability to discriminate trees; see Goloboff 1991b; Farris 1991, for details).

A different approach was used by Farris (1989), with the retention index. In this case, the index is designed to achieve a value of zero when the tree is as bad as possible for the dataset at hand. The *character retention index*, ri, is defined as $ri = (g - s) / (g - m)$, where g is the maximum possible number of steps that a tree may require for the character (easily measured on a polytomy, see Chapter 3). Note that, just like $h = s - m$ (see earlier), then the maximum possible homoplasy is $h_{max} = g - m$. Thus,

$$ri = \frac{g - s}{g - m} = \frac{(h_{max} + m)}{h_{max} + m - m} = \frac{h_{max} - h}{h_{max}} = 1 - \frac{h}{h_{max}}$$

that is, the complement of the retention index measures the fraction of maximum possible homoplasy that the observed homoplasy represents. Unlike the consistency index, ri is a lineal function of the homoplasy for a given character, but a function with different slopes for characters with different values of h_{max} (Figure 7.2b). In Farris' (1989) interpretation, ri measures the fraction of informative variation that is effectively "retained" as synapomorphy on the tree.

Ensemble values of the retention index, RI, result from sums of g, s, and m over all characters, then called G, S and M. This is maximized by the *MPT* (where S is smallest). Note that RI achieves a value of zero on least parsimonious trees. On *MPT*s, RI cannot be strictly zero, although it can get very close (Farris 1991).

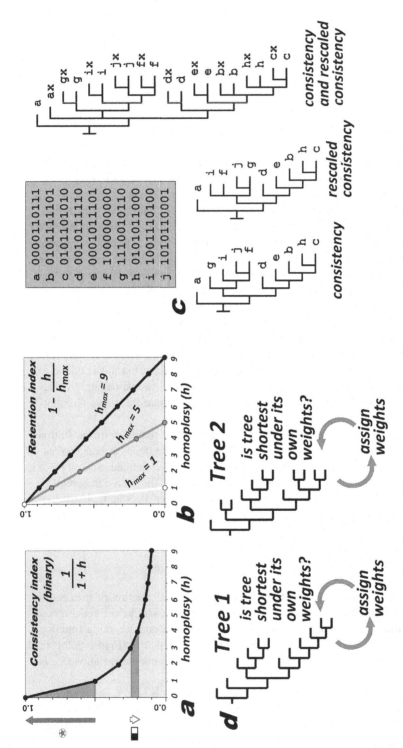

FIGURE 7.2 (a) Plotting of the unit consistency index, c, as a function of the number of extra steps, for a binary character. To the left of the y-axis, the differences in c value for the two conflicting characters of Figure 7.1c. (b) A plotting of the unit retention index, ri, as a function of the homoplasy, for characters with different values of maximum possible homoplasy, h_{max}. (c) An example showing that duplicating all taxa in the matrix may change the results when successively weighting with the rescaled consistency index, c, the results always remain constant when duplicating taxa. (d) Self-consistency (in weight assignments and conflict resolution) is best checked on a tree-by-tree basis. See text for discussion.

With $ri = (g - s) / (g - m)$ and $c = m / s$, their product equals $ri.c = (c - c_{min}) / (1 - c_{min})$, a rescaling of the consistency index that achieves a value of 0 for trees which require as much homoplasy as can be required for the dataset at hand (see Farris 1989 for details). The rescaled consistency index has a shape similar to that of c, but decreases more rapidly, and differently for characters with different values of h_{max}. SAW in Hennig86 used $ri.c$ for reweighting, but this can create problems (Goloboff 1991a). The consistency index measures the homoplasy, but the product $ri.c$ measures the *proportion* of maximum observable homoplasy. Equal amounts of homoplasy lower the retention index (and thus $ri.c$) less if the maximum homoplasy in the character is larger (Figure 7.2b). The maximum homoplasy changes with the number of taxa with different states (e.g. 0's and 1's in binary characters). This reflects an argument that has existed for a long time in the literature. A character where only two taxa have state 1 cannot have more than 1 step of homoplasy on any tree; a character with ten 0's and ten 1's can have ten. Starting with Le Quesne (1969), many authors have tried to take this difference into account in several ways. This was the basis for some of the proposed methods to evaluate data quality (homoplasy excess ratio or *HER*, Archie 1989; permutation tail probability or *PTP*, Faith and Cranston 1991), which permuted states among characters (thus destroying hierarchical correlation but preserving the numbers of taxa with each state; see criticisms of those methods by Farris 1991, 1995; Goloboff 1991b; Carpenter et al. 1998). While consideration of the number of taxa with different states may be useful for specific purposes, it seems clear that what should influence character weights is the observed homoplasy, not the proportion of homoplasy relative to some worst-case maximum (Goloboff 1991a). If a character where only three taxa have state 1 presents just one instance of homoplasy in a tree, then that is because (along evolution) all the remaining 0 states have remained as 0 in the rest of the tree; that continues being less homoplasy than observing two instances of homoplasy in a character with ten 0's and ten 1's (even if it is a higher *proportion* of the maximum possible; 1/2 for the 3:17 partition, and 2/9 for the 10:10 partition). Thus, rescaled consistency is lower for the 3:17 than the 10:10 partition, 0.250 vs. 0.259, while consistency is higher for the 3:17 than the 10:10 partition, 0.500 vs. 0.333.

A consequence of reweighting characters in a *SAW* procedure and rescaling by maximum observable homoplasy is that duplication of the entire matrix—which does not otherwise alter the information in the dataset—may produce different results. An example is in Figure 7.2c. Note that as every taxon is represented by more copies, $ri.c$ approaches c: having more copies of every taxon increases g, but affects neither m nor s. Therefore, as g increases, $(g - s) / (g - m)$ approaches unity, and m / s remains constant. That is why representing every taxon in the dataset of Figure 7.2c with two copies and reweighting with rescaled consistency index changes the result, and produces the same tree as reweighting with the consistency index. The consistency index, which reweights by the absolute instead of the proportional homoplasy, is not affected by taxon duplication.

7.3.2 PROBLEMS WITH SAW

The previous section showed that one needs to be careful with the selection of the function to reweight characters. That problem is at least soluble. The mechanics of *SAW* creates other problems, which can be solved only by substantial modifications of the method.

The first and most obvious problem is that the method may depend on the starting weights. Different sets of initial weights may lead to different final stable solutions (Farris 1969), and there is no obvious criterion to choose from among such solutions, or to choose a set of initial weights. A related problem is then that *SAW* does not in itself provide an optimality criterion by which to compare two given trees; you either take or leave the final trees, but there is no way to evaluate them independently of the process that produced them.

Given that *SAW* assigns character weights on the basis of a set of trees, it is possible that a tree in the final set is most parsimonious, not under the weights it implies itself, but only under the weights implied by the *other* trees. An example is in Figure 7.3, where a dataset produces (by weighting with the consistency index, rescaled between 0–100, starting from all trees produced under equal weights) a set of two trees, trees 0 and 1, stable under the weights they determine together (see Figure 7.3). Note that the weights determined by the set are a composite ($Wts_{(0,1)}$ in Figure 7.3), using the maximum weights that result from either tree 0 or 1. If the weights are set from tree 0 alone, that tree is (the only one) of minimum length. But if the weights are set from tree 1 alone, tree 1 is *not* of minimum length under those weights—another 5 trees are. In other words, tree 0 is acceptable according to its own amounts of homoplasy, but tree 1 is not. Tree 1 is acceptable only according to the amounts of homoplasy on tree 0!

That paradoxical situation only arises from using multiple trees to estimate character weights. As stated, the weights determined by the set are a composite, and result from amounts of homoplasy *that cannot coexist on any single tree*. In the composite weights, the 2nd character is assigned weight 50, and the 7th is assigned 33 (see white arrows in Figure 7.3); however, there is no tree on which the corresponding amounts of homoplasy (1 and 2 extra steps) for the 2nd and 7th character can occur simultaneously; either the 2nd and 7th characters have 1 and 3 extra steps (as in tree 0), or both have 2 (as in tree 1). Likewise for the 4th and 11th characters.

As a tree search during a round of *SAW* proceeds, it may move away from the trees that had been used to set the current weights. This begs the question: why choose between trees (say) X and Y on the basis of the weights determined from trees A, B? Trees X and Y should be compared considering the weights (=amounts of homoplasy) they imply themselves. The weights implied by trees A, B are entirely irrelevant for comparing trees X and Y.

More practical problems with *SAW* are that it is more time-consuming (i.e. it requires several rounds of tree searches) and that it makes measuring group supports difficult. It is clear that one cannot use the final weights and just evaluate group supports on the basis of the final weights, as doing so would distort the relative amounts of favorable and contradictory evidence for each group. Consider the simple case where there are 11 characters in favor of a group, and 10 against:

```
A 0000000000  00000000000
B 0000000000  11111111111
C 1111111111  11111111111
D 1111111111  00000000000
  ╰──────╮──────╯  ╰──────╮──────╯
    10 chars.      11 chars.
```

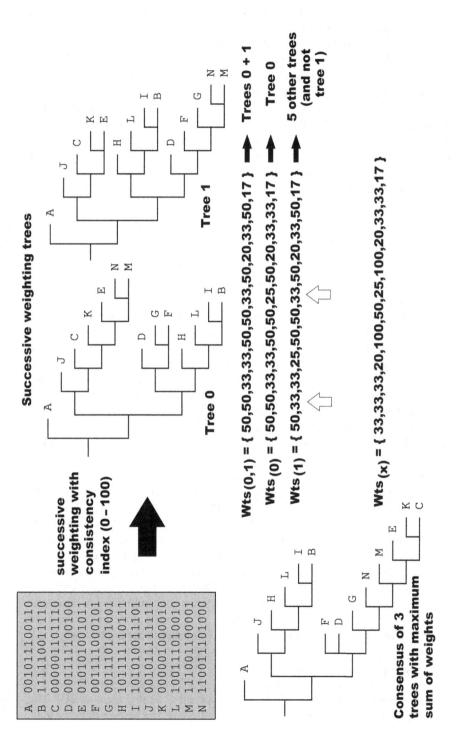

FIGURE 7.3 A case where successive weighting (starting from the results under equal weights) produces two trees when character weights are assigned from both trees combined, but only one of the trees is shortest under the weights it implies. See text for discussion.

It is evident that the group B+C, favored by 11 characters, is supported, but only mildly so (it is contradicted by the 10 characters supporting C+D, for an almost exact tie). The bootstrap frequency of B+C with all weights equal is 58%, correctly showing that the group is supported but weakly so. A *SAW* analysis of the dataset (with consistency index rescaled between 0–100) produces obviously the tree (A(D(BC))), with each of the characters supporting B+C with a weight of 100, and each of the characters supporting the alternative a weight of 50. Bootstrapping now under those weights produces inflated frequencies for the group B+C, 93%, as the characters supporting B+C are upweighted.[4] Measuring support in combination with *SAW* then requires completing *SAW* (until stabilization) for every resampled matrix. This is of course possible, but time-consuming. In the case of Bremer supports, the weights resulting from *SAW* cannot be used (as this would also inflate support values for the groups in the resulting trees), but it is unclear to me what modification would allow properly measuring supports.

7.3.3 POTENTIAL SOLUTIONS

The considerations in the previous section suggest that self-consistency can be better checked one tree at a time. Any possible tree can in principle be checked for self-consistency (as in Figure 7.2d), to establish whether the tree is of minimum length under the weights it implies. Note that this does not require a full search under the weights determined by every tree (cf. the previous discussion, on choosing trees *X* and *Y* on the basis of weights from trees *A*, *B*). All that is needed is establishing whether a tree is of minimum length under its own weights. Any tree which is not of minimum length under its own weights can be discarded. The remaining trees are acceptable under the criterion of resolution of character conflict in favor of more reliable characters embodied in *SAW*. Of course, different trees will imply different weights for the characters; every tree is associated with its own set of weights.[5] The procedure of selecting all possible self-consistent trees does not determine a unique set of weights. That is in fact desirable: given ambiguity, the data may be insufficient to prefer a single tree, and therefore the data are also insufficient to determine a unique set of weights.

Although using the sets of weights determined from (a sample of) every possible tree solves some logical problems of *SAW*, it leaves open the question of whether every one of those trees is an equally acceptable hypothesis. A tree *A* might be self-consistent by virtue of implying that most characters are very unreliable, and resolving conflict *against* those characters. Another tree *B* might be self-consistent by implying that most characters are highly reliable, and resolving conflict *in favor* of those characters. It is clear that tree *B*, self-consistent and implying higher character weights, is a better choice than tree *A*, self-consistent but implying low character weights. Further consideration of the possibility of choosing among self-consistent trees on the basis of their weights leads directly to the criterion for implied weighting.

7.4 IMPLIED WEIGHTING (*IW*)

Implied weighting (*IW*, Goloboff 1993a) was proposed as a refinement of *SAW*. The previous section raised the possibility of selecting trees on the basis of their weights.

Farris (1969) had argued for using concave decreasing functions of the homoplasy (such as c) for reassigning weights. Goloboff (1993a) noted that then a tree-by-tree checking of self-consistency (as described in the previous chapter and schematized in Figure 7.2d) becomes unnecessary if the trees are selected just on the basis of the weights they imply—trees of maximum weight are expected to be self-consistent, just as a byproduct of the weight maximization.

Consider the case of conflict between two characters shown in Figure 7.1c. Tree 1 has 3 steps of homoplasy for the character "black/white", a character which (other things being equal) supports the monophyly of group B+C. The monophyly of B+C is contradicted by the character "star", which on tree 1 itself has a single step of homoplasy. On tree 1, then, character "star" is better correlated with groups than "black/white" and is thus more reliable, but tree 1 resolves the conflict between "black/white" and "star" in favor of the least reliable character. A better tree, according to tree 1 itself, would be tree 2, and indeed starting *SAW* from tree 1 leads to tree 2. But if trees 1 and 2 are compared on the basis of their own weight implications, under a concave function of the homoplasy, tree 2 is chosen automatically, with no need to iterate searches. Consider how the "black/white" and "star" characters differ in their fits to the trees, according to the consistency index (Figure 7.2a). Going from tree 1 to tree 2 decreases the fit, from 3 to 4 steps of homoplasy; this is translated on the y-axis into a small decrease in fit (white down-arrow in Figure 7.2a). Because the shape of the curve used to measure fit is concave, the difference along the y-axis between 0 and 1 steps of homoplasy is much larger than the difference between 3 and 4, and then when going from tree 1 to tree 2, the fit for the "star" character is improved much more (gray up-arrow in Figure 7.2a) than "black/white" is worsened. That is, the differences in homoplasy in the relevant characters are automatically taken into account when maximizing sums of fits measured with a concave decreasing function of the homoplasy, so that the same difference in steps is less significant if occurring in a character with more homoplasy. A tree that maximizes the fits (=weights) is therefore self-consistent in the general sense of resolving conflict between characters in favor of those characters which—on the tree itself—have less homoplasy.

Let us return to the example of Figure 7.3, where *SAW* produced a set of two trees, only one of which was shortest under its own weights. The sum of weights for each of the two trees is 444 (we leave it to the reader to verify this); the sum of weights of the five trees produced by finding the *MPTs* for the weights of tree 1 is 486, and then those trees (each of which is shortest under its own weights) are preferable to either tree 0 or 1. But yet even higher sums of weights are possible, as shown at the bottom of Figure 7.3. Those three trees were selected to have the maximum weights (summing up to 497), and each of them also is—as expected—shortest under its own weights. The advantage of selecting trees so as to maximize a concave decreasing function of the homoplasy is that it automatically takes into account self-consistency, without the need for multiple rounds of searching *alla SAW*. Tree searches, as a consequence, can proceed using the exact same algorithms described in Chapter 5 for fixed weights, thus searching and weighting simultaneously.

The approach just described has several important implications. The weights of the characters are not fixed; every pairwise comparison of the score between two trees will be done according to the homoplasy the characters have on the trees being compared—the "implied weights". For example, when comparing trees X and Y, a given set of weights will be used implicitly, but the comparison between trees Y and Z will be made using a *different* set of weights. This is desirable: when comparing Y and Z, the weights implied by tree X are irrelevant. The problem of attempting to determine a unique set of weights (as done in *SAW* or other methods) is misguided: if the data are insufficient to select a single tree, why should they be sufficient to select a single set of weights? Under *IW*, weights and trees are inextricably tied (via the function chosen to derive weights from homoplasy); multiple trees as the result of an analysis imply multiple alternative sets of weights. All that is required is that when comparing trees Y and Z the *possibility* that one is the correct tree can be taken into account—weighting the characters accordingly.

IW is often misunderstood, confusing both the mechanics and the justification of the method. For example, Congreve and Lamsdell (2016) claim that "[t]he problem [with *IW*] is that the fit is entirely contingent on tree topology; in this way implied weights just acts as a successive weighting system (not a concurrent weighting system as it claims) and is circular in its methodology" (p. 453). It is true that the "fit" is contingent upon tree topology, but so are numbers of steps (in equally weighted parsimony) or individual character likelihoods (in maximum likelihood methods[6]). If the fit of data were not contingent upon tree topology, there would be no way to choose from among different trees! Despite Congreve and Lamsdell's claim, the differences between implied and *SAW* are clear in the description just given; and if the accusation of circularity made no sense in the case of *SAW*, it makes even less sense in the case of *IW*. As the method works without multiple iterations, simply allowing every possible tree to be scored, it is hard to see where Congreve and Lamsdell (2016) imagine the circularity is. *Some* tree is chosen by the method, and it may of course be the wrong tree, but the possibility of incorrect results has nothing to do with circularity (almost the opposite; a truly circular method would choose always the same tree). Perhaps this is due to Congreve and Lamsdell (2016) failing to grasp the actual meaning of "circularity" and "tautology"—after all, they claim as well that their "results also corroborate the accusations of tautology that have plagued implied weights" (p. 457). A "tautology" being "a statement that is always true" (Cambridge dictionary), there is no way in which external evidence (be that the evidence produced by Congreve and Lamsdell, or anyone else) could falsify or corroborate it. Congreve and Lamsdell's claim of what their results corroborate is a contradiction in terms. Many other problems with Congreve and Lamsdell's (2016) treatment are discussed in Goloboff et al. (2017).

7.4.1 WEIGHTING FUNCTIONS

Obviously, the function to weight characters on the basis of homoplasy needs to be chosen carefully. The preceding sections introduced the idea of *IW* by reference to maximization of sums of character consistency indices, c, in binary characters. However, c measures the (complement of the) fraction of change on the tree that represents homoplasy. A single step of homoplasy represents a smaller fraction of the change in a character with many states, and vice versa. Therefore, c will decrease more quickly in characters with fewer states, so that in the case of conflict between characters with different numbers of states and similar amounts of homoplasy, choosing trees so as to maximize Σc will prefer saving homoplasy in characters with fewer states. Consider the conflict between a 7 and a 2-state character, and two trees A and B. Tree A has 1 and 0 steps of homoplasy in each of the characters (respectively), and tree B has 0 and 1; the rest of the tree is well defined by other characters, so that the choice boils down to trees A and B. The sums of c for the 7 and 2-state characters are:

	7-state character		2-state character	total Σc
Tree A	6/7	+	1	1.857
Tree B	1	+	1/2	1.500

Thus, for equal amounts of homoplasy in both characters, the character with fewer states has a larger difference in fit and wins over the other; therefore, tree A is preferred.

Maximizing sums of character retention indices (Σri) would also be a bad choice. The relative weight of a step-difference in a comparison between two trees depends on the slope of the curve (as shown in the discussion of Figure 7.2a). The retention index is always a straight line but with a different slope for characters with different values of $g - m$. Therefore, maximizing Σri will prefer saving homoplasy in characters with smaller $g - m$, and regardless of the homoplasy in the characters. Consider the conflict between two binary characters (i.e. $m = 1$), one of which is an 8:8 partition, and the other a 4:12. Values of g are $g_{(8:8)} = 8$, and $g_{(4:12)} = 4$. Assume most groups are well supported by other characters, and tree choice boils down to two trees, X and Y, where the characters with $g = 8$ and $g = 4$ have 1,2 and 0,3 steps of homoplasy respectively. The sums of ri for the relevant characters are:

	character with $g = 8$		character with $g = 4$	total Σri
Tree X	(8–2)/(8–1)	+	(4–3)/(4–1)	1.190
Tree Y	1	+	(4–4)/(4–1)	1.000

Thus, even if the 4:12 character has more homoplasy than the 8:8 on trees X–Y, the character with the partition 4:12 wins over the 8:8 partition, and tree X is preferred. Other methods where the numbers of taxa with state 0 or 1 influence the choice between trees are Salisbury's (1999) "strongest evidence", and Faith and Trueman's (2001) "profile parsimony" (see discussion at Section 7.7.2).

To avoid such problems, Goloboff (1993a) proposed maximization of a simpler function, one which varies only with homoplasy, and regardless of the possible minima or maxima of the characters. Recall that homoplasy $h = s - m$. The function $1 / (1 + h)$ will behave as expected in the previous examples, implicitly weighting only as a consequence of the amounts of homoplasy, not being influenced by number of distinct states or the number of taxa with different states.

7.4.1.1 Weighting strength

Although $1 / (1 + h)$ is not influenced by the irrelevant factors just discussed, it has the problem that it weights too strongly. If the cost of the first step of homoplasy is defined as unity, then the cost of adding a step to a character with h steps of homoplasy becomes,

$$\frac{2}{h+1} - \frac{2}{h+2} - \frac{2}{2+h^2+3h}$$ [Formula 7.1]

(note numerator is 2 because the absolute cost of the first step of homoplasy is 0.5). Then, the cost of adding a step to a character with 3 steps of homoplasy is 0.100, instead of 0.333 when using c in a *SAW* context. That is, a character with only 3 steps of homoplasy will be 10 times less influential than a character with no homoplasy. Thus, the function can be made to weight less strongly against characters with homoplasy by adding a concavity constant, k, which makes the function less concave (i.e. approaching linearity). Character fits are then measured as

$$f = k / (k + h)$$ [Formula 7.2]

With larger values of k, the function approaches linearity, and the analysis approximates the results under equal (or prior) weights.

The function f measures a goodness of fit, as it is to be maximized, and can be interpreted as measuring the weights implied by a tree. Parsimony analysis is generally described in terms of minimizing homoplasy, and this is measured by the complement of f, or $w = 1 - f = h / (k + h)$; w can be interpreted as "weighted homoplasy", and is an increasing function that asymptotically approaches 1 with increasing homoplasy. Note that since w is the complement of f, the trees that minimize w are exactly the same ones that maximize f. Figure 7.4a illustrates the shape of the curves for various k values. The cost of adding a step to a character with h steps of homoplasy is determined as in Formula 7.1; adding the constant of concavity k and operating algebraically, the relative cost a of adding every step of homoplasy is

$$a_{(b)} = \frac{k^2 + k}{k^2 + h^2 + 2kh + k + h}$$ [Formula 7.3]

This is the default weighting formula in TNT; Table 7.1 shows the resulting values for $k = 1$–20 and the first 10 steps of homoplasy.

In addition to the default weighting function, TNT also allows defining any user-weighting function. A user-defined weighting function is specified as lists of costs for different numbers of extra steps; if the cost of adding a step to a character with n steps of homoplasy is defined as $a_{(n)}$, then the user-defined weighted homoplasy for h steps of homoplasy is $W_{u,h} = \sum_{i=0}^{h} a_{(i)}$ The $w_{u,h}$ values are precalculated and stored in memory, and used to convert h into u for tree-scoring. Although self-consistency cannot be guaranteed for some shapes of the weighting function, the method properly applies the user-defined weights dynamically, during the search. This allows defining weighting functions of any shape (including functions that increase weights with homoplasy). Only functions that are exclusively a function of

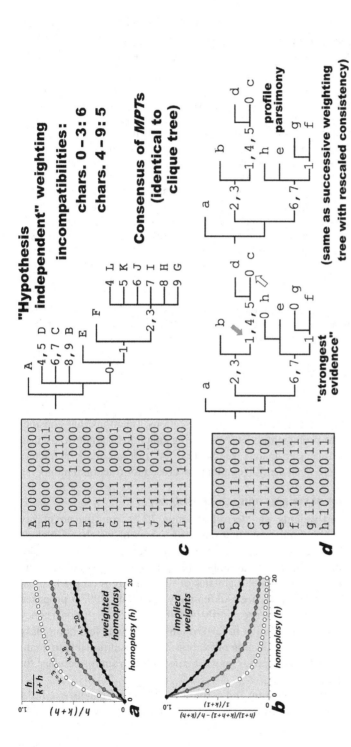

FIGURE 7.4 (a) Shape of the weighting curve under implied weighting, for various values of the concavity constant k. (b) Relative costs (y-axis) of adding a step to a character with different numbers of extra steps (x-axis), rescaled so that the first step of homoplasy has relative cost = 1, for different values of the concavity constant k. (c) A matrix where the *MPT* (or the maximum compatibility tree) displays no homoplasy for the character(s) with more incompatibilities, indicating that weighting (or setting initial weights for successive weighting) based on compatibilities is problematic. (d) A matrix where "strongest evidence" (Salisbury 1999) or "profile parsimony" (Faith and Trueman 2001) produce results that cannot be justified. In both cases, the homoplasy occurring in a terminal (character 0, white arrow) lowers the "weight" of the character more than a homoplasy that occurs in an internal node (character 1, gray arrow). In addition, strongest evidence (due to its dependence on tree shape) produces a group (e+h) for which there are no possible synapomorphies.

TABLE 7.1

Relative Cost of Adding a Step to a Character with Different Numbers of Extra Steps, for Different *k* Values

k	h = 1	h = 2	h = 3	h = 4	h = 5	h = 6	h = 7	h = 8	h = 9	h = 10
1	0.3333	0.1667	0.1000	0.0667	0.0476	0.0357	0.0278	0.0222	0.0182	0.0152
2	0.5000	0.3000	0.2000	0.1429	0.1071	0.0833	0.0667	0.0545	0.0455	0.0385
3	0.6000	0.4000	0.2857	0.2143	0.1667	0.1333	0.1091	0.0909	0.0769	0.0659
4	0.6667	0.4762	0.3571	0.2778	0.2222	0.1818	0.1515	0.1282	0.1099	0.0952
5	0.7143	0.5357	0.4167	0.3333	0.2727	0.2273	0.1923	0.1648	0.1429	0.1250
6	0.7500	0.5833	0.4667	0.3818	0.3182	0.2692	0.2308	0.2000	0.1750	0.1544
7	0.7778	0.6222	0.5091	0.4242	0.3590	0.3077	0.2667	0.2333	0.2059	0.1830
8	0.8000	0.6545	0.5455	0.4615	0.3956	0.3429	0.3000	0.2647	0.2353	0.2105
9	0.8182	0.6818	0.5769	0.4945	0.4286	0.3750	0.3309	0.2941	0.2632	0.2368
10	0.8333	0.7051	0.6044	0.5238	0.4583	0.4044	0.3595	0.3216	0.2895	0.2619
11	0.8462	0.7253	0.6286	0.5500	0.4853	0.4314	0.3860	0.3474	0.3143	0.2857
12	0.8571	0.7429	0.6500	0.5735	0.5098	0.4561	0.4105	0.3714	0.3377	0.3083
13	0.8667	0.7583	0.6691	0.5948	0.5322	0.4789	0.4333	0.3939	0.3597	0.3297
14	0.8750	0.7721	0.6863	0.6140	0.5526	0.5000	0.4545	0.4150	0.3804	0.3500
15	0.8824	0.7843	0.7018	0.6316	0.5714	0.5195	0.4743	0.4348	0.4000	0.3692
16	0.8889	0.7953	0.7158	0.6476	0.5887	0.5375	0.4928	0.4533	0.4185	0.3875
17	0.8947	0.8053	0.7286	0.6623	0.6047	0.5543	0.5100	0.4708	0.4359	0.4048
18	0.9000	0.8143	0.7403	0.6759	0.6196	0.5700	0.5262	0.4872	0.4524	0.4212
19	0.9048	0.8225	0.7510	0.6884	0.6333	0.5846	0.5413	0.5026	0.4680	0.4368
20	0.9091	0.8300	0.7609	0.7000	0.6462	0.5983	0.5556	0.5172	0.4828	0.4516

Note: In All Cases, Values Rescaled so the First Step of Homoplasy Has a Cost of Unity.

h, however, can be defined; as discussed in other sections, functions that consider *m*, *g*, or other factors are problematic. User-weighting functions cannot be used in conjunction with continuous, step-matrix (see Section 7.4.2), or landmark characters (since the function is stored in an array, it cannot be accessed for fractional numbers).

7.4.1.2   Maximization of weights and self-consistency

The very idea of implied weights was derived (Section 7.3.3) from the possibility of choosing among self-consistent trees that maximize sums of weights. Goloboff (1993a, 1995b) argued that choosing the tree that maximizes the overall character weights amounts to choosing the tree that implies that evidence is most reliable, and that this agrees with the general idea of cladistics of selecting trees on the basis of how well they allow explaining evidence. Additionally, Goloboff (1993a) proposed that trees of maximum weights would necessarily be self-consistent. Self-consistency in the abstract sense of resolving conflict in favor of characters with less homoplasy does occur (as discussed in the introduction of Section 7.4), but is it possible to quantify it in a more formal way? This depends on what constitutes an appropriate test for self-consistency.

De Laet (1997, p. 185 ff.; see erratum and correction on p. 184) demonstrated that with the weighting function of Formula 7.2 (and regardless of the value of *k*) any tree A for which $F = \Sigma f$ is higher than in other tree B must imply a lower number of homoplastic steps, weighted according to f—that is, the trees maximizing F are self-consistent, as expected from discussion in the previous section. More precisely, De Laet (1997) demonstrated that the sum of steps for tree A weighted according to the homoplasy of tree A must be less than (or equal to) the sum of steps of tree B weighted according to the homoplasy of tree B, if $F_{(A)} > F_{(B)}$ (or if $F_{(A)} = F_{(B)}$). This is one of the possible ways to test self-consistency.

An alternative (and perhaps more appropriate) test for self-consistency would determine whether an analysis under prior weights fixed at the f values of a tree that maximizes F necessarily leads to that same tree; this is how self-consistency would be checked in a *SAW* context (i.e. as discussed in Section 7.3.3). Note that, in implied weighting, the cost ratios between different amounts of extra steps during a comparison between two trees correspond to ratios for Formula 7.3, not to ratios of f values themselves (Goloboff 1993: 88). The f values on optimal trees can subsequently be interpreted as if they were weights, but they are not exactly the same weights dynamically used during a tree search. Thus, finding a tree of optimal F and then running an analysis under prior weights fixed at those f values may on occasion produce a different tree. Obtaining the same tree under fixed weights is much more likely if the prior weights are fixed at the a values of the tree maximizing F, instead of the f values. However, not even then seems self-consistency warranted: e.g. a variation in 5 steps would be a difference of $5 \times a$ with fixed prior weights, but this would not happen under implied weights (where there would be a smaller difference than $5 \times a$ if adding steps, or a larger one if subtracting); this could perhaps lead to different trees in some cases. And, if setting fixed prior weights from a values rather than f values, then those "weights" are not necessarily being maximized (implied weighting searches trees of maximum Σf, not trees of maximum Σa; although both are often the same trees, that is not the case for all datasets). Self-consistency in this alternative sense and a strict weight maximization can then be mutually exclusive for some datasets. Therefore, Goloboff's (1993a) idea that implied weighting maximizes overall character reliability while at the same time providing self-consistency is only an approximation—a close one, but without mathematical guarantees.

7.4.2   BINARY RECODING, STEP-MATRIX CHARACTERS

In the case of additive characters, the weighted homoplasy in TNT is calculated for the recoded binary variables. Such decomposition had been proposed by Farris (1969) and Carpenter (1988), in *SAW*, and by De Laet (1997), in *IW*.

In the case of step-matrix characters, the value of m is determined by TNT with approximate algorithms (see Chapter 3), and (as before) $h = m - s$. The weighted homoplasy for step-matrix characters is then calculated taking into account the minimum (prior) transformation cost between all possible states, t_{min}:

$$w_h = (h / t_{min}) / (k + h / t_{min})$$

In this way, the scores under *IW* for step-matrix characters are calculated as if the smallest transformation cost was defined as unity. Note that h / t_{min} can be a fractional value, and user-defined weighting functions are stored in indexed tables (which can only be accessed via integers). Therefore, step-matrix characters can only be analyzed with the default weighting function.

7.4.3   Tree Searches

Given that the score is calculated directly from the h values of any tree, IW allows any tree to be evaluated, effectively working as a refined parsimony criterion. Tree calculations proceed exactly as described in Chapter 5, with the only proviso that differences in score are real values instead of integers. For example, values of acceptable suboptimality for processes such as tree-drifting or searches with suboptimal trees need to take into account that score differences may be smaller than unity. The method was first implemented in the program Pee-Wee (Goloboff 1993c, now obsolete); in addition to TNT, the method is implemented also in PAUP*. Keep in mind that PAUP* does not recode additive characters in binary form, so that the fit values reported may show some discrepancy with those reported by TNT (also keep in mind that PAUP* always adds 1 to the k value, so that $k = n$ in TNT equals $k = n - 1$ in PAUP*).

7.4.4   Prior Weights

The IW for a character is independent of prior weights. The user can also apply differential prior weights. Those prior weights are taken into account, simply multiplying character scores by the corresponding prior weight. Prior weights could be used, for example, to represent several independent characters with identical observed distribution with a single one, or for any other legitimate reason. Keep in mind that the reliability of the character itself, as evidenced via inferred homoplasy, is what IW attempts to detect; therefore, in an analysis under IW, the prior weights should *not* be meant to express preconceived notions about the propensity for homoplasy.

The other side of the coin of the previous point is that, when using IW to analyze the matrix, characters which are believed to be highly homoplastic should not be excluded from the analysis. Of course, this is not to say that characters believed to be highly homoplastic should be excluded from analyses under equal weights. But any argument for excluding such characters is even weaker in the case of IW: it is the analysis itself that should identify poor characters, and not prior considerations of reliability. If the character turns out to be very poorly correlated with others, and displays much homoplasy in optimal or near-optimal trees, it will exert only a minor influence on the results.

7.4.5   *IW* and Compatibility

In the 1970s and early 1980s, a method alternative to parsimony analysis, known as compatibility or maximum clique analysis, enjoyed some popularity (especially in botanical circles). The method started with work by Wilson (1965), Camin and Sokal (1965), and Le Quesne (1969), and was further developed by G. Estabrook and C. Meacham (e.g. Estabrook et al. 1977; Meacham 1983). The idea is finding the tree where the maximum number of characters is free of homoplasy at the same time. This tree can be computed by first evaluating the pairwise compatibility among all possible pairs of characters, and then finding the largest set of mutually compatible characters or "clique"; in addition to this phylogenetic approach, clique analysis is also used in several applications of graph theory. Enumerating all largest cliques (given a matrix of character compatibilities) is an NP-complete problem (Karp 1972). Criticism of clique analysis in phylogenetics was provided by Farris (1983) and Farris and Kluge (1985, 1986); the main problem with the method is that it assumes that characters are either absolutely reliable or absolutely unreliable, with no middle ground. Once a character has been observed to display a single instance of homoplasy, the method assumes every instance of similarity in that feature must be due to a separate independent origination (Farris 1983), and there is thus no point in trying to explain similarities in the feature beyond the first separate origination.

Felsenstein (1981) discussed some aspects of character weighting, from the point of view of maximum likelihood. One of the approaches Felsenstein discussed is intended for the situation when one knows that some characters are entirely reliable, others entirely unreliable, and we do not know which is which. Under such circumstances, unsurprisingly, the clique tree(s) are also maximum likelihood trees. One of the points of Felsenstein (1981) was that his weighting approach, which gives characters either a very high or a very low weight, provides a link between parsimony and clique analysis. Felsenstein (1981) also claimed that a very strong weighting function will make *SAW* equivalent to a clique analysis; however, given the mechanics of *SAW*, that claim is incorrect. Taking any tree and reweighting the characters with such a strong function will always lead to the tree being "self-consistent". The actual equivalence (Goloboff 1995b) is with an *IW* scheme, where the weights themselves are to be maximized, and the weights are either 1 (for homoplasy-free characters) or 0 (for characters with any number of steps of homoplasy). Such weights can be defined by the user in TNT (see Section 7.11), as an approach to find sets of mutually compatible characters in datasets with many taxa, nonadditive multistate characters, or missing entries, where exact algorithms for cliques cannot be used. For binary data with no missing entries, TNT also includes J. S. Farris' (2003) implementation of Bron and Kerbosch's (1973) algorithm, which is the fastest available implementation of phylogenetic cliques.

One of the practical problems of clique analysis was that it often produced poorly resolved trees (not many characters are completely compatible with each other in empirical datasets). One of the ways to solve this problem was defining primary and secondary (or higher order) cliques. The primary clique is as just described. The secondary clique would determine the groups that resolve the polytomies of the tree resulting from the primary clique, but that have no more than one additional step of homoplasy (and so on, for higher order cliques). These multi-level clique analyses can also be emulated in TNT, with a user-defined weighting function where the cost of every additional step of homoplasy is sufficiently large, relative to the cost of previous steps. Given the unrealistic assumptions of clique analysis, cliques cannot be recommended for inferring phylogenies, but the method can be useful in some other contexts (e.g. creating supertrees; see Chapter 6).

7.5 WEIGHTING STRENGTH, SENSITIVITY, AND CONSERVATIVENESS

Logic requires that characters with homoplasy are given lower weights, but the strength of downweighting is yet to be determined. That is exactly the purpose of the concavity constant: allowing analysis under different weighting strengths. *IW* has sometimes been criticized (e.g. Turner and Zandee 1995; Congreve and Lamsdell 2016) on the grounds that the results depend on the value of the concavity constant k, but that criticism is poorly justified. It is true that there is no obvious criterion by which to decide exactly how strongly should characters with homoplasy be downweighted, but that is not a problem with *IW* itself—rather, *IW* incorporates a way to take into account that uncertainty via the k constant. Opting for not downweighting homoplastic characters in any way—as done in an analysis under equal weights—is also a decision on how strongly to weight against homoplasy ("not-at-all"), and a decision without rational justification.

Values of k that are too low will approximate a clique analysis, and values of k that are too high will approximate analyses under equal weights, making the different amounts of homoplasy irrelevant for character weights. There can be different arguments for empirical evaluation of differentially vs. equally weighted analyses; one such argument is to prefer the approach which produces more stable or better supported groups. Goloboff et al. (2008a) analyzed 70 morphological datasets under a range of concavities ($k = 5$ to $k = 16$) and estimated the expected stability and group supports. Within that range of concavities, Goloboff et al. (2008a) showed that for *IW* "the results outperformed equal weights regardless of the concavity constant chosen", thus diminishing the relevance of the idea that application of *IW* is problematic because one does not know the exact k value.

Wheeler (1995; see also Giribet 2003) has proposed that when the results depend on a certain parameter whose value cannot be established exactly, a range of values must be tried, to determine whether the results are sensitive to what is assumed. This has been called *sensitivity analysis*, and it can be applied to analyses under *IW*. Although sensitivity analysis has been criticized on the grounds of being unscientific (e.g. Grant and Kluge 2005), such criticisms stem from an overly narrow philosophical perspective (Giribet and Wheeler 2007). Variating the values of the parameter can be used to maximize resolution, congruence between datasets, or support. This has been proposed as a way to select "optimal" values of k (e.g. by Ramírez 2003; see also Mirande 2009). Even if some "optimal" value of k can be determined on some grounds, it seems obvious that any group absent in the results for a slightly different k cannot be considered as firmly established. Therefore, leaving aside the possible utility of such attempts to identify a unique value of k maximizing some criterion, the most conservative approach is to examine a range of possible k values and retain the groups that are common to all the concavities explored. Obviously, concavities that are too low must be avoided, for the analysis then resembles a clique method more than parsimony, thus producing more questionable results. This was used by some critics of *IW* to their advantage: O'Reilly et al. (2016) and Puttick et al. (2017b) intended to belittle *IW* and in their initial experiments $k = 2$ was the worst performing out of six k values evaluated, so they chose $k = 2$ for all their subsequent comparisons, which allowed them to be more critical of *IW* (see discussion in Goloboff et al. 2017). With $k = 6$, the cost of adding a step to a character with one step of homoplasy is already 25% lower than in the absence of homoplasy (from Formula 7.3), so that this is a reasonable lower limit for k. On the other hand, for most morphological datasets, concavities that are too high simply select some of the trees shortest under equal weights, thus suggesting $k = 10$ or $k = 12$ as the upper limit (for $k = 12$, it takes *two* steps of homoplasy for the implied cost of an additional step to be 25% lower; see Table 7.1). In the case of group supports (see Chapter 8), the lowest values for each group can be used; this can be done either manually (repeating support calculations for a range of concavities) or with simple scripts (see Chapter 10).

One of the points repeatedly discussed in relation to *IW* (Goloboff 1993a, 1995b; Goloboff et al. 2008a) is that datasets with different numbers of taxa may require the use of different concavities. In datasets with few taxa, the maximum homoplasy that a most parsimonious optimization may require is limited; as datasets have more taxa, there are ampler opportunities for homoplasy to manifest. The lower limit is the case of datasets with 5 or fewer taxa, where application of *IW* with any concavity produces the same results as equal weights (for such a small dataset, a character can either have or not have homoplasy; there are no degrees of homoplasy). That by itself suggests a milder concavity for larger numbers of taxa. Goloboff et al. (2008a) also considered the case of molecular datasets, which commonly have hundreds or thousands of taxa. In such datasets, some characters can have hundreds of extra steps on near-optimal trees, and—under normally used values of concavity—that means that such characters will be almost completely uninfluential relative to characters with just a few steps. *IW* in such a case comes down to an almost complete elimination of those characters. A possibility (Goloboff et al. 2008a) is to select the value of k such that the ratio r for the weights of a character with no homoplasy and a character with the maximum possible homoplasy in the matrix equals a certain value (all other weight ratios for the matrix will necessarily be within that range). This can be achieved by solving[7] Formula 7.3 for k, when the ratio is r $(r > 1)$ and the reference homoplasy (e.g. the maximum possible) is h:

$$k = \frac{1 - \left(\frac{2h+1}{2}\right) - \sqrt{\left(\frac{2h+1-r}{r}\right)^2 - 4\left(\frac{1-r}{r}\right)\left(\frac{h^2+h}{r}\right)}}{\frac{2}{r} - 2} \qquad \text{[Formula 7.4]}$$

This possibility is incorporated into recent versions of TNT. Note that (like the rescaled consistency index) this determines weighting strength from the maximum possible steps of the characters in the matrix, but (unlike the rescaled consistency index) does so uniformly for all characters in the matrix. Limiting the range of weights in this manner will rarely be important for morphological datasets, where the differences in homoplasy between the worst and best characters will usually be smaller than in molecular datasets with thousands of taxa. However, as the exact weighting strength of a given k value is hard to intuit, some users may prefer to set the value of concavity for a specific implied weight to occur at a given value of homoplasy.

Without invoking a specific evolutionary model, it seems unlikely that the problem of determining the exact value of k can be solved. In fact, it cannot even be logically posed, because k itself is not a parameter of the evolutionary model, but simply
a numerical device used to let the weighting method be more or less flexible in the
penalizing against homoplasy. Variables that are part of the inference model more
than part of the evolutionary model itself also occur in maximum likelihood methods. An example is the gamma distribution; as put by Felsenstein (2004: 219), "there
is nothing about the gamma distribution that makes it more biologically realistic than
any other distribution, such as the lognormal. It is used because of its mathematical
tractability". The α parameter of the gamma distribution determines its shape, but
there is no biological correlate of the "α-parameter". Back to weighted parsimony,
which assumes no specific evolutionary model, there is no reason to think that a
"true" value of k exists, particularly one that will remain constant across time and
throughout the tree. The best way to take into account such type of uncertainty is to
determine to what extent the results depend on assuming a specific value of k—as
done in sensitivity analysis, discussed earlier.

7.6   PRACTICAL CONSEQUENCES OF APPLICATION OF *IW*

The differences between applying equal or implied weights to empirical datasets
follow certain general patterns. An obvious one is that the strict consensus tree for
IW tends to be more resolved than for equal weights. A tree-scoring function (like
the one in *IW*) that is much more finely grained than equal weights makes exact ties
between trees more unlikely, and so such scoring function will produce on average
more resolved consensus trees (this is even truer of maximum likelihood). Therefore,
the more resolved consensus trees cannot be claimed as a legitimate reason to prefer
IW (even if they are a nice byproduct of it). The advantage claimed for *IW* is that it
better takes into account character reliability, and if this works indeed as intended,
it will be so even when *IW* produces a less resolved tree than equal weights (which
may eventually happen).

Evaluating the strength of support for groups is always important, but in light
of the preceding discussion, even more important for *IW*. Under *IW*, there is the
possibility that a tree differing in only a tiny difference in score will lack some
of the group(s) present in the consensus, and it is clear that such groups cannot be
considered as well supported. The problem of group supports then must always be
addressed when using *IW*, either by considering suboptimal trees, or by resampling
methods (see Chapter 8).

In empirical datasets, it is commonly observed that the resampling values of group
support are higher under *IW* (Goloboff et al. 2008a). Methods that produce higher
supports are often interpreted to be preferable (e.g. Källersjö et al. 1999). Although
such interpretation is not undisputable (see Section 7.10.2), methods to measure support based on resampling provide a proxy for the expected stability. Support is not
exactly the same thing as stability, but the two concepts are somewhat related. Thus,
results under *IW* could be expected to have a somewhat greater stability than results
under equal weights, based on such considerations. Of course, it is almost self-evident
that a method that produces better trees should also produce more stable results. For

resampling measures more directly aimed at estimating expected stability, Goloboff et al. (2008a) also observed that *IW* fares better than equal weighting on empirical datasets. Obviously, the ideal test for stability is the comparison between results for empirical datasets as they evolve through time and with work by different taxonomists, but not many such multi-layered datasets are publicly available, and direct comparison between them is often made difficult by incompletely overlapping taxon sets or other factors.

Another pattern that emerges quite clearly when comparing implied and equal weighting analyses is that the two methods tend to produce the same results for datasets with few taxa. For fewer than 30 or 20 taxa, the two methods produce either the same strict consensus tree, or *IW* selects some of the trees that are shortest under equal weights, thus providing a somewhat more resolved consensus than equal weights (but with no contradicting groups). Therefore, application of *IW* is less important in datasets with too few taxa. Only for larger datasets, above 30 or 40 taxa, can the differences in homoplasy for the characters be large enough as to lead *IW* into preferring trees that are not shortest under equal weights—that is, into producing a strict consensus with some groups that contradict the groups in the strict consensus under equal weights. An important point to keep in mind is that, whatever the differences between the results for equal and implied weights, both are parsimony methods. Any group found by *IW* and not equal weights will be supported by some alternative characters acting as synapomorphies, characters that are there, in the matrix, scored and sanctioned by the investigator. *IW* only can emphasize those characters over others, but never creates groups *ex nihilo*.

The tree searches when using *IW* can proceed at almost the same speed as searches under prior (equal) weights. Depending on the type of search to be performed, the ease with which optimal trees can be found influences the times needed to complete a search (e.g. under exact solutions, or driven searches where a pre-specified number of hits to the optimal trees is to be effected). Sometimes, *IW* makes the tree-score landscape smoother, and those search algorithms proceed faster than under equal weights; however, this is by no means universal. Note that the method of indirect calculation of tree lengths (see Chapter 5) is applicable under *IW*. The only minor modification required is that the decreases in score when clipping the tree are calculated as the difference in the values for the fitting function (from the h values before and after the clipping). The same is done for every reinsertion point of the clipped clade. As floating point operations are not exact, the score calculated in this manner may not be exactly the same as the score that would result from first calculating h for each character and summing up all the corresponding values of weighted homoplasies. For coherence in reported values, TNT uses a tolerance of 10^{-3}, recalculating the scores from scratch on the basis of h values only for those trees within the 10^{-3} error margin. *IW* in PAUP* is much slower than in TNT; e.g. for the dataset of Pei et al. 2020, TNT completes a *RAS+TBR* saving up to 10 trees with $k = 8$ in about 0.5 sec., but PAUP* takes 109 sec. to do the same under $k = 7$ (for a speed ratio of about 200:1). PAUP* apparently calculates scores from scratch (from h values derived via indirect length calculations, applying Formula 7.2 to every character) for every rearrangement tried. This is many more floating point operations (with the method used in TNT, only characters for which h changes relative to the divided tree must be checked when reinserting the clipped clade).

7.7 PROBLEMATIC METHODS FOR EVALUATING DATA QUALITY

Several methods aimed at detecting unreliable datasets have been proposed. Methods like *permutation tail probability* (Faith and Cranston 1991) or *homoplasy excess ratio* (Archie 1989) were based on permutation (see critiques by Goloboff 1991b; Farris 1991, 1995; Farris et al. 1994; Carpenter et al. 1998). The *skewness* test (Hillis 1991; Huelsenbeck 1991) was based on the shape of the length distribution of all possible trees (see critique by Källersjö et al. 1992). RASA or *relative apparent synapomorphy analysis* (Lyons-Weiler et al. 1996) attempted to determine "whether a measure of the rate of increase of cladistic similarity among pairs of taxa as a function of phenetic similarity is greater than a null equiprobable rate of increase", but the measure used was plagued with problems (see critiques by Farris 2002; Faivovich 2002). Their problems at distinguishing good from bad datasets aside, all those methods assess datasets as a whole. What is really needed is methods that can identify better or worse *characters*. Some of the methods proposed for assessing character quality have been inspired in methods for evaluation of whole datasets, and they all seem problematic.

7.7.1 TREE-INDEPENDENT

For identification of less reliable characters, other methods not based on homoplasy have been proposed. Sharkey (1989) proposed to derive weights (or initial weights, followed by *SAW*) from the number of character incompatibilities (see comments by Farris et al. 1990; Goloboff 1993a; Wilkinson 1994b). A problem with Sharkey's method is that a character with fewer incompatibilities may well have more homoplasy on the maximum compatibility tree (or, more obviously so, on the most parsimonious tree). The maximum compatibility tree may be that determined by characters which (individually) have more incompatibilities, because the characters with fewer may in turn be incompatible among each other. That is why finding the clique tree is an NP-complete problem after all; just calculating the number of character incompatibilities is a quick calculation, but insufficient for finding maximum compatibility trees. An example is shown in Figure 7.4c, where the first four characters (each with 6 incompatibilities) jointly determine the maximum clique (as well as the *MPT*); each of the other six characters in the matrix (with only 5 incompatibilities) contradicts the first four, but those six are also incompatible with each other. There is thus little reason to set weight *a priori* on the basis of incompatibilities.

Other methods that attempt to determine character reliability from the dataset alone, prior to inferring any trees, are *observed variability* (OV; Goremykin et al. 2010) and *tree-independent generation of evolutionary rates* (TIGER; Cummins and McInerney 2011). Simmons and Gatesy (2016) point out that these methods "can favor convergences and reversals over synapomorphy, exacerbate long-branch attraction, and produce mutually exclusive phylogenetic inferences that are dependent upon differential taxon sampling".

7.7.2 PROBABILITY-BASED

Both Salisbury (1999, 2000) and Faith and Trueman (2001) have proposed methods that resemble *IW* in searching trees optimal under a scoring function that penalizes additional steps of homoplasy more mildly. However, the weighting strength used is determined from probabilistic considerations, and is different for different characters. Yet another attempt (which takes most of the tree fixed) to weight characters on the basis of probabilities of homoplasy under some randomization scheme is by Berger and Stamatakis[8] (2010).

Salisbury's (1999) *strongest evidence* (*SE*) uses as penalty the probability of obtaining the homoplasy observed on a tree if the character is permuted. Obtaining few steps when permuting is of course less probable than obtaining larger numbers of steps; the tree for which the observed number of steps is least probable is to be preferred. The probabilities for every number of extra steps in each character can be converted into minus log likelihoods, and then their sum is to be maximized (just as *IW* maximizes Formula 7.2). Farris (2000) provided a criticism, focusing mostly on Salisbury's (1999) philosophical interpretations. Another problem with *SE* is that the probability of obtaining different numbers of steps with permutation depends on the numbers of 0 and 1 states in every character. Therefore, at equal amounts of homoplasy, characters with more even partitions are then generally more influential (because they have lower probabilities). What is even worse, the probability of a given number of steps for a fixed number of 0's and 1's depends *also* on the topology of the tree. For example, for a symmetric 8-taxon tree, there is no permutation of a 3:5 partition with fewer than two steps; therefore, a symmetric tree with 2 steps is penalized the same as a pectinate tree with 1. Figure 7.4d illustrates this method; the conflict is in the first two characters, with the rest of the tree determined by the remaining six characters. The single tree preferred by *SE* displays the group based on the second character; both the first and second character have an independent transformation in the group BCD. For *SE*, the fact that the first character had a parallel transformation in a terminal, C, instead of having changed in the common ancestor of CD, is taken as a reason to prefer the tree with the group FG. Distinguishing whether the parallelism occurs in a terminal or in an internal node does not make sense; a homoplasy is a homoplasy, regardless of where it occurs. The other problem with *SE*, that it depends on tree shape, manifests in the method preferring a tree with the group EH. The reason for that preference is that the permutation on trees of other shapes [e.g. (E(H(FG)))] would confer a higher probability to the observed numbers of steps. Thus, *SE* leads to conclude the existence of a group not supported by any possible synapomorphies, EH. The method is something entirely different from parsimony, for it distinguishes among trees that have the same numbers of steps for all characters. One just wonders why the numbers of steps (observed and permuted) are calculated by means of parsimony, if the final results can in fact contradict the most basic tenet of parsimony.

Faith and Trueman (2001) proposed *profile parsimony*, *PP*, a method similar to Salisbury's (1999), except that the probability is that of obtaining the observed number of steps by generating a random tree (produced by examining the lengths for the character among all possible binary trees). Arias and Miranda (2004) have shown that *PP* produces results relatively similar to those of *IW*. *PP* at least scores trees based on their numbers of steps, instead of their shapes, but it continues depending on the numbers of 0's and 1's. Just like *SE*, *PP* prefers (see Figure 7.4d) trees with less homoplasy in the character which independently originates at an internal node, instead of a terminal. Note that *SAW* using the rescaled consistency index would have the same preference in this case.

Both *SE* and *PP*, to be correctly calculated, would require that the length distributions used in each case are calculated exhaustively (in the case of *SE*, this must be done for every tree[9] examined!). Arias and Miranda (2004) note that the distribution (in *PP*, but their reasoning extends to *SE*) could be calculated by a random sampling, but that this is likely to be inaccurate (as the length distribution is typically very asymmetrical, very large samples are needed). For *PP*, Carter et al.'s (1990) formula (with Goloboff's 1991a correction for missing entries) for the number of trees with a given number of steps could be used (as done by Arias and Miranda 2004 in their comparisons), but this is applicable only to datasets with binary characters—multistate characters are simply not amenable to analysis by this method.

7.8   IMPROVEMENTS TO *IW*

7.8.1   INFLUENCE OF MISSING ENTRIES

In any weighting method based on homoplasy, missing entries are problematic, in that they can conceal homoplasy. As a consequence, when different characters have different numbers of missing entries, the characters with more missing entries will (on average) be assigned higher weights by *IW*. The fact that uneven numbers of missing entries make it difficult to evaluate homoplasy on a uniform scale is sometimes invoked as an argument to prefer equal weights over *IW*. Missing entries are a legitimate concern, particularly in paleontological datasets (where missing entries abound). However, it is less than obvious that equal weights is preferable. Preferring equal weights simply amounts to trading a method with known problems, for another method with different—and possibly worse—problems.

The caveat about missing entries extends to other methods, like *SAW*, but taking into account the problem in that context is easier. A character i with observed homoplasy h_i and o_i observed entries for the t total taxa can be inferred to have a fraction of the homoplasy. If the missing entries hide, on average, as much homoplasy as is contained in o, then the expected homoplasy if all missing entries were scored would be $e_i = h_i t / o_i$. If using, for example, the consistency index c to reweight, one simply gives every character a weight of $1 - e_i / m_i$ instead of $1 - h_i / m_i$ as usual.

In an *IW* context, taking into account missing entries is more delicate, because in such context a weighting function that drops faster for some characters has the opposite effect than under *SAW*: it will make the characters with the faster-dropping function more influential (cf. the discussion for the retention index in Section 7.4.1). The tempting approach for taking into account missing entries would be, instead of minimizing $\Sigma h_i / (k+h_i)$ as usually done, to minimize $\Sigma e_i / (k+e_i)$. But this would not work as intended, as $e_i > h_i$, it would make characters with missing entries more influential, not less.

A simple solution to this problem was offered by Goloboff (2014a), with *extended implied weights*, and is based on using (for the character with missing entries) a scoring function that lowers the weight more rapidly as homoplasy occurs; in the absence of homoplasy, the weight is equivalent to that of a homoplasy-free character with no missing entries. *IW* works by adding successively smaller steps as homoplasy increases. Then, the cost of adding a step to a character i where h_i steps of homoplasy occur in the tree but the expected homoplasy is e_i should be the same as the cost of adding a step to a character with no missing entries and e_i steps of homoplasy—which is less than the cost of adding a step to only h_i steps of homoplasy. The first step (from $h_i = 0$ to $h_i = 1$) is defined to be the same for all characters; in the absence of homoplasy, all characters are equally influential. That is, instead of the cost of adding a step for h_i being as in Formula 7.3, it is

$$a_{(h)} = \frac{k^2 + k}{k^2 + e_i^2 + 2ke_i + k + e_i} \qquad \text{[Formula 7.5]}$$

where (as earlier) $e_i = h_i t / o$. The values for every step of homoplasy h_i could be precalculated (separately for every character i) and stored in an array, so that the weighted score for h steps of homoplasy is $w_{i,h} = \sum_{j=0}^{h} a_{(k,j)}$. The problem with storing the values in an array is that a dataset may include continuous and landmark characters (see Chapter 9), for which the values of "homoplasy" can be fractional numbers, and arrays can only be accessed by means of integers. For a simpler approximate solution of the problem, Goloboff (2014a) took advantage of one of the consequences of the default hyperbolic function for two concavity values k_1 and k_2, where $k_1 > k_2$. The cost of a step addition for h steps of homoplasy under k_2 approximates the cost of adding a step to hk_1 / k_2 steps under k_1, divided by the ratio of costs of the first step under k_1 and k_2. That is,

$$a_{(k_1, hk_1 / k_2)} \approx a_{(k_2, h)(k_2 + 1)/(k_1 + 1)}$$

Note that the factor $(k_2 + 1) / (k_1 + 1)$ is a constant ϕ, used to rescale all values of the weighting function for the character with missing entries. In this way, for a character with o observed entries out of t taxa, the concavity k_m for the character with missing entries can just be set to $k_m = k_0 o/t$ (where k_0 is the reference concavity, for a character with no missing entries), and the values rescaled by ϕ. For values of k_0 above 10, and relatively large values of o, the correspondence is quite close. For example, for $k_0 = 12$, and $k_m = 4$ (i.e. $o / t = 1/3$), the cost of adding the first 15 steps of homoplasy to the character with only a third of the entries observed is off by 2.50 to 10.34% (see Table 7.2). Given the uncertainties (e.g. in determining how much homoplasy is expected to be concealed by the missing entries; see next), Goloboff (2014a) considered this as an acceptable approximation. Recent versions of TNT include the option to adjust k_m to a value such that the deviation in the cost of adding a step for the first 15 steps of homoplasy is minimal; this is found heuristically and produces smaller values of k_m than $k_0 o / t$. When applying this method, even for smaller concavities, the deviation from the correct values is negligible (see Table 7.2; for $k_0 = 6$ and $o/t = 1/3$, the deviation is within 1% for at least the first 100 steps of homoplasy). Figure 7.5b shows the correspondence between $k_0 = 12$, and concavities corresponding to $o/t = 0.333$ and $o/t = 0.666$; the lines are not perfectly horizontal, but the error being within 0.10% for the values shown, the deviation is imperceptible.

When a user-weighting function is in effect, TNT needs to use arrays to store the values of weighted scores $w_{u,h}$ (see Section 7.4.1.1). In such cases, the same weighting

TABLE 7.2
Comparison Between Different Concavities

h	K4.000	K8.000	Dif4/8	K12.000	Dif4/12	K3.703	K7.855	Dif4/8	K12.000	Dif4/12
1	0.66667	0.65455	+1.82%	0.65000	+2.50%	0.64929	0.64992	-0.10%	0.65000	-0.11%
2	0.47619	0.46154	+3.08%	0.45614	+4.21%	0.45555	0.45608	-0.12%	0.45614	-0.13%
3	0.35714	0.34286	+4.00%	0.33766	+5.45%	0.33727	0.33763	-0.11%	0.33766	-0.12%
4	0.27778	0.26471	+4.71%	0.26000	+6.40%	0.25976	0.25999	-0.09%	0.26000	-0.09%
5	0.22222	0.21053	+5.26%	0.20635	+7.14%	0.20622	0.20635	-0.07%	0.20635	-0.06%
6	0.18182	0.17143	+5.71%	0.16774	+7.74%	0.16768	0.16776	-0.04%	0.16774	-0.04%
7	0.15152	0.14229	+6.09%	0.13904	+8.24%	0.13903	0.13906	-0.02%	0.13904	-0.01%
8	0.12821	0.12000	+6.40%	0.11712	+8.65%	0.11714	0.11714	+0.00%	0.11712	+0.02%
9	0.10989	0.10256	+6.67%	0.10000	+9.00%	0.10004	0.10002	+0.02%	0.10000	+0.04%
10	0.09524	0.08867	+6.90%	0.08638	+9.30%	0.08643	0.08640	+0.04%	0.08638	+0.06%
11	0.08333	0.07742	+7.10%	0.07536	+9.57%	0.07542	0.07538	+0.05%	0.07536	+0.08%
12	0.07353	0.06818	+7.27%	0.06633	+9.80%	0.06639	0.06635	+0.07%	0.06633	+0.10%
13	0.06536	0.06050	+7.43%	0.05882	+10.00%	0.05889	0.05884	+0.08%	0.05882	+0.12%
14	0.05848	0.05405	+7.57%	0.05253	+10.18%	0.05259	0.05255	+0.09%	0.05253	+0.13%
15	0.05263	0.04858	+7.69%	0.04719	+10.34%	0.04725	0.04721	+0.10%	0.04719	+0.14%

Note: The table shows the differences in cost between adding a step to a scoring function with $k = 4$ relative to adding twice that many steps under $k = 8$ (Dif4/8) or adding three times that many steps under $k = 12$ (Dif4/12), in all cases rescaling so that first step of homoplasy has cost unity. Left part of the table shows Goloboff's (2014a) approximation, of multiplying by the proportion of observed entries. Right part adjusts the concavities so that the average error (relative to the cost of $2x$ and $3x$ times the steps under $k = 12$) is minimized for the first 15 steps. The error for subsequent steps of homoplasy (not shown) increases with homoplasy, but for 100 steps under $k = 3.703$ is still within 0.39% (relative to $3x$ times the steps under $k = 12$).

function is used for all characters, but modified for every character i so that the addition of a step for h steps of homoplasy costs the same as for the expected $e_i = ht / o_i$ steps. If e_i is fractional, an interpolation is used (to save time, the interpolator is used only once, when calculating the corresponding $w_{u,h}$ values for each character, and reused during a search). As long as t/o is not a fractional number, this produces exact values of extrapolated costs for the observed homoplasy.

Finally, TNT allows assuming that the unobserved entries u only have a proportion, p, of the average homoplasy in the observed entries. For this, all the preceding considerations apply, with no substantial changes, and $e_i = h_i(u_ip+o_i)/o_i$. The p constant can be used to obtain results intermediate between those just described (when $p = 1$), and a standard *IW* (when $p = 0$). In addition, as all the calculation machinery is already in place, TNT also allows the user to define different concavities for different characters. In that case, TNT will always normalize the values by the cost of the first step of homoplasy relative to a reference concavity. However, arbitrarily defined weighting strengths should be used with extreme caution. In the case of missing entries, the goal of using a different concavity is to make the treatment more uniform—that is, by diminishing the distorting effect of missing entries. Otherwise, the use of different weighting functions for different characters seems very difficult to justify.

7.8.2 UNIFORM AND AVERAGE WEIGHTING OF MOLECULAR PARTITIONS

Another two options implemented in TNT are of little use in morphology-only datasets, but may be important in datasets combining morphology and molecules. Differentially weighting the individual positions within a gene is harder to justify (Section 7.10.5), and then it may be desirable to use a uniform weighting for all the positions in the gene. That uniform weighting may well be a constant defined by the user, or depend on the average homoplasy of the gene. Of course, uniformly weighting a set of morphological characters by their average homoplasy is also possible—given the implementation in TNT—but it seems that this will be applicable only to exploratory approaches, as it seems impossible to define meaningful partitions in morphology. In the case of genes, sets of characters to be uniformly weighted can be defined non-arbitrarily (e.g. genes; first, second, and third positions; segments; etc.). In addition, Goloboff and Arias (2019) showed that uniform weighting by average homoplasy of partitions (approximated by likelihood; see Section 7.9) seemed preferable to standard likelihood models (as judged by the Akaike information criterion; Akaike 1973) in about one fourth of the 37 molecular datasets examined.

Adapting *IW* for uniform weighting by average homoplasy of a set of characters is, like consideration of missing entries, more complicated than in a *SAW* context. Using the total homoplasy H for the entire set and the standard formula, as $H / (k + H)$, will tend to lower the influence of sets with more characters (as larger sets have larger values of H, adding a single step will usually cost less). This is exemplified in Figure 7.5c,d. The correct weighting (Goloboff 2014a) should consider that, in a set with n characters, the average homoplasy is $\bar{H} = H / n$. The score for a set of n characters with uniformly distributed homoplasy $\bar{H}$ is, clearly, $n (\bar{H} / (k + \bar{H}))$ (as in Figure 7.5e). This allows the proper comparisons of amounts of homoplasy.

Using this approach, it is possible to, for example, weight morphological characters individually, and every gene uniformly according to its average homoplasy (as in Figure 7.5c). Adapting algorithms for speeding up tree searches for usage with set weighting requires some precautions (see Goloboff 2014a for discussion). In the case of weighting sets, where some taxa may be scored only with missing entries (as in gene 2 in Figure 7.5c), it is important to take into account missing entries, as discussed in the previous section. Otherwise, the effect of entire blocks of missing entries may be pronounced. In that case, calculation of the set score would use the corresponding concavity.

7.8.3 SELF-WEIGHTED OPTIMIZATION AND STATE TRANSFORMATIONS

Another modification of *IW* is to consider the cost of transformations between states. This was proposed by Goloboff (1997), with self-weighted optimization. The idea was hardly new; Williams and Fitch (1990) had already adapted *SAW* for this purpose, calling the method *dynamic weighting*, and Sharkey (1993) had proposed some alternative indices for weighting transformations. Those previous contributions, however, ignored some of the potential problems with weighting character-state transformations in a *SAW* context.

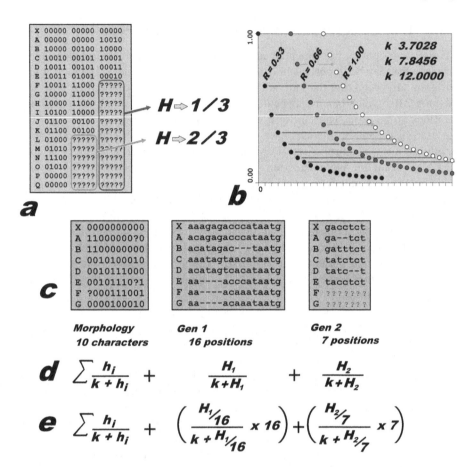

a

b

$H \Rightarrow 1/3$

$H \Rightarrow 2/3$

k 3.7028
k 7.8456
k 12.0000

c

Morphology
10 characters

Gen 1
16 positions

Gen 2
7 positions

$$d \quad \sum \frac{h_i}{k+h_i} + \frac{H_1}{k+H_1} + \frac{H_2}{k+H_2}$$

$$e \quad \sum \frac{h_i}{k+h_i} + \left(\frac{H_{1/16}}{k+H_{1/16}} \times 16\right) + \left(\frac{H_{2/7}}{k+H_{2/7}} \times 7\right)$$

FIGURE 7.5 (a) Characters with missing entries usually receive higher weights than fully scored characters. To remedy this, in a character with u unobserved and o observed entries, the cost of adding a step to a character with X steps of homoplasy can be set as equal to the cost of adding a step to a character with no missing entries and $X(u+o)/o$ steps of homoplasy. (b) The simplest way to approximate the cost proportions expected for the matrix in (a) is to assign different concavities to the characters (rescaling so that the first step of homoplasy always has the same cost). (c) In a matrix with three partitions, combining morphology and two genes, the morphological characters can be weighted individually, and the molecular characters can be weighted collectively (based on the average homoplasy of each gene). (d) Attempting to weight genes collectively by using the standard formula for implied weighting (with the total homoplasy of the gen) would tend to make genes with fewer positions more influential (adding a step to a few steps will generally be more significant than adding a step to many). (d) The proper collective weighting is accomplished by modifying the formula for implied weighting, so that it uses the average implied weight (i.e. calculated with the average homoplasy) multiplied by the number of characters in the partition.

Weighting character state transformations on the basis of a reconstruction on a tree begs the question of the transformation costs used to obtain the reconstruction. In this regard, the method must iterate character-state optimizations, not just tree searches (as done by Williams and Fitch 1990). Consider the example of Figure 7.6a, and assume that transformations are reweighted using $1 / (1 + n_{ij})$, where n_{ij} is the number of changes $i \rightarrow j$ in the reconstruction. Finding ancestral states with equal transformation costs produces a reconstruction (left) with $n_{01} = 7$ and $n_{10} = 1$.

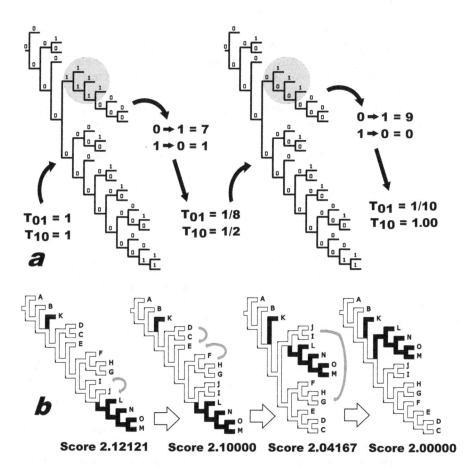

FIGURE 7.6 (a) Using an iterative evaluation of character-state transformation costs would suffer from the same problems of successive weighting, but at the level of character-state optimization. For example, the reconstruction produced by optimizing the character with equal transformation costs implies costs that produce a different reconstruction, which in turn imply a different set of costs. (b) Example showing that weighting characters differently in different branches of the tree cannot be called "parsimony". The cost of a transformation *white* → *black* decreases as similar transformations occur closer in the tree, and this in turn implies that the overall tree cost is decreased if taxa with the *white* state are grouped together—which places independent derivations of the *black* state closer to each other, but amounts to grouping by homoplasy and does not improve the explanation by reference to common ancestry.

Therefore, the corresponding transformation costs must be reset to $T_{01} = 1/8$, and $T_{10} = 1/2$, instead of $T_{01} = T_{10} = 1$. But, for those transformation costs, the reconstruction used is no longer optimal! If ancestral states are calculated (on the same tree) using the new set of transformation costs (producing the ancestral states shown in the right tree), the numbers of transformations between states are now different ($n_{01} = 9$ and $n_{10} = 0$). The transformation costs that correspond to those numbers of changes are then $T_{01} = 1/10$ and $T_{10} = 1$. In the example, recalculating ancestral states with cost ratios 10:1 produces the same set of state assignments, but that need not be so. Finding consistent transformation costs may then require iteratively re-optimizing trees until stability. And, just like in SAW, the final transformation costs implied by a different set of initial costs might have been different: why prefer the final costs that result from initial equal costs? Any set of transformation costs that implies itself is consistent; can there be any grounds to prefer some consistent sets of transformation costs over others?

As the goal of a most parsimonious reconstruction is to find the ancestral states that explain as many similarities as possible by homology, a reasoning similar to that for standard IW can be applied. That would lead to selecting reconstructions on the basis of the summed implied costs for all transformations. That is, for a given tree and character-state reconstruction, the summed weighted differences d in character states are,

$$d = \sum n_{ij} / (k + n_{ij})$$ [Formula 7.6]

Define D as the minimum value of d (among all reconstructions for the given tree). The reconstruction which minimizes d takes into account that the preference for homologizing a state over the other decreases to the extent that there seem to be more repeated transformations into that character in the tree. For multistate characters, Formula 7.6 also takes into account from what state the new one is transforming; see Goloboff 1997 for discussion of less appealing alternatives. Among all possible trees, those with minimum D should be selected.

When applied to the example of Figure 7.5a, the reconstruction shown on the left tree (suboptimal under its implied transformation costs) has $d = (1/8) + (1/2) = 0.625$, and the reconstruction on the right (optimal under its implied costs) has $d = (1/9) + 1/1) = 1.111$. Therefore, evaluation under Formula 7.6 automatically leads to prefer the same reconstruction as dynamic weighting, but without the need to iterate. Goloboff (1997) discussed the need to make the formula lineal for the first two transformations (i.e. there can be zero transformations between two states in a reconstruction, unlike the number of steps in an informative character, and the first transformation does not represent "homoplasy"). The reliabilities of different character state transformations (including asymmetries) are a consequence of the analysis, instead of a premise. A similar approach was independently proposed by De Laet (1997).

The distinction between weighting the entire characters, or weighting the different kinds and directions of transformations, may be important in some datasets. In a dataset (like the one in Figure 7.6a) where a critical character has many $0 \rightarrow 1$ transformations and few $1 \rightarrow 0$, standard IW will downweight the entire character—including $1 \rightarrow 0$ transformations, which seem to be well associated with groups. The present approach will leave $1 \rightarrow 0$ transformations with a high weight, downweighting only the $0 \rightarrow 1$. As in the case of IW, if differential prior costs have been defined for the possible character-state transformations (e.g. in an additive character), those are combined with the implied transformation cost, by multiplication.

The obvious problem with selecting reconstructions to minimize homoplasy under implied transformation costs is that standard algorithms for finding optimal ancestral reconstructions are inapplicable. The calculation of an optimal reconstruction (i.e. one minimizing d) cannot be decomposed into a series of local problems (as discussed in Chapter 3). This is because the cost of assigning a state to a given internal node depends on how many transformations from the state to others are being postulated in other parts of the tree, and that can only be known in a complete reconstruction. Therefore, algorithms working with down- or up-passes are just not possible. Even Sankoff and Cedergren's (1983) algorithm, the most general one for optimization of ancestral states, works by decomposing the reconstruction into a series of local subproblems.

Therefore, finding optimal reconstructions under this criterion requires an exhaustive enumeration of reconstructions. Recall that the number of possible reconstructions increases rapidly with number of possible states, and so applying the method to characters with more possible states will take longer. De Laet (1997: 91–92) also noted that minimization of d would be particularly time-consuming, and proposed that a branch-and-bound reasoning could be applied to save some time. Goloboff (1997) proposed the use of branch-and-bound, but combined with several shortcuts (some of which apply only to characters where the numbers of states is NS or more). These shortcuts reduce the number of nodes for which states need to be tried, and can discard ahead of time many possible reconstructions that would be suboptimal:

a) Nodes which lead to only a single state (or missing entries) are fixed at that state.

b) Nodes ancestral to R or more successive sister branches (default $R = 8$) leading to the same fixed state (as in (a)) are fixed at that state. Lowering R speeds up the search for reconstructions, but too low a value may miss some optimal reconstructions (consider the example of Figure 7.6a).

c) For nodes ancestral to fewer than R successive sister branches fixed at a state β, assume that assigning a state to one node will have the same state as optimal for all the others. The n ancestral nodes can then be assigned a state in tandem; if the ancestral state α is different from β, the assignment implies increasing the number of $\alpha \rightarrow \beta$ transformations by n. That is, a state needs to be tried for just one node, not the R repetitive nodes.

d) Forbid state transformations not occurring in a reconstruction under prior costs, when $NS > n$ (default $n = 5$).

e) When $NS > n$, restrict state assignments to those of the terminals no more than L branches away from the node; if no terminal is less than L branches away, assignment is unrestricted (default $n = 4$, $L = 6$ nodes above the node, or 3 nodes below).

f) When selecting state assignments, select only from those states assigned in an optimization under prior costs (very fast, very error-prone; disabled by default).

The shortcuts (a) and (c) cannot be disconnected (they are error-free); all the other ones may lead to missing the optimal reconstruction and therefore are optional. Shortcuts (b), (e) and (f) were not described by Goloboff (1997), but were added when the method was implemented in TNT (in 2006). Given the complexity of calculating scores under self-weighted optimization, the method cannot be used together with homoplasy extrapolated to missing entries, or weighting sets of characters by their average homoplasy.

For a fast approximation of scores during searches, several additional shortcuts (described by Goloboff 1997) are used. Those are used to quickly reject candidate trees for which the approximate score exceeds the best score found so far by a pre-determined margin, repeating the calculation from scratch (as described in points (a)–(f)) only for trees that are within this limit. By using a combination of all these approximations, it is possible to analyze datasets of a non-trivial size (and even larger when the characters have fewer states). For example, Mirande (2009) analyzed a matrix with 160 taxa and 360 characters. For the matrix of 164 taxa and 853 characters of Pei et al. (2020), including one character with 8 states, a Wagner tree plus *TBR* can be obtained in about 12 hs. on a relatively fast computer (with mulpars off, 2.3×10^6 rearrangements examined, moving or rerooting the clipped clade no more than 4 nodes away, maximum repeat of 6, and forbidding state transformations not occurring under prior costs for characters with 4 or more states).

7.8.4 Weights Changing in Different Branches

As mentioned in the section on Generalities, at the beginning of this chapter, weights constant over all tree branches seem to be a requirement of parsimony. In maximum likelihood, those weights can change in different branches, but they are not determined from any single character—the weights are a consequence of the branch length, determined from the whole set of characters.

Using weights varying over tree branches, without a necessary reference to other characters (as in likelihood), seems difficult. One would tend to think that if the weights are going to be set at a given value over some region of the tree, then that implies that we are defining what the regions of the tree are—and a proper method must allow the tree to vary freely. However, the method for self-weighted optimization of Goloboff (1997) provides a useful framework for evaluating reconstructions, a framework that can be adapted for weighting every state transformation individually, and depending on the distance to other similar transformations. The adaptation was presented at the 27th Annual Meeting of the Willi Hennig Society, and briefly discussed in Goloboff et al. (2009b). To evaluate a given reconstruction, for every combination of different states *x, y*, the $x \rightarrow y$ transformations are numbered in the tree, from 1 to the number n_{xy} of $x \rightarrow y$ transformations in the tree. The number of branches separating the ith from the jth $x \rightarrow y$ transformation is denoted as b_{ij}. Then, the score for the reconstruction is calculated as

$$d = \sum_{xy, x \neq y} \sum_{i=1}^{n_{xy}} \frac{1}{k + \sum_{j=1, j \neq i}^{n_{xy}} \frac{q}{q + b_{ij}}}$$ [Formula 7.7]

where k is the constant for weighting strength (as in IW), and q is a constant for distance to make other changes in the character influence more, or less, the cost of every individual x→y transformation. For every x→y change, every other x→y change is considered (in the summation in the denominator); when all other transformations occur far away, the x→y change contributes less toward the denominator. That is, a given x→y change has a high cost when no similar changes occur in the near vicinity of the tree, and vice versa. For a very large q, other changes have a lower influence on the cost of every individual change, and the method converges to self-weighted optimization.[10] For a small q, the denominator tends to a constant value, k, and the method converges to equal weights (just as with a very large k). Formula 7.7 evaluates an individual reconstruction; for a given tree, the reconstruction minimizing d provides the score to evaluate the tree.

The criterion is called *topological weighting*. All the shortcuts used for speeding up calculation of optimal reconstructions under self-weighted optimization can be used to find reconstructions minimizing Formula 7.7. The method was tested on some empirical datasets, and the trees it produces seem acceptable. However, a more careful consideration of hypothetical examples reveals some disturbing properties (Figure 7.6b). In the first tree, each of the two changes *white → black* has a certain cost; as the tree is changed, grouping some of the branches intervening between the two changes decreases the distance between them (from 5 to 4 branches, then 2, and 1). This has the effect of an apparent "grouping by plesiomorphies": the white taxa are better grouped together because that allows the black taxa to be closer in the tree. The scores shown in Figure 7.6b use $k = q = 6$, but the effect does not depend on particular values for those constants, or even on the specific implementation described by Formula 7.7. Rather, any method that weights a transformation depending on the number of branches separating it from other such transformations will have the same effect. The method might perhaps be justified by reference to specific assumptions about evolution, but cannot be called "parsimony"—the method is considering that the rightmost tree is a better explanation for the distribution of the *black/white* character, even if the degree to which the four trees allow explaining the observed similarities as homology is exactly the same. The method is not scoring trees by reference to genealogical accounting of observed similarities (=parsimony), but instead by something else. Whatever virtues the method may have, the advantages outlined in Chapters 1 and 2 for parsimony cannot be claimed for it, and its use is not recommended. The main interest of the method is theoretical, as it suggests (a) that weights constant over all branches of the tree may be a *sine qua non* requirement of parsimony, and therefore (b) criticizing weighted parsimony methods on the basis that they assume uniformity is misguided.

7.9 IMPLIED WEIGHTS AND LIKELIHOOD

An analog of IW under a Poisson model can be defined, for a likelihood approach that produces results very similar to those of IW (Goloboff and Arias 2019). The method differs from standard likelihood models in considering that, for any given character, change is equally probable along all branches of the tree. When the probability of change is assumed to be the same for all characters (and, for any one tree, fixed at the value that maximizes the likelihood for the entire matrix), the method behaves almost like equal weights parsimony (MP_{lik}; see Chapter 4). The alternative is to let the probability of change (i.e. the branch length) be optimized separately for each character; branch length for characters with low homoplasy will be optimized as short, and vice versa. Such a model mimics *IW*, because the probability of a change is increased less strongly with every additional step. For a single step, all branches are very short; one step is quite unlikely for short branches, so adding one more step makes the probability substantially lower. As many steps occur in the tree and branches have to be made longer to maximize the likelihood, one more step is quite likely, and then postulating another step lowers the overall probability very mildly. This is of course reflected in the log likelihoods. This method was called *implik* by Goloboff and Arias (2019). Figure 7.7a shows the log likelihoods for characters with different number of steps, on a tree with 100 taxa, normalized relative to the difference between the log likelihoods for one or no extra steps (white), and the corresponding costs of addition of another step (*a*, from Formula 7.3) under *IW* with $k = 25$ (black). The *implik* values are obtained by summing over all reconstructions; the correspondence between the single-reconstruction versions (as in Goldman 1990, see Chapter 4; Goloboff and Arias called this single-reconstruction approach *simplik*) and *IW* is closer, but the *implik* model is conceptually closer to standard likelihood models. Figure 7.7a shows that the difference in log likelihood values decreases less and less with every extra step. The curves for the two methods are very similar. Figure 7.7b maps the downweighting resulting from *IW* against the one from *implik*. An exact correspondence would have a straight line with slope unity; the two methods do not have an exact correspondence, but they are very close. This means that *implik* will select almost the same trees as *IW*: likelihood is based on selecting the tree that minimizes the sum of negative log likelihoods. When considering the likelihood from all reconstructions (as in *implik*), the same number of minimum possible steps can have slightly different likelihoods in different topologies (i.e. depending on where the changes occur; see Goloboff and Arias 2019 for extensive discussion), but these differences are minor, especially when there are few changes relative to the number of taxa. Figure 7.8a shows this in a simulated dataset (500 binary characters generated on a 40-taxon model tree where every branch is a random number between 4×10^{-3} and 10^{-1}). Figure 7.8a plots the scores of *implik* and *IW* on 500 near-optimal trees (black diamonds); the diamonds are close to a perfect diagonal. The scores under the Mk model (Lewis 2001, white circles) are instead much more dispersed.

Even if *implik* and *IW* do not produce identical results, they are very close, and *implik* shows that the rationale for implied weights—making every additional step of homoplasy cost less and less—can be justified also from the perspective of likelihood and Poisson models.[11] Just like parsimony (with or without *IW*), the method does not assume that some branches of the tree will have a higher concentration of changes. As it posits change is equally likely in any branch, the method can obviously be inconsistent when changes are concentrated (in long branches). If it is not obvious to the reader that *IW* does not avoid long-branch attraction, consider that *IW* and equal weights become identical for four taxa and that the quintessential example for long-branch attraction is a four-taxon tree (Felsenstein 1978).

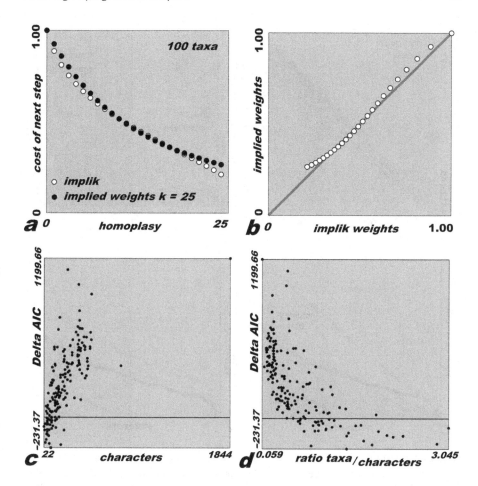

FIGURE 7.7 **(a)** For a model (*implik*, Goloboff and Arias 2019) where length is uniform for all the branches of the tree, but varying for the different characters (and chosen so as to maximize the likelihood), the results approximate those of implied weighting, with the cost of every additional step decreasing with homoplasy. **(b)** Same as in (a), but with cost of every extra step for implied weighting plotted against cost for *implik*. **(c)** Values of Akaike information criterion between the Mk and *implik* models (y-axis), for 182 empirical datasets with different numbers of characters (x-axis). Dots below the horizontal line correspond to datasets where *AIC* prefers *implik* (=implied weighting) over the Mk model. **(d)** Similar to (c), but the x-axis shows the ratio of taxa to characters; there is a strong tendency for the *AIC* to prefer implied weights over Mk when the ratio of taxa to characters is larger.

Although it can be inconsistent when its assumptions are violated, *implik* is a proper likelihood method, and its parameterization is well defined: the method needs to estimate as many branch lengths as characters. In the no-common-mechanism (*NCM*) model of Tuffley and Steel (1997), which produces the same results as equally weighted parsimony, the estimation of a branch length from a single character is based on a single point of observation, and that produces extreme values (branch

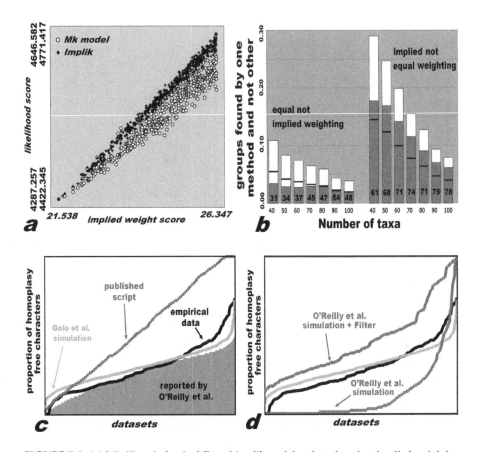

FIGURE 7.8 **(a)** Likelihoods for the Mk and *implik* models, plotted against implied weighting scores, for 500 near-optimal trees in simulated datasets. The similarity in *implik* and implied weighting is reflected in the corresponding dots being placed close to the perfect diagonal, while the plots of Mk and implied weighting scores are much more dispersed. **(b)** Proportions of groups found by one method and not the other (white bars, relative to full resolution), and proportion of groups found by one method and not the other that are correct (gray bars, relative to number of unique groups found by the method), in datasets with different numbers of taxa (and twice as many characters as taxa), simulated with an exponential distribution (as in Goloboff et al. 2017, with λ a random number in the range 0.05–0.15). For each number of taxa, 100 datasets were simulated, analyzed under equal implied weights ($k = 8$), with command xmult=hit 3 (and consensus produced with coll tbr; nelsen*, setting suboptimal to sub 2x0.15 for equal weights, and sub 0.222x0.15 for *IW*). The gray portion occupying less than half of the entire bar represents a case where a group found by the method in question is more likely to be wrong than right (and vice versa). **(c)** The frequencies of characters with no homoplasy, in different datasets (ordered from lowest to highest). Neither the distribution of simulated datasets from Goloboff et al. (2017), nor the one reported by O'Reilly et al. (2018) are exactly identical to the empirical distribution, but both are very close. The values actually obtained with the script published by Puttick et al. (2017b) are far from the empirical distribution, despite the claim by O'Reilly et al. (2018). **(d)** The frequencies of characters with no homoplasy, in the datasets simulated by O'Reilly et al. (2016), with or without filtering by C value (see text), are very different from the empirical distribution. See text for further discussion.

lengths are either zero or infinity; Lewis 2001). This leads to the well-known over-parameterization of *NCM*, and to the impossibility of selecting *NCM* over standard approaches such as the Mk model (Holder et al. 2010; Huelsenbeck et al. 2011) on the basis of the tradeoff between the likelihood and the number of parameters. That over-parameterization does not occur in *implik*—the length is estimated from the many branches of the tree, and the estimation becomes more precise to the extent that there are more taxa. Note that in the classic Mk model, the parameterization is in some sense the opposite: regardless of number of characters, the method must estimate a length per tree branch (with the precision of the estimation increasing with number of characters). Unlike the case of *NCM*, it is therefore possible that *implik* (i.e. *IW* parsimony) is chosen over the Mk model, and that will happen more often when the dataset has relatively few characters and many taxa (and when there is not a concentration of character-state changes in some branches of the tree, which allow the Mk model to increase the likelihood, relative to the uniform branch lengths of *implik*, by making those branches longer). Note how this agrees with the common-sense justification given earlier for *IW* to be used preferentially in cases of larger numbers of taxa: the estimation of homoplasy from few taxa is too imprecise.

As a test of that conceptualization, Goloboff and Arias (2019) compared (by means of the *AIC* criterion of Akaike 1973) the likelihoods and numbers of parameters of *implik* and the Mk model for simulated datasets. Predictably, in datasets where the generating model had the same length for all branches and characters, the *AIC* led to a preference for the MP_{lik} model (where the same length is inferred for all characters and branches; a single parameter). In datasets generated with all branches of the tree of a uniform length (but changing for different characters) the *AIC* led to a preference for *implik*, and in datasets generated with lengths varying for the different branches (but uniform for all characters) the *AIC* led to a preference of the Mk model.

Goloboff and Arias (2019) also compared likelihoods and numbers of parameters in 182 empirical morphological datasets; their results are reproduced here, in Figure 7.7c–d. The horizontal line indicates a tie in the AIC, so that there is no preference for either Mk or *implik*. Points above the line are datasets where Mk is preferred by AIC, and points below the diagonal are datasets where *implik* is preferred. In Figure 7.7c there is a tendency for *implik* to be preferred more often when there are fewer characters. The pattern of preference becomes more obvious when plotting the AIC values against the ratio of taxa to characters: in agreement with theory, *implik* tends to be preferred when there are more taxa relative to the number of characters.

Using *implik* as a substitute for *IW* would be ill-advised; it would just produce comparable results in much longer times. The interest of the *implik* model is theoretical and conceptual. *Implik* provides justification from a likelihood perspective for the main rationale of *IW* (i.e. when calculating tree scores, each successive step of homoplasy must be less costly than previous ones). It also suggests an alternative justification for milder values of *k* when there are more taxa. Uniform weighting for sets of characters on the basis of average homoplasy (as in Section 7.8.2) results naturally as a byproduct of optimizing the branch length as in MP_{lik}, but for sets of characters. The answers to these problems from a likelihood perspective are very similar to the answers obtained from common sense and careful consideration of alternatives, discussed earlier in this chapter.

7.10   TO WEIGHT OR NOT TO WEIGHT,
THAT IS THE QUESTION . . .

Finally, after having discussed possible approaches to character weighting in connection with parsimony, we can reconsider the main arguments for and against character weighting in morphological datasets. The problem has attracted some attention recently, with Congreve and Lamsdell (2016) and a series of papers by J. O'Reilly and M. Puttick (O'Reilly et al. 2016, 2017, 2018; Puttick et al. 2017a, 2017b, 2018; these six papers are by the same core of four authors, in different sequences of authorship) approaching the question almost exclusively by means of simulations. Arguments based on simulations are considered first.

7.10.1   Criticisms of *IW* Based on Simulations

Congreve and Lamsdell's (2016) compared *IW* and equal weights parsimony. They generated datasets with only 22 taxa, and focused on the absolute number of incorrect groups (i.e. relative to the model tree) found by *IW* or equal weighting. For their datasets, *IW* tends to find a larger number of incorrect groups than equal weights, and they conclude from this that equal weights is to be preferred. However, as discussed in several parts of this chapter, *IW* is less relevant in very small datasets; a more useful comparison is against larger datasets. But Congreve and Lamsdell's (2016) comparisons suffer from a more serious problem as well: counting only absolute numbers of incorrect groups is very inappropriate for preferring one of two methods if they produce trees with different degrees of resolution. The relevant comparison is whether the groups found by one method and not the other are more likely to be correct than wrong (Goloboff et al. 2017). Figure 7.8b shows results for simulated datasets; a group found by *IW* and not equal weights is more likely to be right than wrong (average 61% to 78% of the additional groups being correct; the percentage increases with number of taxa). The absolute number of incorrect groups found by *IW* is indeed larger than for equal weights (the white upper parts of the bars are, for the same number of taxa, larger for *IW* than for equal weights), but the number of correct groups more than offsets this. Equal weights also finds some groups that *IW* does not, but in most cases (all numbers of taxa but 90) the majority of the exclusive groups are not found in the model tree (see Figure 7.8b).

The flurry of publications starting with O'Reilly et al. (2016) also performed simulations, using two different models. They concluded that Bayesian analysis (with MrBayes; Ronquist et al. 2012) outperforms parsimony, both with and without *IW*. But in most papers they used a model where characters evolve under a common model and substantial branch-length heterogeneity, which is known to be problematic for parsimony; their experiments were *designed* to make parsimony perform more poorly (Goloboff et al. 2017). And the evidence suggests that the common mechanism used by O'Reilly et al. (2016) and Puttick et al. (2017b) in their simulations does not apply to empirical morphological datasets: Goloboff et al. (2018a) demonstrated that branch lengths inferred from different partitions (defined either on the basis of anatomy, or randomly) are much more different than expected under the Mk model (see Chapter 4, and Figure 4.6b). O'Reilly et al. (2016) claimed to have made their simulations more realistic by generating datasets and then discarding those where the ensemble consistency C (described in Section 7.3.1) was below 0.26 (citing Sanderson and Donoghue 1989 in support); O'Reilly et al. (2016) generated datasets on a tree of a fixed topology with 75 taxa. In their subsequent paper, Puttick et al. (2017b) generated datasets on symmetric and asymmetric trees of 32 taxa; in this case, they made sure that the frequency of datasets with a given C value (between 0.26 and 1, in intervals of 0.05) resembled the frequencies observed in Sanderson and Donoghue's matrices, just by discarding most of the generated datasets. Incredibly, O'Reilly et al. (2018) took Goloboff et al.'s (2017) simulations to task, on the grounds that the resulting homoplasy distribution is unrealistic. Goloboff et al. (2017) had examined 158 datasets (between 50 and 170 taxa), and generated their datasets so that the expected proportions of characters with different amounts of homoplasy are more or less similar to those observed in the empirical datasets. According to O'Reilly et al. (2018), the proportions of characters with no homoplasy in their own simulation is much closer to the one in the 158 empirical datasets than in Goloboff et al.'s (2017) simulation. Figure 7.8c shows the distributions for empirical datasets (black line), matrices simulated as in Goloboff et al. (2017; light gray), and Puttick et al.'s (2017b) simulated datasets (dark grey area). The dark grey area was copied from O'Reilly et al.'s (2018) paper, because the curve they showed cannot be reproduced with the script[12] of Puttick et al. (2017b), which produces the curve shown as a dark gray line in Figure 7.8c. Reproducibility problems aside, the reader can judge in Figure 7.8c whether the light gray curve, or the contour of the dark gray area, comes closer to the black line; frankly, I see a minimal difference, and there is no basis for the claim that the difference is that Goloboff et al.'s (2017) matrices are "dominated by characters with very high consistency and an unrealistically small proportion of characters exhibiting high levels of homoplasy" (O'Reilly et al. 2018). There are more cases where homoplasy-free characters "dominate" (i.e. exceed 50%) in Goloboff et al.'s datasets than in Puttick et al.'s, but fewer than in empirical data (and the proportion of Goloboff et al.'s "dominated" datasets is closer to empirical data). Puttick et al.'s (2017b) interpretation does not fit their own figure.

But there is another question which is of more interest than the tiny differences in those distributions. Sanderson and Donoghue (1989, 1996) had shown that values of C consistently decrease with numbers of taxa in empirical datasets. This is expected from theory: as the maximum possible value of homoplasy is limited by the number of taxa, C cannot reach values below a certain threshold for too few taxa. Sanderson and Donoghue (1989) gathered 42 morphological matrices with 5–68 taxa, and found that plotting C against number of taxa produces a hyperbolic shape, similar to that illustrated in Figure 7.2a for the character consistency indices as a function of increasing homoplasy. Sanderson and Donoghue (1996), with a different set of 50 morphological matrices of 5–56 taxa, confirmed the same trend. Therefore, although Puttick et al. (2017b) generated matrices so that the distribution of C values would match that in all the datasets of Sanderson and Donoghue, the undisputable conclusion of Sanderson and Donoghue (1989, 1996) is that doing so is utterly meaningless: all the datasets of Puttick et al. (2017b) have 32 taxa! A realistic distribution of C values in datasets of 32 taxa can never match the distribution of C values in datasets with different numbers if taxa. Why generate datasets of 32 taxa so that the frequencies of *C* values approach the frequency[13] of *C* values in datasets with varying numbers of taxa? If realism had indeed been a serious concern, Puttick et al. (2017b) should have concentrated on the datasets of Sanderson and Donoghue in the 30–40 taxon range, but there are too few such datasets in the two databases to appeal to empiricism (and what is in there does not justify Puttick et al.'s cutoff of 0.26; Sanderson and Donoghue had three datasets between 30–40 taxa in 1989, with *C* = 0.340–0.430, av. 0.387, and three datasets in 1996, with *C* = 0.415–0.741, av. 0.612). Further, if O'Reilly et al. (2018) were right in expecting that the distribution of homoplasy in their 32-taxon datasets should match the distribution in empirical datasets with 50–170 taxa, and that their 32-taxon datasets are closer to that distribution than Goloboff et al.'s (2017), then the same should apply to O'Reilly et al.'s (2016) own previous simulation of 75-taxon datasets. However, their own simulation of 75 taxa (whether or not the matrices are subject to the *C* = 0.26 filter) produces a distribution very different from the empirical one (Figure 7.8d). Why did O'Reilly et al. (2018) never examine their own previous simulation with the same zeal as they used for Goloboff et al.'s simulation?

Further decreasing the realism of their simulations, Puttick et al. (2017b) generated their datasets on ultrametric trees—that is, their datasets obey the clock. Phenetic methods assume ultrametricity (see Section 1.5.1), and for datasets such as those generated by Puttick et al. (2017b), phenetic grouping produces trees closer to the model tree than either parsimony or likelihood (Goloboff et al. 2018b)! Puttick et al. (2017b) did not consider that their datasets would be adequate to show the superiority of phenetic methods, and rightly so, but then why would one consider those datasets as relevant evidence for choosing among *other* methods? If an asymmetric model tree is ultrametric, then the first splitting branch is the longest, and all successive splits are shorter and shorter. One of the conclusions Puttick et al. (2017b) drew from their simulations was that asymmetric model trees are harder to reconstruct with confidence than symmetric trees, but this is only a consequence of the differences in branch lengths imposed by making the trees ultrametric, not a consequence of differences in symmetry (Goloboff et al. 2018a, 2018b). Puttick et al. (2017b) used as the only measure of tree similarity the Robinson-Foulds distance normalized by maximum possible resolution, which, as they recognize, is influenced by tree-resolution (although they incorrectly stated in Puttick et al. 2018 that "a value of zero indicates two trees that are either identical or that one tree is fully unresolved"; if one tree is fully resolved and the other fully unresolved, the normalized Robinson-Foulds is not maximum, but 0.5, as discussed in Chapter 6). Smith (2019) examined the results of analyzing Puttick et al.'s (2017b) own datasets with *IW*, equal weights parsimony, and Bayesian analysis, focusing on the normalization of Robinson-Foulds distances. The normalization used by Puttick et al.'s (2017b) implies that datasets improve as you eliminate most of the characters (see discussion in Chapter 6, and Figure 6.5d). The same normalization was used by Keating et al. (2020) in their comparison of Bayesian and parsimony analyses, and may account for the apparent superiority of Bayesian methods in those other simulations. Smith (2019) used a more meaningful normalization to measure distance to model tree, normalizing by the number of groups present in the trees (instead of maximum possible resolution). Smith (2019) found that, even with the model trees of Puttick et al. (2017b) having significant differences in branch lengths, the trees produced by *IW* were about as similar to the model tree as the trees produced by Bayesian analysis.

Neither O'Reilly et al. (2016) nor Puttick et al. (2017b) collapsed poorly supported groups for parsimony; they "did not employ bootstrap methods to measure support for parsimony and likelihood analyses because phenotypic data does not meet the assumption that phylogenetic signal is distributed randomly among characters" (Puttick et al. 2017b). But nobody ever proposed (aside from Puttick, O'Reilly, et al.) that the bootstrap needs "phylogenetic signal" to be "distributed randomly" among characters. Brown et al. (2017) and Goloboff et al. (2017) pointed out that Puttick et al.'s (2017b) justification for not bootstrapping is nonsense, and goes against standard interpretation (e.g. Felsenstein 2004). And when Brown et al. (2017) and Goloboff et al. (2017) pointed out that they should have collapsed poorly supported groups for their comparisons with parsimony, Puttick et al. (2017a) then defended their neglecting of support measures on the absurd grounds that most users report support measures but do not actually *eliminate* the groups of low support—as if the bad practices of others could justify their own.

In summary, Puttick et al.'s (2017b) study is plagued by problems and arbitrary decisions. Eventually, they admitted (Puttick et al. 2018) that "data simulated in a framework where characters share branch lengths . . . are biased in favour of an inference framework that explicitly makes the same assumption (e.g. maximum likelihood and Bayesian implementations of the Mk model)", and proposed to simulate datasets with a different model, a model they claim is "non-probabilistic". Their method for generating data consists of allocating character changes to branches chosen at random. In this, their method for generating data is exactly the same as Goloboff et al.'s (2017), where character changes occur equiprobably at any branch; Puttick et al. (2018) do not discuss the similarity. Their results are different from Goloboff et al.'s (2017), however, because they used much larger amounts of homoplasy and focused their comparisons on the standard Robinson-Foulds distance. Their model trees, again, are symmetric and asymmetric 32-taxon trees. They generated 100 datasets, in ten bins of *C* values, from 0.0–0.1 to 0.9–1.0.

They found that "Bayesian inference with the Mk model achieves the highest accuracy in analyses of datasets exhibiting the highest levels of homoplasy". In their opinion

> it is the relative performance of phylogenetic methods in analysis of datasets dominated by homoplasy that is most informative in designing phylogenetic analyses of empirical datasets. . . . In the first bin, CI 0.0–0.1, all methods perform poorly. However, the Bayesian implementation of the Mk model is much more accurate than the next-best method, implied weights parsimony.
>
> **(Puttick et al. 2018)**

They continue using standard Robinson-Foulds values to measure distance to model tree, which tends to favor under-resolved results. Puttick et al. (2018) considered the results in the 0.0–0.1 bin most important, but did not dwell on the implication. For 32 taxa, a dataset with consistency index of 0.096 is expected when every cell in a matrix with very large numbers of characters is assigned states 0 or 1 completely at random, in which case all possible trees have (almost) the same length and the data are maximally undecisive and entirely worthless for phylogenetic inference (Goloboff 1991a, 1991b). Such consistency index is achieved only when the number of characters is very large, but Puttick et al. (2018) generated datasets with just 100, 350, and 500 characters, so the average consistency index in random datasets is somewhat higher. Generating 1,000 random 32-taxon datasets with TNT produces average consistency indices of 0.112 ±0.0013, 0.104 ±0.0005, and 0.100 ±0.0003, for 100, 350 and 1,000 characters respectively. Therefore, the datasets Puttick et al. (2018) consider most important for choosing among methods, those with $0.0 \leq C \leq 0.1$, are *worse* than random. An inference method that can provide correct inferences under such circumstances would be magical indeed.

Of course, neither *IW* nor any inference method can produce meaningful results for random data; those results just correspond to a poor experimental design. Still, an abundance of homoplasy was repeatedly taken by O'Reilly et al. (2017, 2018) and Puttick et al. (2018) as indication that *IW* will perform poorly:

> the presence of large numbers of characters that are congruent with the tree allows implied weights to increase the power of these 'true' congruent characters. This effect will not be possible when increased levels of homoplasy are present or when the true tree is unknown.

(O'Reilly et al. 2017)

But that belief is incorrect; *IW* will make a difference when some of the characters are less homoplastic than others, even if all characters have substantial amounts of homoplasy. This was clear to Goloboff et al. (2017), who discussed that a dataset with 500 1-step and 500 5-step binary characters has the same consistency index as a dataset with 1,000 3-step binary characters; "yet the first data set comprises some characters that are much more reliable than others, whereas the reliability of all the characters in the second data set is exactly the same (thus making a method such as implied weighting simply irrelevant)" (Goloboff et al. 2017). The simulations shown in Figure 4.6c (in Chapter 4) dispel the myth that abundance of homoplasy, by itself, makes *IW* perform worse than Bayesian analysis: in simulated datasets where the majority of characters have 13–14 steps of homoplasy (with only 0 to 2% of characters free of homoplasy), *IW* clearly outperforms Bayesian analysis if the data are generated without a common mechanism.

The only clear conclusion possible from all of this is that the sweeping statements made by O'Reilly et al. (2016, 2018) and Puttick et al. (2017b, 2018) on the superiority of Bayesian methods cannot be justified. The main point of Goloboff et al.'s (2017) simulation was not to claim superiority of parsimony methods in general (a forced interpretation of O'Reilly et al. 2017), but instead that *depending on the method used to simulate data, different inference methods may provide the best results*. That is why Goloboff et al. (2017) concluded that "simulations alone cannot lead to preference for one method over another unless there is empirical evidence in favour of the model used to run the simulations". Goloboff et al. (2018a) showed that the model with a common mechanism does not apply to morphological datasets. Goloboff et al. (2018a) also showed that, while the model where changes can occur with equiprobability over all the tree also fits the data poorly, a model they called "episodic" (where change occurs equiprobably but only within certain regions of the tree), produces datasets very similar to the empirical datasets they examined. Furthermore, since this model lacks a common mechanism, *IW* outperforms Bayesian methods on the generated datasets.

In summary, establishing the absolute superiority of one method over the others only on the basis of simulations and without reference to a specific evolutionary model, is very difficult. The problem of method choice must also consider other lines of argumentation.

7.10.2 SUPPORT AND CHARACTER RELIABILITY

In a now classic paper, Källersjö et al. (1999) analyzed one of the (then) largest existing molecular datasets (2,594 taxa), and estimated the average homoplasy of first, second and third positions. Many authors before had suggested excluding third positions, expected to evolve faster and to have more homoplasy (because of the degeneracy of the genetic code). On Källersjö et al.'s dataset, third positions were indeed much more homoplastic than first and second (average number of extra steps in first, second and third positions on an *MPT* is 21.9, 12.2, and 87.0). That was within expectation, but the unexpected finding of Källersjö et al. (1999) was that the number of well-supported groups for the entire dataset was determined almost completely by the third positions. Reanalyzed with TNT (by Goloboff et al. 2008a), all positions combined determine 1,394 groups; removing the highly homoplastic third positions only 360 well-supported groups remain, and analyzing third positions alone, 1,349 groups.

Källersjö et al. (1999) concluded that perhaps the idea of downweighting characters on the basis of homoplasy alone was misguided. If a group of characters (third positions in this case) determines most of the well-supported groups in a dataset, then indeed it seems to follow that those characters should be considered as most reliable. Yet those were the most homoplastic characters in this case. The reasoning in this case is very different from the one used in simulations, being much more empirically grounded. Several authors who defend the universal use of equal weights (e.g. Grant and Kluge 2003, 2005; Kluge 2003; Miller and Hormiga 2004; Scott 2005; Kvist and Siddall 2013; Congreve and Lamsdell 2016) have cited Källersjö et al.'s (1999) in support of their views on weighting. Ironically, Källersjö et al. (1999) themselves interpreted their results only as against weighting on the basis of homoplasy,[14] not as against every kind of weighting (as shown by the subsequent attempt of one of the coauthors to devise a method that would upweight third positions over first and second; Farris 2001).

Goloboff et al. (2008a) reexamined the problem. The characters in a dataset can be divided on the basis of the number of groups they determine, or on the basis of their reliability. The two divisions do not necessarily agree, as they correspond to different aspects. Consider a dataset for numerous vertebrates comprising a single character—mammary glands. The resulting tree would be very poorly resolved. However, inferring that the character is unreliable because it separates a single group in the tree would be incorrect—the group itself is correct, as far as we know. The results for

Källersjö et al.'s (1999) resemble such situation. Of the 360 groups supported by first and second positions alone, 317 are also well supported by the entire dataset; 325 are supported or uncontradicted by the third positions alone. That is, 88% of the groups determined by first and second positions are supported by the complete dataset, and 90% are compatible with the groups for third positions alone. First and second positions determine few groups, but the groups they determine seem to be (mostly) correct; there is thus no reason to conclude that first and second positions are, individually, less reliable than third positions. The leap from number of supported groups to the reliability of individual characters seems unwarranted. Similar comments regarding Källersjö et al.'s (1999) results were made by Nixon (2008).

   Further strengthening the idea that increased frequencies under resampling do not necessarily indicate that the added characters are more reliable, it is possible to design datasets where adding random characters to a dataset *increases* the average group frequencies (Goloboff et al. 2008a). That is shown in Figure 7.9a–c, where every one of the branches of a poorly resolved 150-taxon tree (e.g. as in Figure 7.9b, but with more taxa) has 10 congruent, homoplasy-free synapomorphies. The average support is rather low, because there are few groups in the tree. This is the first part of the matrix in Figure 7.9a, where it is visually obvious that all characters are nicely nested within each other. Then, half as many characters are added to the matrix, characters where every terminal taxon receives a 0 or 1 state completely at random (with probability $p = 0.20$ for state 1, and $p = 0.80$ for state 0). Any congruence between two of these random characters is mere coincidence; these are the characters in the right side of the matrix in Figure 7.9a, where no discernible pattern is evident. Almost every possible pair of random characters is incongruent, and (on the most parsimonious tree for the full dataset) the random characters have on average over 25 steps (as opposed to an average of about 1.1 extra steps in the original characters). The added random characters are the *quintessence* of unreliability. Yet a strange phenomenon occurs: most of the groups determined by the good characters continue having a good support after addition of the random characters; the random characters are too few, and too incongruent with each other, to substantially lower the support for the preexisting groups. But any fortuitous coincidence in some characters *within* those groups will create an additional group of high resampling frequency (e.g. the synapomorphies for the group also occur in many other taxa, but those other taxa are across some insurmountable barriers). Thus, previous groups remain with high frequency, and some others are added. Figure 7.9c shows the group supports (white circles, relative to a fully resolved tree with all groups having 100% support) before the addition of the random characters on the x-axis, and the supports after the addition on the y-axis, in 200 simulated datasets. With very few exceptions, the support is substantially increased. To make the example more paradoxical, if the characters that increased the group supports are mildly upweighted relative to the initial set of characters (with weights in the 3:2 ratio), the average group supports now decrease (Figure 7.9c, black dots). The upweighted random characters can now overcome many of the groups defined by the initial set of perfect characters, and the whole structure crumbles. In other words, upweighting the very characters that increased group supports, decreases the support for all the groups in the tree.

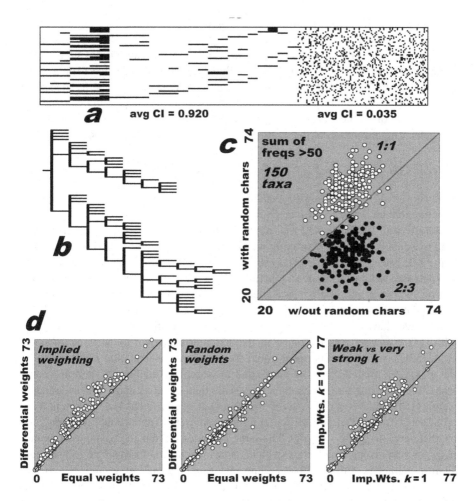

FIGURE 7.9 (a, b) An example of matrix (a) with 2/3 of the characters fully congruent, determining a poorly resolved tree (b) of 150 taxa with many polytomies of up to 8 nodes (and 10 synapomorphies for each branch of the tree), and 1/3 of the characters generated at random (with every cell independently assigned states 0/1, with $P_{(0)}$ = 0.8 and $P_{(1)}$ = 0.2). A black matrix cell represents a 1 state, white represents 0s. The first 2/3 of characters determine a poorly resolved tree, but each of the groups is strongly supported. Adding the last 1/3 of characters (the part of the matrix that looks like a quadcode) does not significantly change the support for the groups defined by the first 2/3 of the characters, but adds (just by chance) some further subdivisions, thus increasing overall supports. **(c)** Plot of the sums of group supports (for groups with frequency above 0.5), with random characters included or excluded, in 200 matrices simulated as described in (a, b). White dots show the supports for perfect and random characters with weight ratios 1:1, black dots show the supports for random characters upweighted relative to the perfect characters, in ratio 3:2. Increasing the weight of the characters that increase the support (the random characters) has the effect of *decreasing* the supports. See text for further discussion. **(d)** Comparison of sum of supports between different weighting methods, for the empirical datasets of Goloboff et al. (2017). Supports are higher, on average, for standard implied weighting with mild values of k; using random weights or very strong values of k does not improve the group supports.

Källersjö et al.'s (1999) dataset is not identical to the example of Figure 7.9a–c, because there the third positions taken in isolation define almost the same groups as the full dataset (while the random characters taken in isolation do not support any groups in the example of Figure 7.9a–c). But the demonstration that supports can be increased with the addition of random characters shows that variation in support measures needs to be interpreted with caution. Measures of support indicate the collective effect of all the dataset on a given group, while the idea of reliability applies when deciding resolution of the conflict between individual characters.

7.10.3 WEIGHTING, PREDICTIVITY, AND STABILITY

Källersjö et al. (1999) assumed that, since third positions determine most of the structure in their dataset, downweighting third positions would lower group supports. Goloboff et al. (2008a) noted that downweighting (as opposed to full elimination) with reasonably bounded maximum weight ratios (i.e. setting k from Formula 7.4 so that $r \leq 15$) need not reduce group supports, since the groups determined by third and first+second positions are compatible. Goloboff et al. (2008a) also showed that, against the prediction implicit in Källersjö et al. (1999), group supports in morphological datasets are generally improved when weighting against homoplasy with IW. Figure 7.9d (left) shows the average group supports (relative to a fully resolved tree with 100% support in all groups) in the 158 morphological datasets of Goloboff et al. (2017), with supports under equal weights on the x-axis, and IW ($k = 10$) on the y-axis. Figure 7.9d (center) shows that the improved supports do not just result from IW measuring tree scores on a finer scale than equal weights, because random weights (where the cost of a step for every number of extra steps is a random number 0.1–4.00, with two decimals, different for each dataset) do not produce any appreciable difference in group supports. The same is true of weights increasing with homoplasy, or very strong concavities (which, as just discussed, downweight too strongly against characters with homoplasy, approximating cliques). Using IW with $k = 1$ or $k = 2$ produces no to minimal improvements in support relative to equal weights (Figure 7.9d right shows $k = 1$ vs. $k = 10$); only at $k = 3$ is a definite improvement over equal weights appreciable.

The general improvement of group supports under IW suggests that (barring pathological situations like shown in Figure 7.9a–c) the method is giving preference to sets of mutually congruent characters, and improving predictivity and stability. The concepts of support and stability, while not exactly the same, are interrelated: groups based on slim evidence can easily be overturned in the future. Although the goal of resampling can be just quantifying amounts of supporting and contradicting evidence, some authors view resampling as a way to predict stability on the basis of introducing perturbations to the dataset (see discussion in Chapter 8, Section 8.5). Therefore, the improved supports when applying IW to morphological datasets would point toward a greater stability. The ideal would obviously be an examination of the actual stability as a dataset progresses through time; however, very few studies have attempted to track evolution of results in this way (but see Murphy et al. 2021). Goloboff et al. (2008a) noted that evaluations intended strictly for stability should proceed differently from methods like bootstrapping or jackknifing, evaluating whether the groups found by only part of the known evidence are supported by all of it (instead of the other way around). Goloboff et al. (2008a) examined several such measures of predictivity (e.g. number of shared groups, number of compatible groups), and found *IW* to outperform equal weights also in this regard. That is reassuring, because a proper method of phylogenetic analysis should produce stable and predictive results. "Predictive" in this context does not mean, obviously, "predicting the true tree". Rather, it refers to observables; attempting a justification for character weighting on this basis is in line with viewing parsimony analysis not just as a way to infer a phylogenetic tree, but also as a way to provide most descriptive classifications, those that allow most efficient diagnoses (see Chapter 1, Section 1.5). From this perspective, *IW* is to be preferred if it better predicts (a) the tree that future observations will uphold, or (b) the distribution of as yet unobserved characters, at least those that have the most consistent distributions—the most reliable ones, the ones that do not defy description (Goloboff 1993a; De Laet 1997). The increased supports under resampling, and under Goloboff et al.'s (2008a) experiments on stability, suggest that taking into account the amounts of homoplasy in the different characters does a better job at producing predictive (and therefore stable) trees than equal weights. It would be naive to expect any phylogenetic method to make perfect predictions, but existing comparisons in empirical datasets suggest that *IW* does better than equal weights in this regard. It is precisely because of this observable behavior of *IW* that one thinks it may be finding trees closer to "true trees" than equal weighting. Paraphrasing Farris (1982), it is not that one wants true theories because they make good predictions, but rather that we believe some of our theories are correct precisely because they make good predictions.

7.10.4 CONVERGENCE BETWEEN RESULTS OF IW AND EQUAL WEIGHTS

The previous section and the comparisons in Figure 7.9d suggest that, in empirical datasets, the results of *IW* are preferable to those of equal weights. Are there any circumstances under which the results between *IW* and equal weights can be expected to converge onto the same tree? Goloboff and Wilkinson (2018) and Goloboff and Arias (2019) have provided independent arguments to believe that *IW* and equal weights can converge for very large numbers of characters. Goloboff and Wilkinson (2018) showed that, for any number of steps $s \leq t / 4$, the dataset with all character distributions of s steps on a tree of t taxa produces that same tree as the unique most parsimonious tree. In such a dataset, as all the characters have the same number of steps, application of *IW* makes no difference; that is, $MPT_{eqwts} = MPT_{IW}$. The same tree is also obtained if a large enough sample of characters with s steps (instead of all possible) is selected. As this holds for any number of taxa or steps (Fischer 2019 demonstrates this for $s = 2$; Goloboff and Wilkinson had proceeded by exhaustive enumeration up to 20 taxa), it is possible to devise datasets where every character has hundreds of extra steps, and yet the MPT_{eqwts} is perfectly well-determined.

The other indication leading to suspect convergence in *IW* and equal weights for large numbers of characters comes from Goloboff and Arias' (2019) simulations to test the MP_{lik} model. In simulations with fixed branch lengths for all the characters, Goloboff and Arias (2019) reported that when the number of characters was very large, the MPT_{eqwts} was identical to the model tree, even when all the branches were very long—so that every character had many instances of homoplasy. The $s \leq t/4$ inequality continues holding for the resulting characters mapped on the model tree, as long as branch lengths do not exceed about 0.5 (much longer than inferred in any empirical analysis; note this is the branch length where parsimony inferences enter inconsistency with constant branch lengths, Figure 4.4c, and discussion in Chapter 4).

Those two interrelated findings suggest that, for all characters having similar amounts of homoplasy (i.e. similar reliabilities), equal weights should find the correct tree when the number of characters is large enough. But—and here comes the interesting part—if n independent datasets where the amounts of homoplasy per character are $h_1, h_2, h_3 \ldots h_n$ all produce the same (correct) tree when analyzed separately, then combining them in a single dataset $1+2+3+ \ldots n$ should *also* produce the same tree. For large numbers of characters, *IW* and equal weighting should always converge to the same result.

The problem is . . . does this not make differential character weighting unnecessary? Yes, but only in the case of datasets with many thousands of characters. In datasets where there are too few characters for a given weight class to produce the correct tree on its own, the resolution of conflict between individual characters becomes relevant. For lower numbers of characters, taking into account the reliability of the conflicting characters (i.e. *IW*) improves results. And, since for characters with amounts of homoplasy exceeding the $t/4$ limit the correct tree is not recovered, making those characters less influential (via *IW*) will also help improve results. Figure 7.10a shows 100-taxon datasets simulated with all branches of the same length, but varying for different characters (as a random number between 0.010 and 1.515). The peak of homoplasy is at 29 steps, and there is a very low frequency of homoplasy-free characters (only 1%). As expected from these considerations, when the number of characters is very large, both *IW* and equal weighting converge to the model tree (i.e. approaching the top and bottom x-axes, using two measures of distance to the model tree), but for lower numbers of characters there is a substantial difference between both methods, with *IW* producing trees closer to the model. When the branch length for different characters varies within a much narrower range (0.010–0.015), so that 70% of the characters are homoplasy-free, the trees recovered by equal weights quickly converge to the model tree, but there is no harm in using *IW*—it converges to the model tree, with growing numbers of characters, just like equal weights. The simulations comparing *IW* and Bayesian analysis (for only 50 taxa, see Chapter 4, Figure 4.6c) show a similar pattern, with *IW* outperforming Bayesian analysis.

Note that, when the length is uniform across branches and characters, most approaches for model choice (e.g. *AIC*) will prefer MP_{lik} (i.e. "equal weights parsimony") over *implik* (i.e. "implied weights"). While MP_{lik} is indeed a simpler model which (under such circumstances) produces about the same likelihood as *implik*, there is no harm in using *implik*—the results are generally identical to those of MP_{lik}. *Implik*, in such a situation, would be making some unnecessary estimates, but those unnecessary estimates do not suffice to make it go astray. Standard model selection is only part of the picture in the choice between methods.

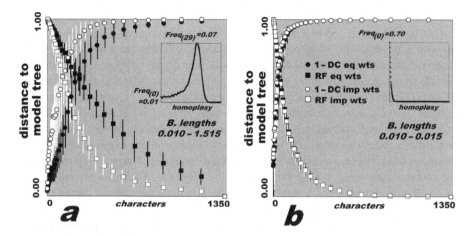

FIGURE 7.10 (a) Distance to model trees, for datasets with 100 taxa and different numbers of characters, generated from a random tree with uniform branch lengths (but changing across characters, so that the homoplasy in the resulting dataset is rampant, as shown in the inset, with only 1% of characters homoplasy-free, and characters with 29 steps being most frequent). Distributions closer to the top and bottom x-axes are closer to the model tree (*RF* is a Robinson-Foulds distance rescaled by number of groups effectively present in the trees, and *DC* is the symmetric distortion coefficient). Equal and implied weighting converge to the same result for very large numbers of characters, but at low or medium numbers of characters implied weighting outperforms equal weights. (b) Same as (a), but with much lower homoplasy (70% of characters in the resulting simulated characters are homoplasy-free). Equal and implied weights produce very similar results, for all numbers of characters, when there is little homoplasy in the dataset.

The preceding considerations indicate points of contact between those authors refusing any possibility of differential weighting, based on the expectation that characters "weight themselves" (as put by Patterson 1982; Kluge 1997; Grant and Kluge 2005), and those who defend character weighting (Farris 1969; Carpenter 1988; Goloboff 1993, 1997, 2014a; Goloboff et al. 2017, 2008a). The idea that, given enough characters, the "signal" will win over the "noise" (e.g. Wenzel and Siddall 1999) is not without merit, and homoplasy-based weighting should produce about the same results as equal weights for very large numbers of characters. The problem with that position is that the number of characters required for that ideal situation to occur is too large, much larger than can be obtained in most morphological studies. When just a few characters can be used to infer a tree, then the relative weights of the (unavoidably) conflicting characters become a serious concern. But the differences between defenders and detractors of weighting is much less a question of irreconcilable principles than usually presented.

7.10.5 Weighting in Morphology vs Molecules

We are now in a position to summarize and reconsider the differences in applications of character weighting to morphological and molecular datasets. It is trivially obvious (at least, to any biologist) that there are important intrinsic differences between molecular sequences and morphological characters. This makes it possible to come up with approaches for weighting in molecular datasets that are simply outside the range of possibilities in morphological datasets. The details of those approaches are outside the scope of this book; in brief, weighting in molecular sequences can take advantage of the fact that sequences form recognizable units—genes, chromosomes, codons, active sites—and that there can be categories of differences—transitions, transversions, stop codons, mismatches in or out of foldable sequences like tRNA. Since these categories span different characters, it is possible to extrapolate from one character to another, for example evaluating the difference in frequency of transitions vs. transversions along *all* the positions of the sequence. Such generalizations are impossible in morphology; a state in one character basically tells us nothing about the states of another (unless the characters are strongly non-independent, a situation that is problematic in itself). This kind of extrapolation forms the very basis for standard maximum likelihood methods, making transformations along some branches of the tree more or less likely.

Additionally, in molecular sequences, it may be reasonable to assume that many changes are of equal "significance"—for example in non-coding sequences, a substitution has no effect on the whole organism. Attempting to evaluate the reliability of an individual position—that is, on the basis of the homoplasy observed for just one character—seems then misguided. In morphology, not only is the possibility of meaningful global assessments of groups of characters unlikely, but the idea that different characters can be of uneven utility is better grounded. Morphological characters have an effect on the phenotype (they would not be "morphological" otherwise!), and this may be subject or not to selective pressures, more frequently than in molecular sequences, in turn determining that the morphological character is better, or worse, correlated with phylogeny.

Some methods for considering different rates among molecular characters have been considered to share similarities with homoplasy-weighted parsimony analysis. For example, Nylander et al. (2004) found similarities in results of a Bayesian analysis with gamma-distributed rates and implied weighting, and explained this with the "intuitive analogy between homoplasy-weighted parsimony analysis and parametric analysis allowing rate variation across sites". It must be noted, however, that this is only an analogy—gamma rates may increase likelihoods when rates are very different, but they still consider common branch lengths and sum up the probabilities of each character belonging to each category (see Chapter 4). Only implied weighting (and its likelihood analog *implik*) evaluates character reliability individually, without other characters coming into play in any way other than jointly determining the tree topology.

In conclusion, the number of usable morphological characters generally being much lower than in the case of sequences makes considerations of character weights more important for morphological analyses. On the other side, the impossibility of establishing categories of change or extrapolating between characters, leaves only the differences in homoplasy (that is, the observed degree of correlation between characters and groups) as the main factor to address the problem of character weights. There seems to be no escaping from this, and the almost superstitious reluctance of many authors to use differential character weighting is unjustified. The frequent "argument" that *IW* can give the wrong tree, while true as far as it goes, would be a real justification only if equal weighting was immune from errors. Of course, it is not.

7.11   IMPLEMENTATION IN TNT

The implementation of *IW* in TNT is, for the most part, transparent to the user. A summary of the commands related to weighting is in Table 7.3. The command piwe (from parsimony and implied weights) controls the use of standard implied weighting as a criterion for tree-scoring and searching. Piwe=k enables the use of implied weights, with concavity k (if no k specified, the default 3 is used; keep in mind that this is a strong concavity, and values of 8–12 are preferable for most morphological datasets). An alternative is to set k indirectly, so that the cost of adding a step to a homoplasy-free character is no more than r times (where $r > 1$) the cost of adding a step to a character with N extra steps. This is set with piwe<r N; if no N specified, the maximum possible value of homoplasy in the current dataset is used. Most of these options can be handled from the menu *Settings > Implied weights > Basic settings*.

If the user has set some character weights as different from unity (with the ccode command), when piwe is in effect, the scores for a character are multiplied by the prior weight (i.e. implied and prior weights are kept different). The ccode command (see Chapter 2 for syntax) can be used (with the = specifier) to define a fixed number of extra steps to be added to the fit calculation for a given character; you can use this when you know that some homoplasy occurs in taxa not included in the matrix, and want to use that knowledge for the analysis of the present dataset.

TABLE 7.3
Summary of TNT Commands Related to Weighting Characters and Character-State Transformations

Command	Minimum Truncation	Action(s)
piwe	pi	Handle implied weights. Sets and displays weighting function
xpiwe	xpi	Handle extended implied weights. Defines sets for uniform weights, and options for handling missing entries
score	sco	Calculate tree scores (if implied weights in effect, reports weighted weighted homoplasy; otherwise, it reports length under prior weights)
fit	fi	Report scores under implied weights (taking into account current options for self-weighted optimization or extended weighting)
cscores	cs	Report scores for individual characters
ccode	cc	Handle prior weights, character status, and internal steps
costs	cos	Handle prior transformation costs between character states
cstree	cst	Define character-state trees
smatrix	sm	Define step-matrices (i.e. prior transformation costs)

When implied weighting is in effect, tree searches report the values of weighted homoplasy, as does the command `score` (which by default reports tree lengths under prior weights). The command `fit` reports the weighted homoplasy, and `fit*` the complement (i.e. Formula 7.2), whether or not *IW* is in effect. The number N of significant digits output in tables reporting tree scores can be changed with `table/N`. In the case of additive characters (or additive character-state trees with the default `cstree+` option), calculation for implied weights decomposes the character in the equivalent binary variables (summing up the scores, or fits, for each of the variables). To consider additive characters as a unit, they need to be set as character-state trees, calculating the scores under `cstree-`, or as continuous characters (see Chapter 9). For step-matrix characters, the scores under implied weighting are $(h / t_{min}) / (k + h / t_{min})$, where t_{min} is the minimum cost of transforming between any two states; for implied weighting h is formally defined as $s - m$, the number of steps minus the minimum possible number of steps (see Goloboff et al. 2021a for discussion of alternative views).

The weighted homoplasies and the fits are reported on the arbitrary scale resulting from the formulae used to weight. Keep in mind that specification of suboptimal values (`subopt` command), for example for searches saving suboptimal trees, must consider decimal values. Near-optimal trees often differ in just a few hundredth units. The values reported are rescaled so that the first step of homoplasy costs unity when defining user-weighting functions (the trick of defining a user-weighting function identical to the standard one, see later, may help better visualize differences in tree scores—the difference between no homoplasy and the first step of homoplasy always costs unity for user-weighting functions). As near-optimal trees can have very small differences in score under implied weighting, thus leading to a single or a few optimal trees only because of the finely graded scoring function, it is especially important to eliminate groups that disappear in trees with a minimal difference in score (either using Bremer supports, or indirectly via resampling; see Chapter 8).

The relative costs of adding a step to different steps of homoplasy is displayed with `piwe[` (or *Settings > Implied weighting > Basic settings*, clicking on the *Show current function* button). A list of step costs can be specified by the user, after `piwe[` (this requires that the dataset be read with implied weights ON, i.e. with `piwe=`). The list needs to define as many costs as extra steps are possible for the dataset; undefined values are considered to be identical to the last value defined (that is why `piwe[1 0` defines a clique-weighting function). TNT reads and stores the user-weighting function, rescaling it so that the first step of homoplasy always costs unity. User-weighting functions cannot be used together with step-matrix, continuous, or landmark characters, and they cannot be converted into `fit*` values (there is no special meaning to the complement of the weighted homoplasy). The option `piwe]` returns the weighting function to the default.

7.11.1 SELF-WEIGHTED OPTIMIZATION

Using `piwe+k` instead of `piwe=k` sets the use of self-weighted optimization. This weights state transformation costs; if prior costs are defined (i.e. with the `cost`, `smatrix` or `cstree` commands) they act as multipliers for the implied costs.

Self-weighted optimization can be used only for relatively low numbers of taxa (no more than 100–200) and character states (about 5–6); characters with more trans-formations in the tree take longer to be optimized, and so exact time predictions are very difficult, depending in part on character congruence. The parameters to control calculations of ancestral reconstructions (for increasing accuracy of the score evaluations, longer times needed) are changed with the slfwt command. The slfwt command is followed by option(s), the most important of which are:

- exops N. Forbid transformations occurring under prior costs for characters with N or more states; default is 5)
- errmargin N. During searches, re-check trees for which initial score estimates are within N units of fit above best score; default is 0.5)
- maxtbr N. During searches using *TBR*, reroot clipped clade no more than N nodes away from original rooting (maxtbr 0 = *SPR*; default is 5]
- setlim N. Set to N the number of states for application of restrictions to state assignments (default is 4)
- maxdist N M. When restricting possible state sets, select from those states present in branches up to N nodes away above a node, and M nodes below (defaults 6–3)
- [no]useminset. Restrict possible state sets by selecting candidates from the state set corresponding to an optimization under prior costs (very fast, very prone to errors; default is no)
- maxrepeat N. When N successive branches have the same (unique) state, consider them as fixed (default is 8)
- [no]verbose. As characters are optimized, report progress; this is useful for testing alternative combinations of parameters, and finding out which individual characters are eating up calculation time (default is no).
- timeout N. Spend no more than N sec (it must be an integer) in optimizing each character on a tree; if timeout reached, the best reconstruction found is used, and calculations proceed with next character. Obviously, the optimal reconstruction may not have been found yet (default is 0, no timeout)

Most of these options can be set, in Windows version, from *Settings > Implied weighting > Basic settings*, selecting *weighting state transformations*, and then the *set params* button. When self-weighted optimization is in effect, displaying (with recons) the individual character-state reconstructions also reports the number of changes of every type of transformation.

7.11.2 Extended Implied Weighting

Missing entries and weighting sets can be taken into account with extended implied weighting (Goloboff 2014a). This is done with the xpiwe command. To enable extended implied weighting, the dataset must be read with implied weighting ON (piwe=), issuing an xpiwe= command after reading the dataset. Extended implied weighting cannot be used together with the piwe+ option (i.e. it applies only to entire characters, not character-state transformations).

The most important function of the xpiwe command in morphological datasets is to determine the handling of missing entries. If combining morphology with molecules, the definition of sets for uniform weighting on the basis of average homoplasy, and the setting of characters with constant weights, can also be important; these are also handled with the xpiwe command. In Windows versions, the corresponding options are changed with *Settings > Implied weighting > extended weighting* (this menu option only becomes available if the dataset has been read with implied weights enabled). From the menus, weighting sets can be applied only to data read in different blocks; however, the syntax of the xpiwe command allows you to define any sets you wish (see later).

7.11.2.1 Missing entries

The option to handle missing entries or different concavities for different characters is xpiwe(. The symbol (can be followed either by *, =, or by a list of characters; it also connects the use of extended implied weights. The options * and = automatically determine different concavities based on proportions of missing entries in every character (see later). Using a list of characters, the user defines the value of concavity for the characters specified. Once the values of the concavities have been defined, either by the user or from the proportions of missing entries in each character, the piwe & command shows the concavities for each character (or the "shifts" in character steps, when user-weighting functions are in effect). With piwe & L (where L is a list of characters), TNT displays a table with the costs of adding a step to each step of homoplasy for every character in list L (keep in mind that the table may include different numbers of values for different characters, depending on their maximum possible values of homoplasy). The simplest command line to read a dataset and take into account missing entries would be:

```
tnt*>piwe=10; p dataset.tnt ; xpiwe(*;
```

The (* and (= option can be followed by the proportion P of homoplasy in observed entries that is to be assumed for missing entries (default P = 0.50); as example, consider a character where only *o* out of the *t* taxa have observed entries. If P = 1, then the cost of adding a step to the Nth step of the character must equal the cost of adding a step to *Nt / o* steps in a fully scored character (remember: this is *not* the same as *t / o* times the cost of the Nth step in a fully scored character). If one wishes to be more conservative and deviate less from observed homoplasy, setting P = 0.5, then the cost of adding a step to the Nth step of the character must equal the cost of adding a step to *0.5Nt / o* steps in a fully scored character. A limit can be set after the proportion P, with <V, for the extrapolation to not increase steps beyond a certain maximum value V (default is 5). This is followed by the reference concavity, /k, to be applied to fully scored characters (if unspecified, the reference concavity is the one set with piwe=k). In summary, xpiwe (*P <V /k. Note that P = 0 or V = 1 are equivalent to standard implied weighting. If weighting sets are defined (with the xpiwe[or xpiwe] options, see Section 7.11.2.2), TNT will honor those sets (extrapolating from the average homoplasy in the observed entries to the total missing entries in the set). When defining the dataset, inapplicable characters (which should not increase

the counts of missing entries!) can be indicated with the symbol * (=set of all states but the largest) instead of ? or −.

The difference between the (* and (= options is that the former approximates the extrapolation into the larger number of steps for the reference concavity just by using k values in the same ratios as *Pt / o*. The = option heuristically adjusts k values for the character with missing entries so that the average cost differences for each of the first 15 extra steps and *Pt / o* times those steps in the reference concavity is as small as possible.

The use of adjusted k values allows extrapolating missing entries also in the presence of continuous, landmark, or step-matrix characters. When a user-weighting function is in effect (in which case continuous, landmark, or step-matrix characters cannot be used), the steps are extrapolated into the reference weighting function (in that case, the * and = options become identical). If the extrapolation leads to a fractional number of steps for the reference (fully scored) character, an interpolation of the user-defined weighting function is carried out (as this requires a substantial number of calculations, it is done only once, when preparing the data for optimization or searching, storing the resulting values for later reuse). The interpolation is done on the basis of the user-defined costs for every additional step of homoplasy (and not from the table of weighted scores, *w*, assembled from the user-defined costs); the interpolator used is the same one invoked with the scripting expression intpol (you can use this to check the exact values of interpolation between given points; see Chapter 10 for how to use scripts).

If xpiwe(is followed just by a list L of characters (without * or =), terminating with /k, all the characters in list L get concavity k. Concavities set with this are always honored by TNT; you can use this to modify a set of concavities established on the basis of proportions of missing entries. Keep in mind that if weighting sets have been defined, all the characters in a set must have the same concavity; attempting otherwise triggers an error message. When user-defined weighting functions are in effect, k is used (as in the previous option) as a factor to extrapolate.

Finally, the xpiwe) option sets a single concavity (or user-defined weighting function) for all characters, returning the system back to standard implied weighting.

7.11.2.2   Uniform weighting of characters or sets

TNT allows defining sets of uniform weighting on the basis of homoplasy, and characters with constant weight. These options can be useful when combining morphology and molecules (application of standard implied weighting to sequences is uncommon). To give some characters a constant weight, the syntax is xpiwe /NxF L, where L is the list of characters for which every step costs the same as F times the cost of adding a step to a character with N extra steps (under the reference concavity or user-defined weighting function). Note that this is "weighting", in a sense—to give some character(s) a constant weight, the user must decide what that weight is. Using xpiwe/0x1 will give the constant-weight characters the same weight as a homoplasy-free character, but if the rest of the characters are being downweighted with homoplasy, this will represent an upweighting of the constant-weight characters over most of the rest.

Character sets are defined with xpiwe[or xpiwe] and weighted with average homoplasy. Set weighting is peculiar in that it makes parsimony-uninformative characters relevant to the analysis (in a set with many invariant characters, most of the characters are indeed free of homoplasy, are they not?). For set weighting, when user-defined weighting functions are in effect, TNT needs to interpolate costs (since it uses average numbers of steps, which can be fractional).

The option xpiwe[L defines characters in list L as a set for homogeneous weighting. The list L can be optionally followed by =N, which divides the set in subsets or chunks of N consecutive characters each. Alternatively, L can be followed by /, and a partition (e.g. xpiwe[{rbcL} /12:3 or xpiwe[{CytB} /1:2:3); then the first, second, and third positions within list L are grouped as specified. Not specifying 1, 2, or 3 leaves those positions outside the subset(s). The option xpiwe] is the same as xpiwe[, but it does all data blocks automatically, without the need to provide a list L of characters. Thus, xpiwe]/12:3 creates a set for 1st and 2nd positions, and another with 3d, for each of the data blocks, while xpiwe] = 10 divides each of the data blocks in chunks of 10 consecutive characters each.

To remove characters from lists of weighting sets and fixed-weight characters, use xpiwe!L (where L is a list of characters). Other options of the xpiwe command serve to report or save current set definitions or status. Xpiwe* writes to log file the list of characters in weighting sets, and the list of characters with fixed weights, in a format that is readable by TNT (this can be used to save current definition for later use). Xpiwe& reports the status of weighting sets for subsequently listed tree(s). If tree-list is preceded by either steps, score, homoplasy, or size, then it reports the corresponding value (default is size). Under different concavities, you can also report the concavity values of each set (kvalue) or cost of adding a step to an average character in the set (cost).

NOTES

1 To be fair, Kluge and Grant's main argument for abandoning Farris' (1983) standard justification was that, in their view, the connection between explanatory power and extra steps collapses in the presence of inapplicable characters. Kluge and Grant's (2006) view was subsequently refuted by De Laet (2015) and Goloboff et al. (2021a); in fact, De Laet (2005) had already provided arguments against that view.

2 I must note here that "empirical" may be a better term than "a posteriori", but it is not an ideal term, either. *A priori* weighting methods, despite being problematic, can also be made by reference to "observations", thus having an "empirical" component. There is no point for arguing in favor of one usage over another, as long as what is being implied is clear.

3 A few years later, Swofford et al. (1996) no longer characterized the method as circular.

4 Note that jackknifing or bootstrapping in PAUP* or TNT on the basis of the implied alignment produced by POY (Wheeler et al. 2006) suffers from analogous problems. Given the computational difficulty of analyzing unaligned sequences, this is often done as an expediency (e.g. Frost et al. 2006; Faivovich et al. 2010), but if alignment itself is ambiguous, the treatment constitutes only an approximation and can overestimate supports (as noted by Faivovich et al. 2005).

5 Evidently, multiple resolutions with zero-length branches may have the same sets of weights, in the presence of ambiguous optimizations.

6 The likelihood for a character is also dependent on the branch lengths (see Chapter 4), which is in a sense part of "the topology".

7 Goloboff et al. (2008a), in their footnote 4, provided an incorrect formula. I thank J. Bosselaers and J. De Laet, both of whom called my attention to the error. The empirical comparisons of Goloboff et al. (2008b) used k values determined heuristically, so the error in the formula did not affect their results.

8 Berger and Stamatakis (2010), as customary in computer-science papers (cf. Goloboff 2015), cite none of the previous work on weighting done by systematists.

9 To be fair, the calculations only need to consider every possible tree *shape* (e.g. as in Goloboff and Wilkinson 2018), not every possible tree. This, however, would make the calculations only slightly easier.

10 For the example in Figure 7.6a, and $k = 10$, the tipping point for preferring three independent derivations or a gain and a reversal is at about $q = 8.39039406$.

11 W. Wheeler notes (personal communication) that the approach could also be justified from a Bayesian perspective: the more changes in a character, the higher the probability that the character is in the class of characters with a high rate of change.

12 I assume they just copied the wrong script to the Supplementary Material. The published script has some obvious errors. For example, it always uses the asymmetric tree as model. As Puttick et al. (2017b) report substantial differences in results between symmetric and asymmetric trees, they must have had another script without this problem. Another inconsistency between the paper and the script is that Puttick et al. (2017b) stated (p. 2) that "for parity, [they] characterized the result of all phylogenetic methods as the majority-rule consensus of resultant tree samples", but the script provided uses the majority rule only for the equal weight parsimony analysis—analyses under implied weights actually use the strict consensus tree in the distributed script (with the `nelsen` command of TNT). All this makes the results reported by Puttick et al. (2017b) wholly irreproducible.

13 Puttick et al. (2017b) stated that their simulation mimics the distribution of C values in Sanderson and Donoghue's (1996) database (which reports three decimals and starts with 0.415, 0.783, 0.459, 0.644, 0.433, 0.833, 0.742. . .). However, in the published R script the "target distribution of homoplasy" is listed as the series 0.79, 0.72, 0.44, 0.63, 0.43, 0.39, 0.31, 0.56, 0.37. . . Except for the first value, 0.79, occurring once instead of twice, the series of values used in the script is *identical* to the sequence of two-decimal values published by Sanderson and Donoghue (1989), not (1996) as cited by Puttick et al. (2017b).

14 Even more ironic, some of the authors (T. Grant, A. Kluge) interpreting Källersjö et al.'s (1999) results as mandating equal weighting, reject in other contexts the use of resampling measures as "unscientific" (e.g. Grant and Kluge 2003). Apparently resampling is acceptable when they view it as reinforcing their unshakeable belief in equal weights.

8 Measuring degree of group support

Previous chapters have discussed the problem of how, given a set of observations, one can establish which tree(s) best fit the data. A group which is present in all optimal trees for a dataset is said to be *supported* by the dataset—providing the best possible genealogical explanation of the observed similarities requires positing the existence of the group. Thus, the support for a group relies on the best achievable tree scores being better with the group than without. That serves to indicate whether a group is supported or not, but the problem is more subtle than just that. The evidence in favor of some groups (e.g. insects, vertebrates) is abundant and almost undisputed, but the evidence for other groups—even if positive—is much leaner and contradicted by other characters. The two situations are very different. Whether the evidence supports a group is determined from the strict consensus of *MPT*s, but how strongly it does so is a different matter, a matter that needs to be properly addressed and quantified.

8.1 THE DIFFICULTY OF MEASURING GROUP SUPPORTS

The strength of support cannot be determined from separate consideration of individual characters; the entire dataset needs to be considered. It must be emphasized, however, that what needs to be evaluated are the consequences of the entire dataset for a particular group of taxa, which is also different from evaluating the dataset in the abstract. The preceding chapter (Sections 7.3.1 and 7.7) briefly mentioned measures for evaluating whole datasets. Those measures evaluate data quality with various degrees of success and justification, but there is a big difference between evaluating the quality of the data and the strength of support for specific groups. The point of whether a dataset is itself "unreliable" does not carry much consequence. If several datasets are available, the principle of globally considering all the evidence still requires that the weakest dataset is combined with others. If the weakest dataset supports conclusions contradicted by the majority of the evidence, then it will be uninfluential—in systematics, there is no way to conclude that some characters are misleading, other than observing that they are contradicted by *other* characters. A worse limitation with measures of dataset quality is that they simply do not address the main problem of interest, which is whether or not the evidence in favor of a given *group* is convincing. Even when appropriate within their domain, measures of data quality are irrelevant for this problem. Some years ago, reporting of consistency or retention indices was almost mandatory, but with the realization that those measures have little to do with group supports, their reporting gradually became much less common.

DOI: 10.1201/9780367823412-8

This chapter is concerned almost exclusively with measuring the amount of evidence in favor of a group, and the term "support" is reserved for that concept. Many authors have attempted to devise measures indicating the probability that a group is correct (e.g. Felsenstein 1985b; Yang and Rannala 2005; Anisimova and Gascuel 2006). Although that probabilistic notion is sometimes called "support", in this book the term support always refers to the evidence alone—how much evidence favors the conclusion, and how is the interplay between favorable and contradictory evidence. Of course, an explicit assessment of the probability of a conclusion being correct depends in part on the amount of evidence supporting the conclusion (i.e. the part that is called "support"). Some empirical relationship between the amount of favorable evidence and the statistical confidence is then to be expected. With very scant evidence, confidence will be low, and vice versa. But the two aspects—support and truth—are separate, from the logical point of view, and this chapter is concerned only with the former. And, despite that general correspondence, the relationship between the two may be far from linear, and specific values of probability given certain values of favorability of evidence (however measured) will depend on specific models of character change—and none seems well justified for morphology. Measuring probability of correctness requires generalizing well beyond the dataset at hand and requires strong assumptions (about evolutionary models, about the dataset being a representative and unbiased sample, etc.); measuring amounts of evidence requires no such assumptions (Farris 2002). But even the simpler goal of properly measuring amounts of evidence becomes complicated by several factors, as this chapter illustrates. Only some sections in this chapter will deal with problems related to measuring confidence in the correctness of a group.

The problem of how to measure strength of support for a group cannot be addressed by just counting its number of synapomorphies or by its branch length. It is often said that groups have "support" from synapomorphies. Only a group for which some characters are mapped as unambiguous synapomorphies can be supported— for every character mapping as a synapomorphy of the group, additional steps are required if we do away with the group; if no synapomorphies, no additional steps when losing the group. It is in this sense that the notion of a synapomorphy "supporting" a group matches the general notion of support outlined earlier, that a group is "supported" if and only if removing the group requires decreasing the fit to the data. The length of the branch provides an upper bound of the score difference to lose the branch—collapsing the branch into a polytomy will increase tree score by no more than the length of the branch.[1] But the length of the branch subtending a group is only an *upper* bound for the support; the support can be much less than the length of the branch if *other* characters can be better explained by alternative groupings. This is so even when the character(s) supporting the group are entirely free of homoplasy (i.e. the tree displaying the group has homoplasy only in the characters contradicting the group). Consider datasets 1 and 2:

1		2			
V	000	A	0000	000	000
X	000	B	0000	111	111
Y	111	C	1111	111	000
Z	111	D	1111	000	111

In dataset 1, the group YZ is supported by three homoplasy-free synapomorphies, and in dataset 2, group CD is supported by four. Yet it is obvious from the examples that concluding that group CD is better supported than YZ would be erroneous: there is no character contradicting YZ, while there are six characters contradicting CD. Thus, a proper measure of support must consider not just the evidence that seems favorable to the conclusions, but also the evidence that seems to be *against*. This is the reason for the failure of some measures of confidence in groups, such as Rzhetsky and Nei's (1992) measure of "Confidence Probability" (which simply tests for whether the length of the branch is significantly more than zero and then can produce absurd evaluations of support in specific cases; see Farris et al. 1996, 1999, 2001).

Although branch lengths or numbers of synapomorphies do not suffice to demonstrate support for a group, simply counting the characters that favor and contradict the group will also be insufficient, for two reasons. The first, most obvious reason why simple counts do not work is that one also needs to consider whether the contradicting characters themselves are congruent. In dataset 2, there are four characters for group CD, and six against. The seemingly contradictory evidence is more than the favorable, yet it is clear from the dataset that it globally upholds CD. That is because the six characters contradicting CD also contradict each other, so that it is not possible for a single tree to save homoplasy for all six characters simultaneously (three is the best any tree can do). Dataset 2 has only four taxa, so the conflict between characters is evident, but with more taxa the maze of possible contradictions becomes difficult to visualize.

The second, more subtle reason, is that the notion of a character "supporting" or "contradicting" a group applies to idealized examples such as datasets 1 and 2, but in more complex datasets the distinction between the two becomes blurred. A character can be said to support a group if removing the group from the tree necessarily increases length for the character, but this may be the case for local rearrangements within the tree—that is, when the rest of the tree is resolved by other characters—yet more global rearrangements that eliminate the group would save steps in the character that is locally "favorable". Consider dataset 3 (from Goloboff et al. 2003a):

	3				
A	0	0	0	0	0
B	1	0	0	0	0
C	1	0	0	0	1
D	1	0	0	1	1
E	0	0	0	1	1
F	0	0	1	1	1
G	0	1	1	1	1
H	1	1	1	1	1

which produces a single pectinate *MPT*, (A(B(C(D(E(F(GH))))))). The group EFGH is supported by a reversal (a $1 \rightarrow 0$ change) in the first character; no other character acts as a synapomorphy of EFGH. But, at the same time, the only character that contradicts the group is the first character, which by itself (i.e. without interactions with other characters) would support the group BCDH instead of EFGH. Furthermore, no other character by itself seems to support EFGH directly (they are all super or subsets of EFGH), but eliminating any of the characters other than the first causes group EFGH to become unsupported by the dataset, and so does doubling the weight of the first character (i.e. upweighting the character that "supports" EFGH removes the "support" for EFGH). This simple example shows that classifying characters into those that are "favorable" or "contradictory" to a group is not always possible, and whether a character "supports" a group may well depend on the interaction with other characters. The notions of favorable and contradictory characters will be used throughout this chapter, but it must be kept in mind that they refer to simple idealizations.

A corollary of the support being defined by character interactions and not the sum of individual characters is that the total support of a group cannot be calculated from summing contributions from individual characters; the support must be calculated from the collective of characters. This makes it necessary to use indirect methods to measure group supports. Two approaches are widely used for this. The first is comparing the optimal tree with other trees; this is known as Bremer support and goes beyond simple counts of synapomorphies because it automatically takes into account whether the characters contradicting a group can all improve their fit simultaneously, on a single tree. The second approach is resampling, taking a subsample of characters. Should contradicting sets of characters exist, subsampling will retrieve characters from the larger set more frequently (Farris et al. 1996), thus indirectly detecting the relative amounts of favorable and contradictory evidence.

8.2 BREMER SUPPORTS: DEFINITIONS

Comparing the scores of two trees provides a natural and intuitive way to decide how convincingly the evidence supports the tree of better score. In the context of parsimony, support measures based on comparing tree scores (either global scores, or scores of individual characters) are generally called Bremer supports. For a specific group, the difference in length between the best possible trees having the group, and the best possible trees lacking the group, is called the *absolute Bremer support BS*:

$$BS = Length_{without} - Length_{with}$$

If the best trees without the group are more parsimonious than the best trees with the group, then *BS* is negative—properly indicating that the group is contradicted by the evidence, with magnitude of *BS* indicating strength of contradiction. This idea was put in practice by Bremer (1988), and formalized in more detail by Bremer (1994). Bremer himself called this just "support", but the measure is widely known as Bremer support. Some authors (Donoghue et al. 1992; Mishler et al. 1992) have called Bremer supports "decay indices"; Goodman et al. (1982) had used the same idea (cf. Grant and Kluge 2008a), but without giving it a name. Some modifications of *BS* consider character-by-character differences in score, instead of the absolute differences, and hence the name *absolute* is used here for *BS*.

BS is similar in form to the likelihood ratio, often used to quantify confidence in groups (e.g. Anisimova and Gascuel 2006); when likelihoods are converted to log likelihoods, ratios become differences, just like *BS*. Within the context of data obeying the assumptions of standard Poisson models used in model-based phylogenetics, the likelihood ratio may have advantages over the absolute Bremer support, because that measure incorporates inferences about branch lengths (i.e. probabilities of change). Very homoplastic or noisy data are expected under that model only for very long branches; then, even if one tree is of better likelihood than the rest, the difference in likelihoods for alternative trees will always be very small. So, likelihood ratios under standard *ML* models take this situation into account better than the plain *BS*. In the non-parametric framework of parsimony, likelihood ratios seem closer in spirit to modifications of the *BS* (such as relative or combined Bremer supports, see later). Of course, *BS* and likelihood ratio are fully equivalent under the *NCM* (no-common mechanism) model of Tuffley and Steel (1997) (see Wheeler 2012: 337 for details).

BS, by virtue of comparing scores on specific individual trees, properly handles cases like the one shown in dataset 2 (see earlier), where the number of characters supporting a group is more than the number of characters contradicting it, but the contradicting characters are in turn in conflict with each other. For dataset 2, the best tree (A(B(CD))) is 16 steps long, while the alternative trees (A(C(BD))) and (A(D(BC))) are 17 steps long; $BS_{CD} = 17{-}16 = 1$. Compare this to dataset 1, which has only three synapomorphies for group YZ in the tree (V(X(YZ))), of three steps; best trees lacking YZ are (V(Y(XZ))) and (V(Z(XY))), of six steps, so that $BS_{YZ} = 6{-}3 = 3$. That is, support for YZ in dataset 1 is substantially stronger than support for CD in dataset 2, even if CD is supported by four synapomorphies, and YZ by three. *BS* is therefore not just the sum of the supports given to the group by individual characters (or partitions), but instead takes into account character interactions.

BS values for a group need to be calculated by comparing scores of the best possible trees with and without the group. Obviously, if *BS* is to be calculated for several groups (JK, LM, etc.) the best trees lacking each of the individual groups must be considered separately. The best trees that lack group JK may well all still display group LM; a tree lacking both groups simultaneously will simply not be the best possible tree without either group. This requires doing appropriate tree searches to evaluate *BS* values, a point to which we return in Section 8.3. A diagram of how to calculate *BS* values (and other variants of Bremer supports) is shown in Figure 8.1.

8.2.1 Variants of Bremer Supports

8.2.1.1 Relative Bremer supports (*RBS*)

While the evaluations of support produced by *BS* are frequently meaningful, they can be problematic in some cases. *BS* only considers the extent to which the favorable evidence exceeds the contradictory, but the *ratios* between favorable and contradicting evidence may also be relevant. Consider dataset 4, and the *BS* values for two possible groups, CD and FG:

```
                           4
A  0  0  0000000000000000000  00000000000000000000000  00
B  1  0  1111111111111111111  00000000000000000000000  00
C  1  0  1111111111111111111  11111111111111111111111  00
D  1  0  0000000000000000000  11111111111111111111111  00
E  0  1  0000000000000000000  00000000000000000000000  00
F  0  1  0000000000000000000  00000000000000000000000  11
G  0  1  0000000000000000000  00000000000000000000000  11
```

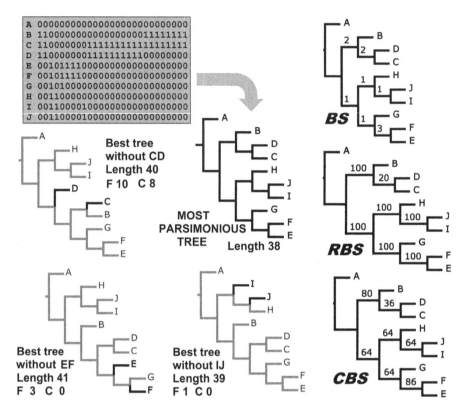

FIGURE 8.1 Basic calculations for Bremer support; *BS* = absolute Bremer supports, *RBS* = relative Bremer supports, *CBS* = combined Bremer support.

BCD and EFG obviously form two (uncontradicted) groups on the *MPT* (the first two characters in the matrix). Within BCD, many characters favor CD, but almost as many favor BC (the actual count is 23 vs. 20). Within EFG, there are only two characters that favor FG, but none that contradicts it. A logical conclusion in this case is that CD is almost "unsupported"—there is almost as much evidence that speaks against the group as there is in favor. Yet the *BS* values for CD (3 steps) are higher than those for FG (2 steps). Taking this into account requires some way to evaluate support that can, if needed, rank the groups differently from *BS*.

Goloboff and Farris (2001) thus proposed the *relative Bremer supports, RBS*, which are based on the relative fit difference, *RFD*, between two trees. Given two trees A and B of a different score (with A more parsimonious than B), amounts of evidence favoring and contradicting the better tree can be approximately quantified by comparing fits of individual characters. Call F the sum of (weighted) step differences in the two trees, for all the characters that fit the more parsimonious tree better. Call C the sum of (weighted) step differences in the two trees, for all the characters that fit the less parsimonious tree better. When $C = 0$, there is no character leading to prefer tree B over A. The ratio C / F expresses the proportion of contradicting evidence, relative to the favorable evidence. The *RFD* is defined as the complement of *C / F*, or *(F − C) / F*. Note that the numerator of this expression, *F − C*, equals the *BS*. An *RFD* of 1 between two trees implies that no evidence in the dataset seems to favor the worse tree. Although *RFD* is not calculated directly from numbers of favorable and contradicting characters (instead taking into account differences in character fits between the two trees), it is possible to visualize it in those terms. The conflict can be assimilated to the conflict of *1 − RFD* contradicting characters per favorable character; for example an *RFD* of 0.8 amounts to the conflict of 2 characters vs. 10.

The relative Bremer support *RBS* of a group is defined as the smallest difference in *RFD* needed to remove the group from the tree. *BS* is unbounded, but $0 \leq RBS \leq 1$. As *RBS* measures proportions of favorable and contradictory evidence, *RBS* is the same (unity) for two groups supported by 1 and 100 uncontradicted characters. Another caveat with *RBS* is that it requires special precautions for calculating the *RFD* value of the suboptimal trees, when multiple *MPT*s exist. Comparison of a suboptimal tree with the (most parsimonious) tree A may produce a given value of *RFD*, different from that obtained by comparing the suboptimal tree with the (most parsimonious) tree B—even if A and B have the same scores. Goloboff and Farris (2001) provided an example of this, and showed (their fig. 1) that the correct evaluations of *RBS* result from assigning every suboptimal tree the *largest RFD* difference with any of the optimal trees. Another problem pointed out by Goloboff and Farris (2001) is that calculating *RBS* from the smallest possible *RFD* without further restrictions can be misleading. Dataset 4 serves as an example. There is a single optimal tree, (A((B(CD))(E(FG)))), for that dataset, with 67 steps. The suboptimal tree (A((D(BC))(F(EG)))), with 72 steps, lacks both groups CD and FG. This is in fact the tree with the smallest *RFD* compared to the optimal tree; on that tree, *F = 25* and *C = 20*, and *RFD = 0.20*. Thus, one could be tempted to conclude that there is character conflict in the support for group FG, equivalent to the conflict of 10 characters vs. 8. But in fact group FG is contradicted by no characters and should properly receive *RBS = 1*. That problem is eliminated in this case if the value of *RBS* for every group is calculated considering only the best trees lacking the group (i.e. trees with a score difference within the *BS* of the group). In the present case, that is either (A((B(CD))(F(EG)))) or (A((B(CD))(G(EF)))), with 69 steps each. For those trees, *F = 3* and *C = 0*, producing *RBS = 1*—as expected. In the implementation of TNT, calculation of *RBS* values considering only trees suboptimal within the *BS* value of each group is optional. However, in more complex cases, it is still possible that considering only trees within the *BS* values produces incorrect assessments of the proportion of characters for and against the group.

8.2.1.2 Combined Bremer supports

Both *BS* and *RBS* measure different aspects of the support. *BS* measures the extent to which favorable evidence exceeds contradictory evidence; *RBS* measures the ratios between favorable and contradictory evidence. Groups can be ranked in different order according to both *BS* and *RBS* because both aspects of the "support" can vary independently—when one is larger the other may be smaller. As both aspects are relevant for quantifying amounts of evidence in favor of a group, it is advisable to report both *BS* and *RBS* (TNT offers facilities to superpose several support measures on tree plots). In many cases, however, it is simpler to have a single measure that combines both aspects. Resampling does exactly that, but it is also possible to combine *BS* and *RBS* in a way that emulates the results of resampling for simple cases. This is the *combined Bremer support, CBS*, proposed by Goloboff (2014c).

To understand the logic behind *CBS*, it is necessary to relate it to the behavior of resampling measures (discussed in more detail in Section 8.4). In an independent-removal jackknifing with probability of deletion p, when a group is supported by r uncontradicted characters, the expected frequency of the group equals $1 - p^r$; if the deletion probability is set to $p = 1 / e$, this is the same frequency expected for the group under bootstrapping when the matrix has a very large number of characters (see Section 8.4). Note that $1 - p^r$ will (unlike *RBS*) vary with the number of uncontradicted characters—approaching unity as r grows larger, taking on the value $q = 1 - p$ (i.e. the probability of retaining a character) when a single uncontradicted character sets off the group. This is sensible—as there is more undisputed evidence in favor of a conclusion, the conclusion is better supported.

The most problematic cases for *RBS* are when comparing groups supported by different numbers of uncontradicted characters, all of which have *RBS = 1*. That problem is remedied by the quantity

$$CBS = \left(q \times RBS\right)^{1/BS}$$

which will behave similarly to resampling measures in the case of different numbers of uncontradicted characters. With a single character supporting a group, *RBS = BS = 1*, and *CBS = q*—just as in jackknifing. As there are more uncontradicted characters in favor of the group, *CBS* approaches unity. If there are many conflicting characters supporting and contradicting the group, *CBS* tends to 0 if the supporting characters outnumber the contradictory ones in one or a few characters. The example of Figure 8.1 shows this behavior for the group CD, with much lower *CBS* support than BCD, even if both groups have the same *BS*; *RBS* evaluates EF and IJ as equally supported (even if EF is supported by three uncontradicted characters and IJ by only one), but *CBS* gives EF a higher support than IJ. Therefore, *CBS* seems to provide a more meaningful assessment of the overall strength of the evidence for groups than either *BS* or *RBS*.

Although *CBS* approximates resampling frequencies in simple cases, no exact agreement can (or should) be expected. *CBS* can never be positive for any group that is not supported by the data, as it is defined by reference to optimal trees. As discussed next, resampling can attribute high "support" to groups that are not present in any of the optimal trees for the data—that is, groups that have no support at all. In those cases, deviating from the values produced by resampling is an *advantage* of *CBS*.

Jackknifing can be performed under different probabilities p of deletion, and *CBS* can use different values of q, with larger values of q $(0 < q < 1)$ producing larger values of support (as in jackknifing). Another (more practical) advantage of *CBS* is that it can be estimated more quickly than resampling (e.g. via branch-swapping, as discussed in Goloboff et al. 2021b). In the case of implied weighting, the difference in score may depend on the weighting function used (i.e. unless the values are normalized in some way, as in Formula 7.3). In that case, *CBS* can be defined as

$$CBS = \left(q \times RBS\right)^{w_0/BS}$$

where w_0 is the cost of the first step of homoplasy. This serves to normalize the case where *BS* itself is smaller than unity (which can happen under implied weights). With this, *CBS* for groups supported by uncontradicted characters is identical to *CBS* under prior (equal) weights.

8.2.1.3 Relative explanatory power

A different kind of normalization of *BS* was proposed by Grant and Kluge (2007), the *relative explanatory power* or *REP*. While the idea behind *RBS* and *CBS* is to possibly change the ranking of groups of *BS*, *REP* is intended to never change the ranking of groups by *BS*, using the same normalizing factor for all the groups in the tree from the same dataset. The normalizing factor changes in this case for different datasets. *REP* is defined as *BS* / (*G* − *S*), where *G* is the length of the worst possible tree (as in the retention index, see Section 7.3.1), and *S* is the number of steps of the most parsimonious tree (i.e. the denominator is the maximum possible *BS*, if all characters were congruent and supported the same split among terminals).

Grant and Kluge (2008b) asserted that *BS* or *REP* provide the only "objective" measures of group support, because *RBS* (as well as the subsequently described *CBS*) can alter the ranking of groups under *BS* (*REP* does not). They considered (p. 1057, and their fig. 5) that *RBS* giving "an uncontradicted group corroborated by a single synapomorphy and an uncontradicted group corroborated by 100 synapomorphies the same support value" is unacceptable, because "the relative strength of the latter hypothesis is clearly greater". They further complained that when "adding a single contradictory synapomorphy to each case . . . the [*RBS*] jumps from scoring both clades equally to ranking the two clades at opposite extremes". But despite their repeated appeals to objectivity,[2] Grant and Kluge (2008b) had not considered the behavior of their own preferred statistic too carefully. Farris and Goloboff (2008) showed (in their fig. 1) that *REP* has the exact same behavior Grant and Kluge (2008b) criticized in the *RBS*, accusing Grant and Kluge (2008b) of a double standard. Grant and Kluge's (2010) subsequent discharge simply engaged in more distortion:

> Farris and Goloboff's (2008) first example (Fig. 1) demonstrates that *REP* and [*RBS*] can exhibit the same behavior. Grant and Kluge (2007, 2008b) never claimed they could not.

It was not Farris and Goloboff (2008) who had claimed that the behavior of *REP* for their first example was objectionable. It was Grant and Kluge themselves, but the reader would never suspect that from Grant and Kluge (2010); they never mentioned their former complaint about variations in support under the addition of a single character. After Farris and Goloboff (2008) showed that *REP* behaves in that regard just like *RBS*, what had been unacceptable in 2008 magically became correct in 2010.

Other examples used by Farris and Goloboff (2008) to indicate problems of interpretation with *REP* are even more damning, and Grant and Kluge's subsequent attempt of defense is equally futile. The normalizing in *REP* seems mostly harmless as long as it is used only to compare supports within datasets; it just shares the same problems as *BS*. But the situation is entirely different when comparing groups in different datasets. As *REP* is normalized by $G - S$, it is influenced not only by the numbers of characters supporting a group, but also by the numbers of 0's and 1's in the characters (and this happens whether or not those characters have anything to do with the group). One of the examples of Farris and Goloboff (2008) is reproduced here, in Figure 8.2a. The group supported by the first dataset has the maximum *REP* value, but adding congruent characters and taxa *decreases REP*. That is because the G value for each character also increases in one step for each new dataset, and

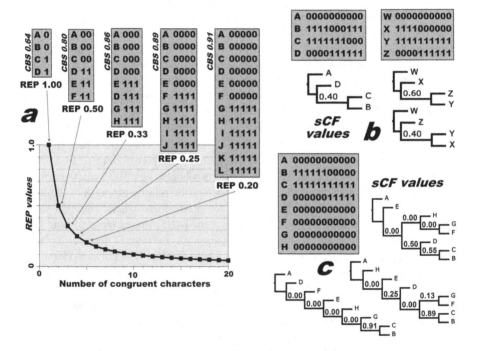

FIGURE 8.2 Problematic measures of group support. **(a)** *REP* (Relative explanatory power, Grant and Kluge 2007) cannot be used to compare supports among groups for different datasets. According to *REP*, a group supported by 20 congruent characters is *less* supported than a group supported by a single character (from Farris and Goloboff 2008). **(b)** *sCF* (Site concordance factors, Minh et al. 2020) only measure the proportion of characters that seem favorable to a group, leading to conclude that the "support" for BC (supported by the data) is the same as for XY (contradicted by the data). **(c)** As *sCF* is calculated on the basis of quartets, the values for different groups depend on the tree onto which *sCF* is calculated. The support for BC is calculated correctly (as 6/11 = 0.55) on the first tree, but incorrectly evaluated as 0.89 or up to 0.91 in other trees. Some groups for which there are no possible supporting characters in the dataset (BCD, GF, BCDFG) are evaluated as having "support" on some of the trees.

REP = BS / (G − S). Therefore, *REP* leads to the conclusion that a group supported by 20 identical characters (evenly partitioned in 0's and 1's) in a dataset with 42 taxa is supported much more weakly than a group supported by a single character in a dataset with 4 taxa. Grant and Kluge (2010) responded by pretending that the behavior shown for *REP* in Figure 8.2a is appropriate. Recall that in (2008b) Grant and Kluge had the tenet that "the relative strength" for an uncontradicted group corroborated by 100 synapomorphies "is clearly greater" than for an uncontradicted group corroborated by a single synapomorphy. The argument used to discard that tenet in (2010) is particularly curious: "for support values to be comparable across datasets they must take into account how strong support could be [*G − S*] . . . the same [*BS*] value is proportionately smaller because [*G − S*] is greater". But how large is *BS* as a proportion of *G − S* is how *REP* is *defined*. That is, according to Grant and Kluge (2010), *REP* indicating that matrices with more congruent characters provide a lower support is correct because *REP* indicates a lower support. That is, to justify the results produced by *REP*, they gave no argument at all—they simply pointed again at the results produced by *REP*.

Some independent criterion of the thing to be measured is needed when designing a new statistic. Although the statistic they proposed did not satisfy their own criterion, at least Grant and Kluge (2008b) had a criterion: a group set off by more uncontradicted characters is better supported. But they abandoned all hope of an independent criterion in (2010), when considering that the only justification needed for *REP* was *REP* itself. Of course *REP* measures *something*; the problem is what that something is, and it does not seem to be anything that could be sensibly called "support". The rationale for *RBS* altering the rankings produced by *BS* is clear and independent of *RBS* values in particular: the relative amounts of evidence for and against a group must be considered. How many more characters favor than contradict the group (what *BS* measures) is not the only relevant factor in determining support; one must also consider the proportions of characters for and against the group, and *RBS* attempts to measure this. *CBS* provides a tradeoff between both aspects, preventing groups defined by different numbers of uncontradicted characters to have the same value of support (as in *RBS*), and the proportion of characters favoring and contradicting a group not being taken into account (as in *BS*). But only by defining an independent notion of adequacy is it possible for a measure of support to be right or wrong; none can be invoked for *REP*.

8.2.1.4 Site concordance factors (*sCF*) and group supports

Minh et al. (2020) recently proposed the *site concordance factors (sCF)*, a support measure that resembles Bremer supports in not being based on resampling. However, any similarity ends there. Minh et al. (2020) proposed the *sCF* with the intent of "capturing underlying agreement and disagreement in the data", but *sCF* cannot be really interpreted as measuring any type of support or character conflict. The measure is calculated by reference to a group in a fully resolved tree, enumerating quartets of taxa *a−d*. Taxa *a* and *b* are taken from the two main splits of the group, taxon *c* is taken from the sister of the group being evaluated, and *d* from the remaining of the tree (see Minh et al. 2020 for details). The individual concordance factor *CFq* for a quartet *q* is the number of characters which support the split *ab|cd*, divided by the number of characters that are parsimony-informative for the quartet. The site concordance sCF is the average CFq over all quartets.[3] In datasets with only four taxa (e.g. as in Figure 8.2b), a single quartet can be defined, and it is clear that the sCF simply becomes the number of characters supporting the group divided by the number of parsimony-informative characters. This is no different from just counting numbers of synapomorphies supporting the group, which (as just discussed in reference to dataset 2) can be misleading because it does not consider whether contradicting characters agree with each other. In Figure 8.2b the dataset for A–C supports the group BC with $sCF = 0.40$ (4 out of 10 characters); that sCF value is the same as for group XY in the dataset for W–Z, a group which is *contradicted* by the dataset. If supported and contradicted groups can receive the same sCF value, it seems obvious that sCF values cannot be interpreted as measuring the degree to which a group is supported or contradicted.

The values of sCF can be misleading for only four taxa, but for more taxa the problems are even worse. For larger numbers of taxa, trees can be different yet contain the same group. Although Minh et al. (2020: 2728) compared sCF to "tree-independent methods of investigating phylogenetic signal", the sCF values are in fact strongly dependent on the tree used to calculate them. Not just suboptimal trees; even alternative *MPT*s can produce widely different values of sCF for the same group. This is because the number of quartets that can be enumerated, and the proportion that supports a given split of the quartet, changes with the numbers of taxa in each of the four groups from where a, b, c and d are taken. Consider the dataset in Figure 8.2c. With no elimination of zero-length branches, as needed for calculation of sCF values, there are 945 *MPT*s of 16 steps, all of which contain the group BC (supported by six characters, contradicted by five). In the first tree shown, $sCF_{BC} = 0.55$ (or 6/11); on that tree, all the quartets are formed by combining B, C and D with every one of {A, E, F, G, H}. For every one of those five quartets, there are six characters supporting the split that matches the group BC, and five contradicting it; 0.55 is a proper assessment of conflict. But on other *MPT*s the quartets that can be formed change, so that sCF strongly depends on the topology used for calculating it. On the bottom tree, most of the decisive quartets that can be formed support the split that matches BC; $sCF_{BC} = 0.91$. The differences in sCF for other groups are even more disturbing, for they imply that there is some "support" for groups that cannot possibly be supported by the dataset. Groups BCD, FG and BCDFG have sCF values of 0.50, 0.13, and 0.25 (respectively), even if the dataset contains no possible synapomorphies for those groups. The sCF was proposed with the aim to quantify support for splits in individual partitions of phylogenomic datasets; in that case, the reference tree will usually be the optimal tree for the combined data, which may well not be optimal for the partition in question, increasing the probability of assigning artificially high values to some groups. Minh et al. (2020) apparently used quartets to bypass the need for optimizing the whole tree,[4] but one would be better off just optimizing the tree and counting the number of characters that appear as synapomorphies of every group; that at least would correctly reveal that no group in any of the 945 *MPT*s for the dataset in Figure 8.2c has any support, with the only exception of BC.

8.2.1.5   Partitioned Bremer supports

Many datasets are formed by joining together different partitions. The problem of
how the group support from the combined dataset is distributed among the partitions
may be of interest. But in this connection, it is rather obvious that just as it happens
with individual characters, the BS of the individual partitions does not need to add
up to the BS of the combined dataset. Consider again the three partitions of dataset
2, discussed previously:

	1	2	3
A	0000	000	000
B	0000	111	111
C	1111	111	000
D	1111	000	111

one partition (the first, with $BS_{1,CD} = 4$) supports the group CD and two (second
and third, with $BS_{2,CD} = BS_{3,CD} = -3$) contradict it. But the Bremer supports of the
individual partitions are not additive—they cannot be added up to produce the total
$BS_{CD} = 1$. Doing so would ignore that the partitions interact with each other—second
and third partition conflict with each other.

   Baker and DeSalle (1997) proposed a measure, which they called the partitioned
Bremer support or PBS, calculated in such a way that the values from each partition
are guaranteed to add up to the total BS of the group in the combined dataset. To
calculate the PBS of a group, all the best possible trees where the group G is present, and all those where the group G is absent, need to be found; call those sets of
trees $\{T_G\}$ (group is present) and $\{T_{-G}\}$ (group is absent). The PBS of partition i for
group G ($PBS_{i,G}$) is then defined as $\bar{L}_{i,-G}$ (the average length of the partition in trees
$\{T_{-G}\}$), minus $\bar{L}_{i,G}$ (the average length of the partition in trees $\{T_G\}$). For partitions 1,
2, . . . N, and group G, these operations guarantee that $BS_G = \sum_{i=1}^{N} PBS_{i,G}$. This equal-
ity results because $BS_G = \bar{L}_{c,-G} - \bar{L}_{c,G}$, where $\bar{L}_{c,-G}$ and $\bar{L}_{c,G}$ are the average lengths
for the combined dataset for the trees with and without group G. But, for partitions
1, 2, . . . N,

$$\bar{L}_{c,G} = \bar{L}_{1,G} + \bar{L}_{2,G} \ldots + \bar{L}_{N,G}$$
$$\bar{L}_{c,-G} = \bar{L}_{1,-G} + \bar{L}_{2,-G} \ldots + \bar{L}_{N,-G}$$

Replacing $\bar{L}_{c,G}$ and $\bar{L}_{c,-G}$ for the sums of partition averages and rearranging the
terms:

$$BS_G = (\bar{L}_{1,-G} - \bar{L}_{1,G}) + (\bar{L}_{2,-G} - \bar{L}_{2,G}) \ldots + (\bar{L}_{N,-G} - \bar{L}_{N,G})$$

which is to say,

$$BS_G = PBS_{1,G} + PBS_{2,G} \ldots PBS_{N,G}$$

The use of averages to calculate the PBS is a significant difference with standard
BS, which can be calculated by just finding a single best tree with the group, and a

single tree without. Using averages makes the *PBS* liable to be affected by different approaches to eliminate zero-length branches (the conclusions in legitimate parsimony procedures are unaffected by elimination of zero-length branches).

The *PBS* is widely interpreted (e.g. Gatesy et al. 1999; Brower 2006; Wheeler 2012) to indicate the contribution to the total *BS* of a group provided by each partition. Schuh and Brower (2009: 165) consider it gives "a means to assess the degree of congruence or incongruence for partitions at individual branches, rather than simply for the tree as a whole". That interpretation, however, can be quite problematic. Brower (2006: 385) mentioned that I had designed datasets to show potential problems with *PBS*, but he gave no further details. The problems arise because the trees used for calculation of *PBS* for a partition are determined by the whole dataset, not by the partition itself. Then, the trees used to calculate *PBS* for the partition may not properly indicate degrees of support (or lack thereof) for the group in the partition. An example is shown in Figure 8.3a, a dataset with three partitions and global *BS* values as shown in the tree to the right of the dataset. For partition 1 and group DE, $PBS_{1,DE} = +5.0$; for partition 3, $PBS_{3,DE} = -3.0$. The values of *PBS* would indicate that partition 1 contributes positively toward *BS*, and partition 3 negatively. However, partition 1 taken by itself *contradicts* group DE, and partition 3 *supports* it—exactly the opposite of what is indicated by the *PBS*.

Figure 8.3b shows an even simpler case of misleading *PBS* values. Consider first a 3-partition dataset, with the third partition comprising only the five black characters. The global *BS* for group BC (supported by 10 characters, contradicted by sets of either 6 or 5 conflicting characters) is 4. Partition 3 contradicts group BC (and quite strongly so), but *PBS* suggests it is indifferent to the group, and that all the conflicting signal is contained within partition 2; $PBS_{2,BC} = -6$ and $PBS_{3,BC} = 0$. The inadequacy of *PBS* is further highlighted when a single (gray) character is added to partition 3. As a single binary character has been added, one would expect the Bremer supports to be modified in no more than a single step, but both $PBS_{2,BC}$ and $PB_{S3,BC}$ vary in *three* units each, now becoming $PBS_{2,BC} = PBS_{3,BC} = -3$. Furthermore, $PBS_{2,BC}$ changed in 3 units but nothing has changed in partition 2; only partition 3 was modified. Brower (2006) defended the use of *PBS* despite these shortcomings, on the grounds that other methods to measure supports are also hard to interpret in specific cases. While it is true that most methods to measure support have problems of interpretation in specific cases, the magnitude of the failure of *PBS* seems much worse, and sufficiently important as to cast serious doubts on its use in general.

The problems just discussed are an unavoidable consequence of the *PBS* for all partitions summing up to *BS* for the combined dataset. The *BS* for the combined dataset just cannot be expected to be calculable from sums of supports in the partitions, because partitions interact. If it were possible to meaningfully define global supports from supports in the individual *partitions*, it would also be possible to meaningfully define global supports as sums of support from individual *characters*—and we have seen at the beginning of this section that that is impossible. The *PBS* manages to produce measures for the individual partitions that add up to the *BS*, and the fact that they lineally add up ipso facto means that the values for the individual partitions cannot be considered as any kind of *support*. A more meaningful assessment of whether individual partitions support or contradict a group, and how strongly they do

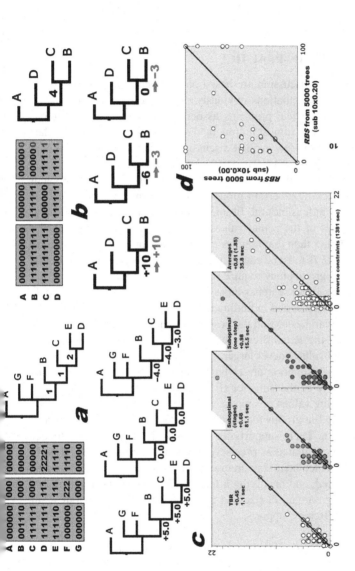

FIGURE 8.3 (a) *PBS* (Partitioned Bremer supports; Baker and DeSalle 1997) may indicate that a partition (the first) contributing a group (DE) contributes a favorable signal, and that a partition (the third) supporting it contributes a contradictory signal. **(b)** For the three partitions (without including the last, gray character), *PBS* implies that all the contradicting signal for group BC is in the second partition, even if all four characters in the third partition contradict BC. When adding the gray character, *PBS* for the second partition changes from −6 to −3 (even if the second partition was unmodified), and *PBS* for the third changes from 0 to −3 (even if a single character was added to the partition). **(c)** Comparison of different methods for speeding up calculation of Bremer supports. Values calculated intensively (with reverse constraints) are plotted on the x-axis, and different approximations (recording score differences for *TBR* moves, saving trees in several stages of increasing suboptimality, saving suboptimal trees in just one stage, and comparing average scores of heuristic searches constrained to have and not have the group). **(d)** For calculation of relative Bremer supports (*RBS*) saving some suboptimal trees, it is necessary to set a value of acceptable relative suboptimality, *RFD* (see text for definition). For Goloboff's (1995a) nemesiid spider dataset (84 taxa), when saving 5,000 trees without taking into account *RFD* (y-axis), most *RBS* values are overestimated relative to saving 5,000 trees but rejecting those with *RFD* > 0.20.

so, is obtained by separately calculating the (standard) *BS* values for each individual partition. For group BC in Figure 8.3b, the supports from the three partitions would obviously be +10, −6 and −5 for the initial dataset (changing to −6 in the last partition after adding the gray character), correctly indicating that partition 1 contains all the evidence in favor of BC, and that both partitions 2 and 3 contradict BC. As expected, the global *BS* of BC does not equal the sum of *BS* values for the individual partitions.

8.3   BREMER SUPPORTS IN PRACTICE

The preceding examples and discussion are based on exact calculations of Bremer supports. In larger datasets, calculations obviously cannot be carried out exactly. This section considers the practical problems associated with calculating Bremer supports. The literature contains almost no practical recommendations for calculating Bremer supports. One of the few explicit discussions of practical aspects is that of Brower (2006), but while generally correct, it lacks many details.

Of course, if the problem of finding *MPT*s is NP-complete, so is the problem of finding the best trees lacking a certain group *G*, and this means that (whenever exact tree calculations are unfeasible) the Bremer supports calculated via heuristic searches are only estimates of the actual values. Often, the effort invested in the search for the *MPT*s is greater than that used for calculating supports; as every group in the strict consensus of *MPT*s must be considered separately, finding the best tree(s) lacking every one of those groups may require repeating independent searches, and it is then necessary to make those searches more superficial in order to save time. If the unrestricted search has succeeded in finding the *MPT*s, then the value estimated from using less intensive searches to find the best trees lacking group *G* will constitute only an *upper bound*—the actual Bremer support may be the same or lower than the heuristically calculated values, never larger.

The precision with which supports are to be calculated may vary with circumstances, and does not need to be the same for all groups. In most analyses, the main interest will lie in calculating values of support with more precision for poorly supported groups, as those are the groups for which finding additional evidence would be a more pressing concern. Then, supports for groups with support 1 or 2 (under equal weights) need to be calculated as precisely as possible. Supports for strongly supported groups can be calculated less precisely—a group with (say) 8 steps of support is strongly supported, and mistaking it for a group with support 10 will not do much harm. Depending on dataset size and characteristics, there are four main approaches to calculating Bremer supports:

a) Performing searches under reverse constraints for every group *G* to be evaluated.
b) Searching suboptimal trees, saving successively longer trees until the group *G* is absent from some trees (if all groups are to be evaluated, until the strict consensus of the suboptimal trees is fully unresolved).
c) Recording score differences during *TBR* branch-swapping, keeping track of the minimum cost of moves that break the monophyly of each group.
d) Calculating average differences in length between independent searches constrained to have group *G*, and searches constrained to lack the group.

Given that the problem is defined in terms of finding, for every group G of interest, the best trees that lack G, most of the discussion that follows applies equally well to all the recommended variants of the Bremer supports—*BS*, *RBS*, and *CBS*.

8.3.1 Performing Searches Under Reverse Constraints

This is the method that allows the most precise estimations of supports. The downside is that it can be time-consuming. For precision in estimations of supports for groups of large supports (in trees where many other groups have low supports) this may be the best method, because saving suboptimal trees that are much worse than the best possible ones may require saving too many trees to fit in memory. Often, a good alternative is using the next method (saving suboptimal trees) to quickly identify the groups with lower supports, and then using reverse constraints to calculate values only for the remaining, more strongly supported groups.

8.3.2 Searching Suboptimal Trees

Calculation of Bremer supports by searching for suboptimal trees can be effective only for relatively small datasets, or datasets that produce low numbers of *MPT*s; otherwise, this method is simply impractical. In principle, one could save large numbers of longer trees; when enough trees, sufficiently suboptimal, have been saved, their strict consensus will be fully unresolved—that is, for every group in the optimal trees, there will be at least one tree where the group is missing, making it possible to evaluate the sacrifice in optimality in trees that lose each group. The maximum value of suboptimality that must be allowed can be determined from the branch lengths (see discussion of `bsupport` and `subopt` commands, in Section 8.6.1), to avoid saving unnecessarily long trees. However, more precautions than just setting a maximum value of suboptimality are needed, because not any set of trees up to a certain length will do. Simply setting the value of suboptimality to a large value, and saving a large number of trees by swapping from optimal trees in just one sitting, is likely to overestimate values of support. To understand why this is so, recall that branch-swapping is performed by effecting rearrangements on preexisting trees (optimal trees, in this case), and saving those within the value of suboptimality defined by the user until the memory buffer is filled. As an example, assume the search for suboptimal trees in a dataset of 150 taxa is to be done under equal weights, saving up to 10,000 trees with up to 30 steps beyond the shortest trees. *TBR* on a single tree for 150 taxa needs to perform about 750,000 rearrangements (depending on tree shape; see Chapter 5), but if trees up to 30 steps are to be accepted, then the first 10,000 rearrangements or so (i.e. about 1.5% of the possible *TBR* rearrangements!) will be accepted. This will fill the memory buffer with trees that are very long, and therefore miss many trees that are just a few steps longer, trees that are harder to identify and would have been found when doing the remaining 98.5% of the swaps. Therefore, saving suboptimal trees in just one round is likely to overestimate the support of poorly supported groups—those with just a few steps of Bremer support.

This can be avoided if the search for suboptimal trees is done in stages, saving successively worse trees, and increasing the size of the memory buffer for every iteration. For example, save up to 1,500 trees 1 step longer; 3,000 more trees 2 steps longer; and continue increasing number of suboptimal steps (1 at a time, or the cost of the first step of homoplasy if doing implied weighting) and trees to save (1,500 at a time), until all groups have been collapsed. While many authors do use this strategy for calculating Bremer supports (e.g. Miller and Hormiga 2004; Mattoni et al. 2011), it is common to see authors who search suboptimal trees in just one round (increasing the chances of miscalculating supports for groups of low support).

When searching for suboptimal trees, the goal is only to find some trees that are slightly longer than the best trees held in memory. Note that this differs from the goal of standard tree searches: here, one does not expect to find shorter trees than found before—quite the opposite. Therefore, once branch-swapping has filled the memory buffer, there is no point in continuing swapping—if the search for optimal trees has been careful enough, we can safely assume no better trees will be found, and any time used in swapping the remaining trees is completely wasted if the tree buffer is already full. TNT has the option to stop automatically as soon as the tree buffer is filled (see Section 8.6.1.1). It must also be emphasized that, to estimate values of Bremer supports, one does not need to find *all* the trees that are one step longer. Brower (2006) noted that, as longer trees are being saved, it becomes "increasingly difficult to perform adequate searches to ensure that all equally parsimonious trees have been discovered", and that insufficiently large numbers of trees found (or saved) may lead to overestimating the Bremer supports; he had in mind calculations for *PBS*, which require calculating length averages of all best trees with and without the group. For calculation of *BS*, *RBS* or *CBS*, finding all the trees that are within a certain level of suboptimality is much less important, as missing some trees does not imply an overestimation, at least in the cases where the approach of saving suboptimal trees is feasible (i.e. medium sized datasets where the number of *MPT*s does not exceed a few hundred trees). The values of *BS*, *RBS* and *CBS* depend on the score(s) of the best tree(s) without the group, not on the number of trees having that score. Therefore, the best course of action is to set the size of the tree buffer so that some thousands of additional trees are saved for every additional step of suboptimality, and abandoning the search as soon as the tree buffer is filled. When that is done, searches for suboptimal trees take relatively little time.

If the suboptimal trees are being saved for calculation of *RBS* or *CBS*, instead of *BS*, it is advisable to also set a value of relative suboptimality, *RFD*. With this, a suboptimal rearrangement is accepted only if it is within both the absolute score difference and the relative score difference. *RBS* and *CBS* are more strongly influenced by character conflict than *BS*, and a rearrangement that produces a tree adding steps in some characters but saving them in none does not identify character conflict. Figure 8.3d exemplifies this, by saving (starting to swap from the 72 optimal trees in Goloboff' 1995a dataset) 5,000 trees up to 10 steps longer (for this example, in just one search), rejecting those with a *RFD* of 0.20 or more (on the x-axis), and accepting trees only on their basis of their absolute scores (y-axis). The *RBS* supports are generally overestimated (as much higher) when *RFD* is not taken into account for saving the suboptimal trees.

To summarize, the recommended strategy for estimating Bremer supports by searching suboptimal trees has three elements: (a) repeat searches, increasing suboptimality in small steps every time, (b) gradually increase the number of trees to save to memory, in a few thousand trees every time, and (c) every time, abandon branch-swapping and move to the next level of suboptimality as soon as the tree buffer is filled. As an example, Figure 8.3c compares the results of applying different strategies for calculating Bremer supports (*BS*) to the embiid dataset of Szumik et al. (2019, with 157 taxa). This dataset was chosen because it represents the limit of what can possibly be analyzed with suboptimal trees; the number of *MPTs* under equal weights is unknown, but (from 20 independent hits to minimum length) TNT found 100,000 equally parsimonious trees (after swapping only 50,000 trees, and it continued adding new trees when stopped), the strict consensus of which had 98 groups with average support 2.91. The different approaches are plotted (on the y-axis) relative to searches with reverse constraints (in the x-axis of all plots); note that every method produced a lower *BS* than searches with reverse constraints for a few groups, indicating that superficial searches with reverse constraints are not, either, guaranteed to produce correct results. The estimations illustrated in Figure 8.3c started swapping from optimal trees (30 trees found in 10 independent hits to the optimal length, 2259 steps). Every stage in the searches for suboptimal trees saved an additional 1,500 trees 1 step longer. Given the large number of *MPTs*, 1,500 trees per search cycle is too few, but it was used to illustrate the differences between increasing suboptimality in stages and searching in one bout. Saving suboptimal trees (30,000+ trees of the maximum level of suboptimality to try, 21) in one stage produced larger overestimations of *BS* than the gradual increase. The average *BS* in this dataset is about 2.50, and searching suboptimal trees in one stage overestimates this by an average 0.98 steps. Searching in stages takes a longer time than searching in one step (81.1 instead of 15.5 sec.), but overestimates *BS* in only 0.68 steps, on average. Note that the example purposefully used relatively low numbers of additional trees per stage; a real analysis for a dataset of this many taxa would need to save larger numbers (e.g. 5,000) per stage, gathering a total of about 100,000 trees, for precise estimates; of course, an alternative is to use searches for suboptimal trees to estimate supports for the groups of lowest support (e.g. up to 8 steps of *BS*, which is a large fraction of the groups), then switching for searches under reverse constraints for the rest of the groups.

A trick that can be used to improve estimations with this method (without greatly increasing time) is to do some (e.g. 100 to 1,000) *RAS+TBR* saving a single tree per replication and keeping the final tree for each replication, and adding all the resulting trees to the pool of suboptimal trees. Most of the trees resulting from those *RAS+TBR* will be suboptimal, which is precisely what is needed for calculation of Bremer supports, and may help pinpoint parts of the tree that are difficult to resolve (particularly in regard to different islands, for which just *TBR* swapping from optimal trees may not be ideal).

8.3.3 RECORDING SCORE DIFFERENCES DURING TBR BRANCH-SWAPPING

This is usually the fastest method, and can be used on any dataset. It takes advantage of the fact that, during *TBR*, tree lengths (or *RFD* values) can be calculated very quickly (due to the shortcuts used by TNT, see Chapter 5, Section 5.1). During *TBR*, if moving a branch to an alternative location produces a tree N steps longer than the original tree, none of the nodes (=groups) between the two locations can have a support greater than N; the same applies to alternative rootings of the clipped clade (i.e. none of the nodes between old and new root can have support greater than N). If a tree better than the input tree is found during this process, TNT will not move onto that new tree, and the support for nodes between old and new location will be negative—in this way, the method serves to properly detect strength of contradiction. This can be applied to several input trees, swapping each—it is of course possible that losing a group *G* costs fewer steps on some of the *MPTs*. Optionally, if some of the input trees to be swapped are more parsimonious than others, the score difference between the input trees themselves can be added to the estimated supports. This method has been used in POY since 1997 (see Wheeler 2012: 330), and has been implemented in TNT since 2008.

To avoid having to find the nodes in the path between origin and destination of every possible move, the user must define limits for absolute and relative score differences. For the latter limit (RFD_{lim}) many rearrangements can be rejected only on the basis of the length increase when reinserting clipped clade, with no need to calculate the resulting *RFD* explicitly. This is because when dividing the tree in two, the decrease in total score is X; when evaluating a reinsertion, the increase in score will be Y; the total increase in tree length for the move is $Y - X$. But $Y - X$ is the same as $F - C$ (the numerator of the *RFD*), and $F \leq Y$ (F can be less than Y when some characters have their steps decreased when clipping and then increased again when reinserting). Therefore, only those rearrangements fulfilling $(Y - X) / Y < RFD_{lim}$ can possibly produce an acceptable *RFD*, which is to say that any reinsertion where the increase in length Y equals or exceeds $X / (1 - RFD_{lim})$ can be rejected without further calculation.

Although it might seem that this is a very crude approximation, it provides relatively good estimates in very short times. This is, in part, because the speed achievable during branch-swapping allows examining (implicitly) larger numbers of trees than could actually be saved to memory. For example, for the dataset in Figure 8.3c, estimating *BS* values with *TBR* produced the most accurate values (average overestimation 0.45, or about 18%) other than reverse constraints (the slowest but most precise method). *TBR* swapping, taking 1.1 sec, is about 80 times faster than a multistage search for suboptimal trees (the next fastest method). The *TBR* swapping (with 30 optimal trees as input) produced 1,230,807 rearrangements within the set limit of 21 steps (and no *RFD* limit).

8.3.3.1   The ALRT and aBayes methods

In the context of likelihood, Anisimova and Gascuel (2006) have proposed *approximate likelihood ratio tests (ALRT)* based on branch swaps. In a subsequent paper, they propose a modification, the aBayes method (Anisimova et al. 2011). Both methods are based on comparing the likelihoods of the three possible trees that result from nearest-neighbor interchanges (*NNI*) around the group of interest (they use further time-saving shortcuts, such as re-optimizing only the branch lengths adjacent to the move). Being based on *NNI* (rather than *TBR*), those tests examine much lower numbers of alternative trees. Given the prohibitive times required for likelihood calculations, the *ALRT* and aBayes methods may be advantageous in the context of likelihood, but they can easily lead to errors. For example, alternative near-optimal locations of a group more than a node away from the location in the optimal trees cannot be detected by *NNI* moves. In the context of parsimony, where distant locations can be tested via *TBR* in negligible times, there is no point in attempting such a type of test.

8.3.4 CALCULATING AVERAGE DIFFERENCES IN LENGTH

When the number of optimal or near-optimal trees precludes estimating *BS* values by saving suboptimal trees, an alternative quick estimation of *BS* can be used with programs that cannot record score differences for *TBR* moves, as long as they implement the possibility of reverse constraints and randomized starting points for searches (e.g. PAUP*; Swofford 2001). Normally, one would do a thorough search to find the optimal trees, and this makes it necessary to use searches with reverse constraints of the same intensity. Superficial searches (e.g. a single *RAS+TBR* with no mulpars) will miss the optimal length by a certain average difference, say X steps. If one can assume that the excess length in such a search is approximately the same when constrained for monophyly and for non-monophyly, then calculating average lengths (e.g. of 10 *RAS+TBR*) allows obtaining estimates of *BS* values. In more detail: consider the case where the best possible trees with group G have a length L_G, but *RAS+TBR* constrained to have G produces on average trees of L_G+X, and the best possible trees without group G have a length $L_{\sim G}$, but *RAS+TBR* constrained to not have G produces on average trees of $L_{\sim G}+X$. Calculating the difference between the average lengths for both positive and negative constraints amounts to $(L_{\sim G} + X) - (L_G + X)$, which is the same as $L_{\sim G} - L_G$—the correct Bremer supports.

While repeating 10 *RAS+TBR* with different random seeds for every group to be tested might seem like a lot of work, each replication can search and save a single tree. This is much less work than doing intensive searches under reverse constraints—with an intensity that needs to approximately match the intensity used to search for optimal trees. For the dataset of Figure 8.3c, using this method (with a TNT script) produced much faster evaluations than multi-stage searches for suboptimal trees (35.8 vs. 81.1 sec, more than twice as fast). The average values of *BS* supports differed in 0.51 from the values calculated with intensive searches under reverse constraints. However, note that this process (unlike the other methods described) can either over- or underestimate *BS*; the average unsigned difference with the intensively calculated values is 1.85 (and a standard deviation of 1.87). The fact that the average is higher than the actual *BS* indicates in this case that the assumption of searches with and without constraints missing the optimal length by the same difference may not be met exactly; a single *RAS+TBR* in TNT seems to be missing the best possible trees by a greater difference under reverse constraints—half a step more. This may have to do with details of how the initial trees are made to obey the reverse constraints in TNT (see Chapter 5, Section 5.5), and it may be implementation-dependent. In any event, the difference is relatively small, suggesting that the method may be useful in specific cases.

8.4   RESAMPLING METHODS

Besides Bremer supports, the other major approach to evaluating group supports is by means of resampling. In resampling (summarized in Figure 8.4), some of the characters from the original dataset are used to form new, *pseudoreplicate* datasets, just as if one were creating a new matrix from scratch, except that the characters are chosen from those in the original dataset. Optimal trees are calculated for every pseudoreplicate dataset, under the corresponding criterion (*MP, ML*). Finally, the trees produced by the analysis of each pseudoreplicate are compared and summarized, producing group supports. The general idea is that groups occurring more frequently for the pseudoreplicate datasets are better supported. Inclusion or exclusion of a character in the pseudoreplicate dataset is most easily handled via character weights (i.e. with weights indicating number of copies of the character, zero indicating exclusion).

The exact goal of resampling can vary, depending on the interest of the researcher. The earliest proposals of resampling methods in phylogenetics (Felsenstein 1985b; Lanyon 1985) attempted to estimate statistical properties of the data. Numerous authors (e.g. Hillis and Bull 1993; Efron 1979; Efron et al. 1996; Berry and Gascuel 1999; Felsenstein 2004; Yang 2006; Anisimova and Gascuel 2006) have discussed phylogenetic resampling (especially bootstrapping) from this perspective. But even from the statistical perspective the resampling may be intended for different purposes. Some authors (e.g. Felsenstein 1985b; Siddall 2002; Hovenkamp 2009) expect resampling to answer questions of stability and repeatability: how likely is it that a new phylogenetic analysis will continue supporting this group? Other authors (e.g. Hillis and Bull 1993; Anisimova and Gascuel 2006; Hoang et al. 2018) use resampling as means to estimate the probability of a clade being correct. A proper method to assess this probability would then have groups of frequency X% being true in X% of the cases. Last, some authors (e.g. Sanderson 1989, 1995; Efron et al. 1996) have discussed using the bootstrap as a frequentist type I error rate: a bootstrap proportion of F indicates that the probability of sampling a dataset that supports the group, if the observed dataset was taken from a universe of characters which does not support the group, is $1 - F$ (Efron et al. 1996 themselves indicate difficulties with this interpretation). For most of those statistical interpretations, it is necessary to make certain assumptions; for example that the characters are independently and identically distributed, that the frequencies of the different character types in the observed dataset properly reflect those frequencies in all the yet unseen characters, and so on. Further, the "correctness" of a clade could mean either historical accuracy, or the idea that the universe of characters from where the observed dataset was sampled includes the group in the optimal trees. The first—historical accuracy—also requires that the characters evolve in such a way that the method used to reconstruct phylogeny is itself statistically consistent; the second—analyzing all possible characters would produce the same result—requires more spartan assumptions.

In addition to the statistical justification, it is possible to view resampling as a means to assess *only* support—that is, only to quantify how much evidence favors or contradicts the conclusions. In that way, resampling can be used to indicate properties of the dataset itself, properties not easily detectable by other means. Table 8.1 summarizes the effect of the alternative resampling and summary methods discussed

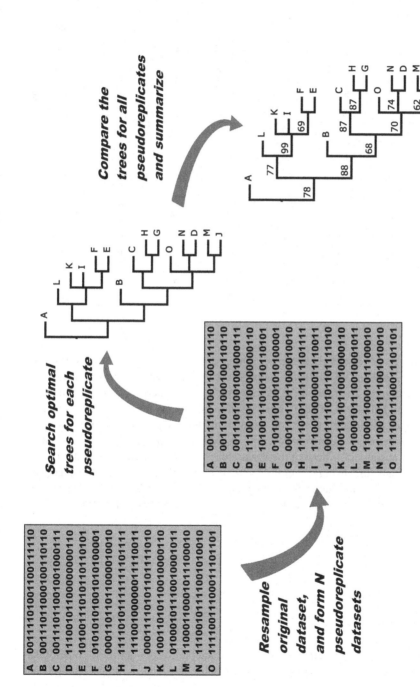

FIGURE 8.4 General summary for resampling. The three stages (resampling, analysis of the pseudoreplicates, and final summary of results) influence the results in different manners.

TABLE 8.1

Summary of Different Methods Discussed in this Chapter, for Resampling and Summarizing Pseudoreplicates

		affected by irrelevant characters	*distorted by character weights/costs*	*values affected only by character conflict*	*can measure strength of contradiction*	*can measure support below 50% frequency*	*Can suggest support for unsupported groups*
Resampling	*Boot*	yes	yes	no	y/n	y/n	y/n
	Jack (fixed 50%)	yes	no	no	y/n	y/n	y/n
	Boot(Poisson)	no	yes	no	y/n	y/n	y/n
	Jack (indep. p = 0.36)	no	yes	no	y/n	y/n	y/n
	Sym.Resampling	no	no	no	y/n	y/n	y/n
	No-zero-weight	no	no	yes	y/n	y/n	y/n
Summary	*Freq. within repls.*	y/n	y/n	y/n	no	no	often
	Strict consensus	y/n	y/n	y/n	yes	yes	rarely
	Frequency	y/n	y/n	y/n	no	no	yes
	FreqDiffs (GC)	y/n	y/n	y/n	yes	yes	yes
	Frequency slope	y/n	y/n	y/n	yes	yes	no

in this chapter. As discussed at the beginning of this chapter, even the modest goal of measuring amount of character support is far from simple, and resampling may help in that regard. The first authors to advance such an interpretation of resampling methods were Farris et al. (1996). No special requirements for the dataset to be representative, or the inference method being consistent, apply in this case. The only question is whether the results of resampling are indeed apt for measuring the quantity of interest, support—much of the discussion that follows focuses on that problem. The basic idea behind using resampling to detect strength of support is that character conflict—an important factor decreasing support—necessarily results in decreased resampling frequencies. Consider the case of matrix 5, with two conflicting character patterns,

```
        5
M  000000 000000
N  111111 111111
O  111111 000000
P  000000 111111
   typeA  typeB
```

Matrix 5 contains the same number of characters of types A and B. When resampling, the chances of a character of type A being included in the pseudoreplicate dataset are exactly the same as for a character of type B; therefore, some pseudoreplicates will have more characters of type A (thus favoring group NO), some pseudoreplicates will have more of type B (favoring NP instead), and some pseudoreplicates will result in ties. As both the numbers of type A and B are increased, the probability of obtaining a tie in a pseudoreplicate decreases, and the frequency tends to 50% for group NO, and 50% for NP. Thus, in principle, group frequencies can be used only to indicate the support of groups occurring in 50% or more of the pseudoreplicates (see later, Section 8.4.3.2, for alternatives). If the number of characters of type A is increased, then every pseudoreplicate is likely to contain a few more characters of type A than B, increasing the frequency of group NO. Therefore, the frequencies of groups NO and NP will generally reflect the relative numbers of characters supporting one or the other group.

Two related points few authors have discussed (an exception being Goloboff et al. 2003a) is that all the different goals may be legitimate and—more importantly—that different goals may require the resampling to be performed and summarized in slightly different ways. Most discussions take bootstrapping to be a monolithic, unmodifiable method, and discuss the problem of how its results should be interpreted. A more profitable discussion should consider that, depending on what is to be measured or estimated, the resampling can—and should—be carried out in different ways. The standard bootstrapping emulates the situation where all the existing data are discarded and a dataset is formed again, by an independent investigator oblivious of our previous dataset. A group may be very likely to show up in the optimal trees for that new dataset, even if unsupported by the current dataset (an actual example is in Figure 8.6a). In such circumstances, a method intended to provide a measure of repeatability really should, like the standard bootstrap, attribute high value to those unsupported groups likely to show up again. But attributing such a high value to a group that is not supported by the data is undesirable if the goal is to measure support—the method would ideally need to be modified to eliminate that behavior, or used with the understanding that it is only a fallible approximation.

8.4.1 PLOTTING GROUP SUPPORTS

Resampling could in principle be used to evaluate different aspects of the results (Li and Zharkikh 1994; Efron et al. 1996)—for example, confidence in the tree as a whole. For group supports, the interest is only in evaluating the frequency of clades (Felsenstein 1985b; Farris et al. 1996). Given that the group frequency is commonly used to summarize results, and only groups with frequency above 50% are supported in simple cases, it is common to simply take the trees resulting from analysis of all pseudoreplicates and calculate their majority rule consensus tree. This is usually not ideal: which groups are supported is indicated by the strict consensus of the optimal trees for the original dataset; resampling is intended to evaluate how strongly they are supported.

In most cases, instead of using the majority rule tree for all pseudoreplicates, it is convenient to calculate the support for pre-established groups—the groups present in the strict consensus of optimal trees. A group present in the consensus may have a frequency well below 50%—this would, in principle, indicate a very low support.

Different programs for resampling work differently; some allow plotting support values on pre-defined groups, others simply construct a majority rule tree and do not give the user control over the groups to evaluate. In TNT, both options are available. Having the possibility of doing both is important for testing alternative methods for measuring supports, allowing the comparison of values for supported and unsupported groups—a perfect measure of support should correctly distinguish both, and always assign higher values to the former.

8.4.2 DIFFERENT RESAMPLING METHODS

This section discusses the implication of different ways to form the resampled matrices. This section leaves aside the problem of how the search for every pseudoreplicate dataset is carried out; just assume for the time being that the actual *MPTs* are found without errors for all pseudoreplicates. The influence of approximate search algorithms is discussed separately, in subsequent sections. Similarly, the final results are assumed to be summarized with group frequencies; some implications of using frequencies, as well as some alternatives, are also discussed separately.

8.4.2.1 Bootstrapping

In bootstrapping (Felsenstein 1985b), every pseudoreplicate dataset is formed by sampling characters from the observed dataset, with replacement, until a new dataset with the same number of character is formed. Therefore, the frequency of a group G supported by r uncontradicted characters in a dataset with n characters is,

$$F_G = 1 - (1 - \frac{r}{n})^n \qquad \text{[Formula 8.1]}$$

For a matrix with a large number of characters and a group supported by a single uncontradicted synapomorphy, $F_G = 0.63$. Note that the frequency of G is affected not only by r, but also by n. For a 2-character dataset where $r = 1$, $F_G = 0.75$, and for a single-character dataset, $F_G = 1.00$. If bootstrapping is to be used to measure group supports, that is problematic (Carpenter 1996; Farris et al. 1996): invariant or irrelevant characters (i.e. characters that do not directly define or contradict G) neither add nor detract from its support. Harshman (1994) considered that the effect of irrelevant characters was weak enough as to not be a cause for concern, but also noted that the effect is mitigated by adding a large number of (imaginary) invariant characters to the dataset, which stabilizes frequencies; in that case, Formula 8.1 tends to

$$F_G = 1 - e^{-r} \qquad \text{[Formula 8.2]}$$

Farris et al. (in Horovitz 1999) noted that for large numbers of characters, the resulting character weights exactly follow a Poisson distribution with mean 1 (see also Goloboff et al. 2003a). Therefore, the weight of a character in the pseudoreplicate dataset can be determined by rolling a multifaceted die, where weights 0, 1, 2, . . . n have the same probabilities as in a Poisson distribution. As the weight of a character in this Poisson bootstrapping is determined independently of the weight of other characters (i.e. a die is rolled separately for every character), F_G is influenced only by the number of characters supporting group G (or contradicting it, if any), but not by autapomorphic and invariant characters.

The perturbation introduced by the bootstrap is quite strong; Felsenstein (1985b) considered this as preferable over the milder perturbation introduced by other possible resampling schemes where a single character is dropped in every pseudoreplicate dataset (e.g. Lanyon's 1985). However, in datasets where most groups are poorly supported, such a strong perturbation may produce very low supports for most groups. That would properly reflect the conflict in the dataset, but it may not be too useful to discriminate among groups—all are relatively poorly supported, but still there can be better and worse groups. Using a milder resampling may better separate those. As discussed in Kopuchian and Ramírez (2010), the degree of perturbation in bootstrapping can be varied by creating datasets with more characters than observed in the original dataset—as the number of resampled characters grows, the frequency of every character type in the pseudoreplicates will approach their frequency in the original dataset, thus producing results more similar to those for the original dataset (and then higher frequencies for groups with some actual support, however weak). In the case of a Poisson bootstrapping, the same is accomplished by drawing character weights from a Poisson distribution with a larger mean. Evidently, the values of group support under a milder resampling cannot be directly compared to those under a stronger one, so if you use this option you need to be explicit about it.

8.4.2.2 Jackknifing

The original statistical technique of jackknifing consists of leaving out a single observation and recalculating the parameter of interest. In the first phylogenetic application, Lanyon (1985) proposed to drop one character at a time. Felsenstein (1985b) discussed the jackknife, and considered that dropping a single character would produce too little variation. He proposed leaving out half of the observations, which would generate much more variation (for n characters, $n! / (n / 2)!^2$ possible datasets, instead of just n). However, the delete-half jackknife will (a) introduce too much perturbation, so that only very strongly supported groups will survive, and (b) also be affected by uninformative characters (Farris et al. 1996). The effect of adding uninformative characters is in the opposite direction of bootstrapping, causing a group supported by r uncontradicted characters to have a *higher* frequency than before addition of the irrelevant characters (the frequency increases because the supporting characters are less likely to be eliminated if there are now more characters to choose from). The problem of irrelevant characters cannot be solved just by eliminating autapomorphies from the dataset; characters that neither favor nor contradict a group can be characters supporting *other* groups, and those cannot be eliminated. As a simple solution to the problem of irrelevant characters, Farris et al. (1996) proposed an independent-removal jackknife, where the exclusion of every character is decided by separately rolling a die. If the probability of excluding a character is $p_{(del)}$, the frequency of a group G supported by r uncontradicted characters is $F_G = 1 - p^r$. Farris et al. (1996) proposed using $p_{(del)} = e^{-1} = 0.368$ and then the results will approximate those obtained with bootstrapping corrected for autapomorphies (as in Formula 8.2).

Note that (just as in the Poisson-corrected bootstrap) the numbers of characters in the pseudoreplicate matrices created by this process is not constant. Some pseudoreplicates will have more characters than others, and the probability of creating a matrix with either all or none of the characters is (if small) more than zero.

8.4.2.3 Symmetric resampling

Goloboff et al. (2003a) discussed further complications that can arise when some characters, or some character-state transformations, have different weights. Consider dataset 6,

```
            6
_____
A      0      0000...0001
B      1      1111...1111
C      1      0000...0000
D      0      1111...1111
    one char.    N chars.
  of weight N  of weight 1
```

with the weights as defined, there is an exact tie between the characters that favor BC and those that favor BD. However, neither bootstrapping (in their standard or Poisson version) nor independent-removal jackknifing (for any $p_{(del)} < 0.50$) produce BC and BD in equal frequencies. Consider jackknifing first, where probability of retaining a character is $p_{(ret)} = 1 - p_{(del)}$; group BC will be absent in the pseudoreplicate only when the first character is removed, or when none of the other characters is. As it is very unlikely that none of the other characters is eliminated, group BC will appear with a frequency that approaches $p_{(ret)}$ as N grows. The exact frequency is

$$F_{BC} = 1 - \left(p_{(del)} + p_{(ret)}{}^{N} \right) \qquad \text{[Formula 8.3]}$$

when $p_{(del)} = e^{-1}$ and N is large, $F_{BC} \approx 0.63$. This effect is because the average weight of a character in a pseudoreplicate is 0.63; the probability of a weight lower than the average (i.e. zero) is 0.37, and the probability of a weight higher (i.e. one) is 0.63. There is thus an asymmetry of $0.63/0.37 = 1.70$. In jackknifing, the only case where this asymmetry does not occur is when $p_{(del)} = 0.50$; in that case, $F_{BC} = F_{BD}$. For $p_{(del)} < 0.37$, the asymmetry grows and F_{BC} shows up even more frequently (although eventually, when $p_{(del)}$ approaches zero, F_{BC} decreases again, because ties become more likely).

In the case of bootstrapping, the asymmetry is milder, and it goes in the opposite direction—a character has a higher probability of being downweighted than upweighted, relative to the average. For a Poisson-corrected bootstrap, with mean 1, a character has a probability 0.367 of being excluded (sampled zero times), and a probability 0.264 of being sampled more than once. The asymmetry is thus $0.367/0.264 = 1.390$, and it pushes in the opposite direction of jackknifing. Therefore, a Poisson bootstrapping of dataset 6 produces (approximately) $F_{BD} = 0.52$—it is weak support, but support nonetheless—and $F_{BC} = 0.45$. For a Poisson of mean > 1, the asymmetry changes, resembling (or exceeding) that in jackknifing with $p_{(del)} = 0.36$.

For dataset 6 which supports neither group (BC or BD), an option (implemented in PAUP*) to attempt fixing the problem of an unsupported group being incorrectly identified as supported, is to resample as if the characters of higher weight occurred several times in the matrix. This does not produce satisfactory results (e.g. randomly generated matrices where all characters have high weights produce high "supports"), and is impractical for several reasons. Of those reasons, the most important ones are that (a) it would not be applicable to characters where the difference is in some state transformation costs and not others (e.g. additive or Sankoff characters), and (b) it would not be applicable, either, to analyses under implied weights, because in that case the character weights vary during the search, and cannot be predicted ahead of time. Both cases (unequal transformation costs, use of implied weights) are common in analysis of morphological datasets.

To visualize a solution, consider first the effect of using an upweight resampling, where the weight of a character can be doubled if the character is chosen, or left unmodified otherwise. A character is chosen to be upweighted with probability $p_{(up)}$, and retained as is with $p_{(ret)} = 1 - p_{(up)}$. This produces the opposite effect of jackknifing; BD will be absent only when the weight of the first character is doubled, or when none of the other character weights are doubled. That is,

$$F_{BD} = 1 - \left(p_{(up)} + p_{(ret)}^{N} \right) \qquad \text{[Formula 8.4]}$$

An up-resampling procedure produces the exact mirror effect of jackknifing (compare Formulae 8.4 and 8.3). Therefore, a procedure where characters can be eliminated or duplicated with the same probability, will produce pseudoreplicate datasets that support BC and BD with equal frequency. Goloboff, Farris, Källersjö et al. (2003) called this *symmetric resampling*, and is intended for use in cases where character or state transformation weights are not uniform. That is, in symmetric resampling, $p_{(up)} = p_{(down)}$, $p_{(change)} = p_{(up)} + p_{(down)}$, and $p_{(ret)} = 1 - p_{(change)}$.

8.4.2.4   No-zero-weight resampling

It is possible for a group to be very poorly supported even if the group is not directly contradicted by any character. This can happen when there are missing entries, and when whether a character acts as a synapomorphy of a group depends on how the rest of the tree is resolved. An extreme example is shown in Figure 8.5a (using one of Goloboff's 2014b datasets with no homoplasy but optimal trees hard to find with Wagner trees and branch-swapping). The first two characters of the dataset delimit two groups, B–I and J–L. The resolution within B–I is determined by the characters in the second block; those characters can be plotted without homoplasy on a tree (and only one tree), but elimination of most combinations of one or two characters produces a poorly resolved tree—all six characters are needed *simultaneously* for a single optimal tree to be defined. The characters resolving B–I are completely homoplasy-free, but the supports as calculated with either the standard bootstrap or symmetric resampling (see Figure 8.5a, values above branches) are very low. The frequencies for all of the groups supported within B–I are 4–7% with bootstrapping, and 10–14 % with symmetric resampling. In the other part of the tree, the group JL is supported by five characters and contradicted by four; the evidence contradicting JL

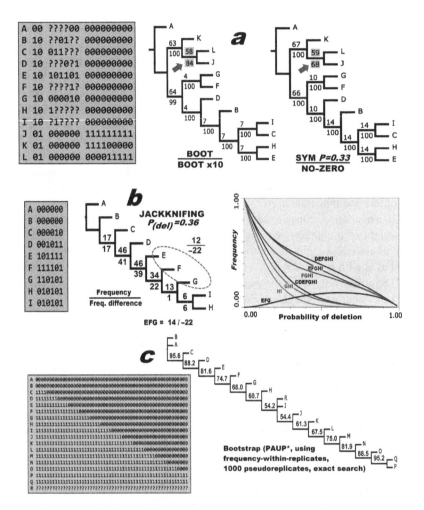

FIGURE 8.5 (a) Groups favored and contradicted by almost equals amounts of evidence (JL) may have much higher frequency than uncontradicted groups (all subgroups of B–I) under standard bootstrap or symmetric resampling. Milder perturbations of the dataset (e.g. bootstrapping datasets with x10 the number of characters of the observed dataset) increase the frequencies of uncontradicted groups, but also produce a relatively high frequency (84%) for near-unsupported group JL. A symmetric resampling (no-zero-weight) where no character is deleted (i.e. only down—or upweighted) better serves to pinpoint character conflict, assigning frequencies of 100% to any group that is supported by uncontradicted characters, but lower frequency (68%) to the near-unsupported group. (b) A dataset where a supported group (HI) has a much lower frequency than a contradicted group (EFG). The difference in frequencies (*GC*) better separates supported and contradicted groups. (c) The use of frequency-within-replicates (*FWR*) may attribute support to entirely unsupported groups. As taxon R is scored only with missing entries, it can make any group in the rest of the tree non-monophyletic at no cost. However, more uneven partitions (e.g. PQ, or all taxa but AB) are made non-monophyletic by fewer possible locations of R, and thus are assigned a high support when using *FWR* (TNT does not implement *FWR*; these results were obtained with PAUP*, using an exact search for each of 1,000 bootstrap pseudoreplicates).

is almost as much as the evidence favoring it, yet the frequency is much higher than for any of the groups within B–I, approaching 60%.

In some contexts, a measure indicating the degree of *conflict* involved in defining the group may be useful. Such a measure should lower the value for a group if and only if there is character conflict concerning the group—some characters support it, some characters contradict it. Such a measure should indicate any of the subgroups of B–I as "perfect"—no conflict or contradiction is involved, only scant evidence. That can be easily done with a symmetric resampling where the weight for the characters is either down– or upweighted, but without never going to zero; this is called no-zero-weight symmetric resampling. With such a resampling scheme, the frequency for groups contradicted by some characters is about as low as it is in standard resampling, but the uncontradicted groups receive values of 100% (as in Figure 8.5a, values below branches on the right tree). It must be emphasized that the no-zero-weight resampling is *not* a measure of support; the idea is instead to identify only one of the *components* of the support, character conflict—only actual character conflict can lower the values of groups. This can be especially useful for paleontological data (e.g. Pei et al.'s 2020 analysis of dromaeosaurids), where numerous missing entries often make the support of groups to depend on several characters simultaneously (as in Figure 8.5a).

One might think that the same effect is achieved by using a milder bootstrap (where more characters than observed are resampled, see Section 8.4.2.1) or a milder symmetric resampling (where $p_{(up)} = p_{(del)}$ are small). That milder resampling would indeed give higher values to the subgroups of B–I, as shown in Figure 8.5a (values below branches, left tree): resampling 10 times more characters than in the original dataset, bootstrapping produces frequencies of 100% for all subgroups of B–I. However, there is a catch: the milder bootstrapping would also give higher frequencies to groups that are almost as contradicted as they are supported, such as JL. The mild bootstrapping gives JL a frequency of 84%, and while lower than subgroups of B–I, this seems inappropriate. The no-zero-weight resampling, instead, gives all subgroups of B–I a frequency of 100%, but JL a much lower frequency (68%) than the mild bootstrapping, correctly identifying conflict. This allows separating groups with low support because of lack of evidence, and groups of low support because of character conflict. The contradiction can either be a direct contradiction to the group (as JL in Figure 8.5a), or indirect, as a contradiction of a character needed for the group to be supported. An example of the latter is in Figure 8.5b, where the first character supports the group HI as a reversal; no character in the dataset directly contradicts HI, but several characters contradict the characters needed for HI to be supported. Therefore, the frequency of HI is appropriately low (51%), even under no-zero-weight resampling.

8.4.2.5   Influence of number of pseudoreplicates

In principle, group frequencies could be calculated exactly, by exhaustive enumeration of all possible pseudoreplicate datasets. The number of possible datasets varies with resampling method. In the case of jackknifing, where characters are only absent or present, the number of possible outcomes for n characters is 2^n. This amounts to different strings of 0's and 1's in the resulting character weights. The dataset for each weight combination can be analyzed (ideally, with an exact search), determining whether the result displays every group of interest. As long as $0 < p_{(del)} < 1$, all possible strings will be eventually sampled if large enough numbers of pseudoreplicates are generated. The probability of obtaining a given string depends only on its number of 0's and 1's—for example, a string like 011011 will be obtained with probability $P_{011011} = p_{(del)}^2 \times (1 - p_{(del)})^4$, because two characters are deleted and four retained. Once the presence or absence of every group of interest has been determined for every possible string of weights, the exact probability of obtaining the group equals the sum of probabilities of all the weight strings containing it—this allows calculating the expected group frequency under any value of $p_{(del)}$. The calculations would be similar for symmetric resampling, but requiring the enumeration of 3^n weight strings (as the weights of a character can take three values) instead of 2^n. While much more laborious, the same could be calculated for bootstrapping.

The exhaustive enumeration of weight strings is feasible only for very small numbers of characters, and then resampling is almost universally done by creating some number of pseudoreplicate datasets with random number generators. Obviously, group frequencies being estimated, the larger the number of pseudoreplicates, the greater the precision of the estimation. As an extreme example, consider doing a single pseudoreplicate of jackknifing in a 1-character dataset—only two outcomes are possible, a support of 100% (which will be obtained in a proportion of $1 - p_{(del)}$ of the attempts) or 0% (obtained in a proportion of $p_{(del)}$). The correct support $(1 - p_{(del)})$ will never be obtained in runs of one-pseudoreplicate; the estimate is not biased, but there is a lot of variance. The larger the number of pseudoreplicates, the smaller the variance.

Some studies (e.g. Hedges 1992; Hoang et al. 2018) have argued that very large numbers of pseudoreplicates must be done. According to Hedges (1992), to estimate supports within 1% for groups with support 95% or less, up to 4,000 pseudoreplicates may be needed. However, other factors come into play—most datasets are large enough to require the use of only approximate tree searches (rather than exact ones), and the estimation of supports may itself be distorted by several factors even if using an exact search (see Section 8.4.4). Controlling for those two additional factors is very difficult. Furthermore, it can be observed in practice that support values for most groups often become stable much earlier than the thousands of pseudoreplicates expected by Hedges (1992), often within a few hundred replicates (Pattengale et al. 2010).

Therefore, running very large numbers of pseudoreplicates may be futile (Goloboff et al. 2021b)—the measure of supports via resampling is only an approximation, with many sources of error that cannot be corrected just by doing more replications. Since the numbers themselves are only a rough proxy for group supports, there seems to be no point in estimating those numbers with great precision (Goloboff et al. 2021b). The idea of not using excessive numbers of pseudoreplicates can be further justified when the resampling upweights or removes characters independently (as in Poisson bootstrapping, jackknifing, or symmetric resampling). For example, if characters 0–4 support a group AB, and do not interact with characters 5–9 supporting another distant group XY, then there is no need to examine all the combinations of weights for characters 0–9 for determining frequencies of AB and XY; frequencies of AB and

XY are independent, and as long as the sampled pseudoreplicates contain enough combinations of weights for characters 0–4, and for characters 5–9 independently, this is enough (Goloboff et al. 2021b).

8.4.3 FINAL SUMMARY OF RESULTS

This section focuses on how to summarize the trees for the pseudoreplicate datasets for estimating group supports. As mentioned before, the results for the pseudoreplicates can be used to make estimates other than monophyly of groups (e.g. Efron et al. 1996); such an application is not considered here.

8.4.3.1 Frequency-within-replicates (*FWR*) or strict consensus

Tree calculations for the pseudoreplicate datasets often produce multiple *MPTs* (or most-likely trees, if using *ML* as optimality criterion). Since Felsenstein (1985b), the usual approach to summarize the multiple trees is by the so-called *frequency-within-replicates* approach, *FWR*. In this case, every tree in a pseudoreplicate that found n trees is weighted $1 / n$; this means that a group found in k out of n trees is passed to the final calculator of supports with a weight of k / n. This approach was first implemented in Phylip (Felsenstein 1993), and incorporated into PAUP* (Swofford 2001) as the only option for handling multiple trees. The approach is also used by MPBoot (Hoang et al. 2018); other programs (e.g. MEGA, Kumar et al. 2018; RAxML, Stamatakis 2014; PhyML, Guindon and Gascuel 2003) approximate in practice the results of *FWR* as a side-product of saving a single tree per pseudoreplicate (although bias in tree search may add distortions in that case; see Section 8.4.4). The rationale advocated for the *FWR* approach (Felsenstein 2004) is that otherwise the pseudoreplicates finding many trees would be overemphasized in the final calculation of frequencies. This is true, of course, only if the final calculator of frequencies is unable to tell what pseudoreplicate each of the trees came from; keeping track of that provenance would allow correcting for the "overemphasis".

Numerous authors (Goloboff et al. 2003a; Goloboff et al. 2021b; De Laet et al. 2004; Goloboff and Pol 2005; Simmons and Freudenstein 2011; Simmons and Goloboff 2013) have pointed out that group weighting by *FWR* can produce serious distortions of the supports. An example from Goloboff and Pol (2005) is in Figure 8.5c. All the taxa except R form a well-supported skeleton, as a pectinate tree; taxon R has only missing entries for every character, and thus can be equally parsimoniously placed in any of the 31 branches of the skeleton formed by taxa A–Q. More uneven taxon partitions offer fewer branches for R to be placed so that the monophyly of the group will be violated, and thus groups including only few taxa (e.g. PQ), or excluding only few taxa (i.e. the basal nodes of the tree) will have higher frequencies. The effect is very obvious and pronounced, with an almost exact symmetry in values moving away from the middle of the tree.

Needless to say: none of the groups shown in the tree of Figure 8.5c has any support—every one of those groups can equally parsimoniously include or exclude taxon R. The figure illustrates values under bootstrapping, but any other resampling method would produce similar values if results are summarized with *FWR*. The high values for some groups might be justified perhaps from some perspective (e.g. if we use resampling as a proxy for stability and we assume that when finding data for R it will land on any of the branches with equal probability). However, there is nothing in the *data* to define the position of R as done by *FWR*; from the perspective of measuring support, these results do not make any sense.

In general, the *FWR* approach will produce distorted measures of support when a group that is supported weakly (or not at all) is nonetheless present in a high proportion of optimal or near-optimal trees. The common way to summarize the results of MCMC (Monte Carlo Markov Chains, see Chapter 4) is by calculating frequency of groups in the final sample of trees, and that method produces results very similar to those of *FWR* (Goloboff and Pol 2005)—the results seem to be mutually reinforcing, but they are in fact a consequence of a similar bias! This can also involve cases with no missing entries (for an example, see Goloboff and Pol 2005, Figure 8.2). In resampling, the problem is avoided, or at least significantly decreased, if a strict consensus tree is used to represent every pseudoreplicate. This is actually simpler than calculating *FWR* (requiring less programming effort), and avoids the problem of "overemphasizing" pseudoreplicates that produce many trees—every pseudoreplicate produces just one tree, a strict consensus. As the purpose of resampling (when used for measuring supports) is to find out how frequently a group is supported among pseudoreplicates, using the strict consensus of all the *MPTs* for the pseudoreplicate is the logical, obvious choice: it only includes supported groups. When a strict consensus is used to summarize pseudoreplicates, the correct conclusion—a complete bush—is obtained for the dataset of Figure 8.5c. If that is considered too uninformative, the use of pruned trees may be preferable, calculating the supports for all the groups that result from excluding R from consideration. In that case, all frequencies tend to 100% (note that the *FWR* results do not serve even that purpose: all groups in the R-less skeleton tree are equally supported, not unevenly as the *FWR* tree would indicate).

Another consequence of the *FWR* approach is that the criterion used to collapse zero-length branches will affect the resulting values of group supports. Recall that the different criteria to collapse zero-length branches do not (or should not) affect the groups that are perceived as supported, as long as all trees distinct under each criterion are considered. But, in the case of *FWR*, the collapsing criterion affects the number of trees that are considered as distinct (see Chapter 6), and therefore can change the weights assigned to every tree found in a pseudoreplicate. This can happen even if doing large numbers of pseudoreplicates and an exact search for every resampled dataset—the effect does not result from calculation errors, but is built into the method. An example is in Figure 8.7c.

8.4.3.2   Frequency differences (*GC*) track support better than absolute frequencies

Even with the corrections for irrelevant characters and using strict consensi, frequencies are not always interpretable as group supports. The notion of resampling was introduced (at the beginning of Section 8.4) together with the idea that it can properly measure supports only for groups with frequency above 50%. The expectation was then that supported groups would turn up with frequencies above that value, but that is not always the case. Some supported groups can have frequencies well below 50%, even if uncontradicted by any data. An example was discussed already, with the data-set of Figure 8.5a. Another example is shown in Figure 8.5b; group HI is actually sup-ported by the data (with $BS = 1$), but has a very low frequency under any resampling method. The tree in Figure 8.5b shows group frequencies under jackknifing with $P_{(del)} = 0.37$, because exact frequency of HI can be calculated easily—eliminating any character, or combination of characters, the group disappears. The probability of eliminating none of the six characters in the dataset is $(1-0.37)^6 = 0.06$. In con-trast, group EFG is contradicted by the data, but its frequency (empirical, with 1,000 pseudoreplicates, an exact search for each) is 0.12, twice the frequency of group HI. Below 50%, that is, there can be groups *contradicted* by the data with higher frequency than groups *supported* by the data. An even more misleading comparison would be for HI (supported, with frequency 0.06) and an unsupported group XY, in a 4-taxon dataset for taxa W–Z with 100 synapomorphies for XY and 100 for XZ. The frequency of such unsupported group XY would approach 50%, much higher than for supported group HI.

The fact that the actual support of groups and resampling frequencies can be transposed for frequencies below 50% (as in the examples just discussed) also casts doubts on the use of frequencies to measure supports for groups with frequency above 50%. If the frequency does not correctly indicate degrees of support *below* 50%, why would it indicate support correctly *above* 50%? The logical conclusion is that absolute frequency can *always* be misleading as an indication of support.

The example of groups XY and XZ supported by 100 characters each serves to illustrate the basis for an alternative to absolute frequencies. Both groups XY and XZ occur with the same frequency, ≈50%. Their support is exactly zero, and so is the *difference* in their frequencies. This suggests that a better indication of support can be obtained by calculating the difference in frequency between the group, G, and the most frequent contradictory group, C. Goloboff et al. (2003a) called this difference the GC value. For the example with taxa W–Z and 200 conflicting characters, and barring sampling error, $GC_{XY} = GC_{XZ} = 0$. GC seems better correlated with support than absolute frequencies, and absence of support is not indicated by 0.50 (as in absolute frequencies), but instead by 0. In the example of Figure 8.5b, HI has a much lower frequency than EFG, but no character directly contradicts HI; $GC_{HI} = 0.06$. EFG is directly supported by the first character, but it is also contradicted by the sec-ond, third and fifth; among pseudoreplicates, EFG is contradicted more often than it is supported, so that $GC_{EFG} = -0.22$. This latter value illustrates another important advantage of GC relative to absolute frequencies: not only does it reach values of 0 in the absence of support, but the negative values indicate strength of contradiction; that is, GC varies between −1 (maximum contradiction) and +1 (maximum support). A group being contradicted in every pseudoreplicate is clearly different from a group that is indifferent for every pseudoreplicate; both situations are perceived as the same by absolute frequencies,[5] but clearly different from the perspective of GC.

The GC values can measure support for groups of low support more meaning-fully. Although the general recommendation is to use resampling for measuring sup-ports for the groups present in the strict consensus of most parsimonious trees, it is common practice to just construct a summary tree with the trees resulting from all pseudoreplicates. When building that tree with frequencies, the trees are typically very poorly resolved; the frequency difference consensus (i.e. the tree that displays

all the groups of positive GC) is normally much more resolved and a better heuristic
guide for further inquiries into the structure of the dataset.

Although the previous section showed that the strict consensus tree is a better
summary than FWR for each pseudoreplicate, using the strict consensus in the case
of GC is only an approximation. A group may be *unresolved* in the strict consensus
for a pseudoreplicate, but *contradicted* in every one of the individual trees used to
construct the consensus. In such a situation, we would be incorrectly counting an
actual contradiction as indifference. An example (from Goloboff et al. 2003a) is dataset 7 (the first character is nonadditive):

7
A 00
B 00
C 10
D 11
E 21
F 20

When collapsing zero-length branches (with rule 4), dataset 7 produces six distinct
trees of four steps, the strict consensus of which is fully unresolved—there is no
support for group DE, because the two characters in the matrix are in a tied conflict.
In a pseudoreplicate where the first character is eliminated or the second character is
upweighted relative to the first, the group DE is supported. But, in a pseudoreplicate
where the second character is eliminated or the first character is upweighted relative to the second, the resulting dataset produces (under rule 4) three distinct trees,
(A(B(EF(CD)))), (A(B(CD(EF)))), and (A(B(CD)(EF))), all of which contradict DE,
but which produce an unresolved consensus. Therefore, if using the strict consensus to summarize results, DE is supported in some pseudoreplicates, and appears to
never be contradicted, thus having a positive GC (the exact value varies depending
on how resampling is done, but all resampling methods produce a positive GC for
DE). This problem would be avoided if the individual trees for each pseudoreplicate
were examined, calculating the frequency with which each group is present among
some of the *MPT*s. Although this would correctly recognize the lack of support of
group DE, it would require more complex calculations (and a lot more RAM); it has
not been implemented in TNT (or any program, to my knowledge). TNT only implements the approximation with strict consensus trees for each pseudoreplicate.

Goloboff et al. (2003a) report other simple cases where GC would produce an
incorrect assessment of supports, even if the individual trees were used. They showed
dataset 8 as an example (the first character is nonadditive):

8
A 000
B 001
C 101
D 110
E 210
F 210

When the last character is removed or the first character is upweighted the dataset determines a single *MPT*, (A(B(C(D(EF))))); the second character unites DEF, and by placing C as sister to DEF, the state 1 becomes "plesiomorphic" at that level, so that EF are joined by sharing state 2. But when the third character is upweighted, C cannot go together with DEF, but instead with B, and the optimization of the first character is ambiguous, no longer providing a potential synapomorphy for resolving DEF as (D(EF)). In the absence of any potential synapomorphies, the resolution of DEF with third character upweighted is a full trichotomy—any resolution is equally parsimonious. Being preferred in some cases and never contradicted, DE has a positive *GC*, even if unsupported by the original data (and even if using individual trees instead of consensi). Goloboff et al. (2003a) reported the exact value $GC_{DE} = 0.22$ for symmetric resampling with $p_{(up)} = p_{(down)} = 0.33$; the results for other ways to resample the data are similar.

8.4.3.3   A death blow to measuring support with resampling

The previous sections showed cases where the resampling frequency of supported groups is below 50%. This alone suffices to cast doubt on the use of frequencies, but it still seems rectifiable by the use of *GC*. Most authors (e.g. Felsenstein 1985b; Farris et al. 1996; Goloboff et al. 2003a) assumed that (curable effects of asymmetry aside, as described in Section 8.4.2.3) groups with resampling frequency above 50% would be necessarily supported. It is possible, however, to contrive examples where groups contradicted by the data have arbitrarily high frequencies—given enough taxa and characters, possibly approaching frequency of 100%. This can happen regardless of whether pseudoreplicates are represented by strict consensus trees or not. In such cases, even if the correction of *GC* usually produces better assessments of support than raw frequencies, it does not suffice to eliminate the problem of unsupported groups with raw frequency above 50%.

Figure 8.6a shows an example constructed to show that behavior (a similar example was provided in the supplementary material of Goloboff et al. 2021b; Figure 4.2c discussed in Section 4.7.3 shows results for datasets simulated to mimic this same behavior under likelihood or Bayesian methods). The dataset has three types of characters; consider first the behavior under resampling for the first two types of character (in black). The first type supports the group BC; the second type includes series of characters (each type is one less character than the first type), alternatively grouping C with D, E, . . . O. Given the black characters only, group BC is supported, but there is almost as much evidence for each of CD, CE, . . . CO. Resampling such a dataset produces a low frequency for the supported group BC; as C can go with almost the same probability with any of 13 taxa (B, or D–O), the frequency of BC will tend to 1/13, particularly as there are more characters of each type (but with the first type always having one more character than the characters that group C with any of the alternatives, to keep BC supported). If the dataset included more taxa (for a total of *T*) and C could go almost equally with any of the *T* − 3 taxa other than B, frequency of supported BC would tend to *1 / (T − 2)*. That is a case of a very low frequency for a supported group, such as the cases discussed in the previous section; the tree is shown to the left of the dataset.

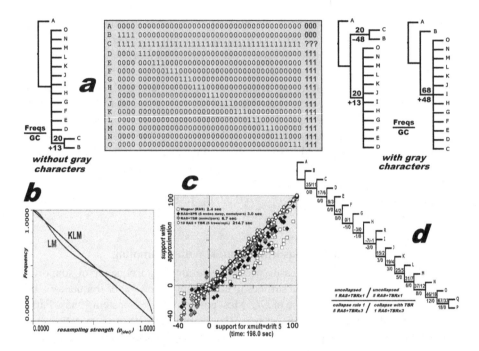

FIGURE 8.6 (a) When the gray characters are not included, supported group BC has a very low frequency (expansion of the pattern shown makes frequency of BC approach $1/(T-2)$). When the gray characters are added, contradicted group C–O has a high frequency (expansion makes frequency of C–O approach $(T-3)/(T-2)$). (b) Trajectory of group frequencies as resampling (jackknifing) perturbates more mildly, calculated by exhaustive enumeration ($p_{(del)}$ varying in intervals of 0.0125), for an example dataset from Ramírez (2005). Uncontradicted group KLM is supported by a single character, and LM is supported by six and contradicted by eight (four characters for KM, and four for KL). Ramírez (2005) showed a sudden change of the trajectory of the frequency, but his results are empirical; exact calculations show no sudden changes. (c) Calculation of supports for 300 pseudoreplicates of symmetric resampling (summarized with GC) with $p_{(up)}=p_{(del)} = 0.33$, for the groups in the strict consensus of $MPTs$ for the embiid dataset of Szumik et al. (2019), with different search intensities (y-axis), relative to very intensive calculations (x-axis). The effect of more superficial searches is to make most supports appear lower than if calculated more exhaustively. For search algorithms that saved multiple trees, a single tree was picked at random to represent the pseudoreplicate; trees were not collapsed. (d) As ambiguity in the pseudoreplicate datasets is taken into account more carefully for the dataset in Figure 8.5c, the artificially high supports obtained for unsupported groups when using FWR or a single uncollapsed tree tend to zero (GC values shown on tree).

Now consider the third type of character (bold, gray); those characters delimit a split for taxa D–O, relative to A and B; as C has missing entries for those characters, it can go in or out of the group D–O with no cost for the characters of the third type. For every pseudoreplicate where C goes with a taxon other than B, the group C–O is also supported—by the gray characters. As C goes with D with frequency approaching $1/12$ (or $1/(T-2)$ in the general case), and the same with E, F, or any of the other taxa, the frequency for contradicted group C–O becomes the sum of the frequencies for CD, CE, . . . CO, or $(T-3)/(T-2)$ in the general case. It is easy to see that, for increasing numbers of taxa and characters with an almost exact tie, the frequency of the contradicted group that excludes A and B quickly approaches 100%.

Figure 8.6a shows the results for symmetric resampling (1,000 pseudoreplicates), with frequencies above branches, and GC values below. Not only does the supported group BC have a low frequency (and a negative GC value), but the contradicted group C–O has a high frequency (and a strongly positive GC value). The same happens, of course, for any type of resampling (even no-zero-weight symmetric resampling). The previous section discussed cases where groups with some support can have low frequencies, but the existence of cases where contradicted groups can have arbitrarily high frequencies (such as the example in Figure 8.6a) makes it impossible to use any correction based on frequencies.

Perhaps the supports for a particular empirical dataset will not be affected by this kind of interaction—but that is almost impossible to determine on a case-by-case basis. A good measure of group supports should ideally be immune to such problems, and the possibility of situations such as this makes the supports measured by resampling always dubious, and best considered as merely an approximation.

8.4.3.4 Frequency slopes

The previous sections showed that frequency differences provide a better assessment of group supports than plain frequencies, yet can be misleading in some cases, and no correction based on frequencies under a given resampling strength seems possible. Goloboff et al. (2003a) discussed the possibility of an alternative approach to summarizing the results of resampling. Consider the consequences of using resampling with different probabilities for modifying the original dataset; as the perturbation is milder and milder, the pseudoreplicate datasets will be more similar to the original dataset, and then the frequency will tend to 100% for any group that is supported, and 0% for any group that is unsupported (or contradicted). This is shown in Figure 8.5b, in the right diagram, which shows exact calculations of jackknifing group frequencies (in the y-axis) plotted against different values of $p_{(del)}$ (in the x-axis). The frequencies follow different trajectories, but the slope at which a supported group intersects the point of 100% support (if supported) or 0% support (if unsupported) in the left part of the diagram may be giving information on how favorable is the evidence for the group. A well-supported group (none exists in Figure 8.5b) would approach frequency of 100% already at moderate values of resampling strength, but a poorly supported group will do so only as the resampling is very mild. Therefore, the slope of poorly supported groups as the resampling strength approaches zero will tend to a large negative value. Likewise, the slope of an unsupported group favored and contradicted by large and equal amounts of evidence will tend to a large positive number, approaching frequency 0 only when resampling strength is very close to 0.

Goloboff et al. (2003a) discussed the possibility of using the slopes, and proposed an approximate method to calculate the slope at low values of resampling strength; the method is prone to error and needs large numbers of pseudoreplicates to achieve some precision (the method uses an extrapolation based on the exact calculation of frequencies described in Section 8.4.2.5; see Goloboff et al. 2003a for details). Since then, there has been almost no exploration of the possibility of using frequency slopes to evaluate supports; an exception is Ramírez (2005), who suggested that the frequency trajectories may have sudden changes of slope when resampling strength tends to zero (his figs. 1–2). However, Ramírez (2005) used empirical estimations of frequencies at different resampling strengths to calculate trajectories, which is prone to errors; exact calculations for his example under jackknifing (Figure 8.6b, group LM) show only smooth transitions along all values of $p_{(del)}$.

8.4.3.5   Rough recovery of groups

As discussed in Chapter 6 (Section 6.1.2.2), a group may be non-monophyletic on a tree because just one or two (out of numerous) taxa have been incorrectly excluded from the group, or included inside. In that case, the group is "close" to be recovered by the tree; that situation is different from the case where nothing resembling the group exists in the tree. The "closeness" to the group can be measured by the number of taxa incorrectly placed in the group. Sanderson (1989) had already discussed the possibility of taking into account how close the tree is to contain a group if interest; Goloboff et al. (2009a) and Lemoine et al. (2018) used a similar approach to measure degree of recovery of groups in large trees. When groups similar to the group of interest are found in a large number of replicates, this would indicate that excluding just a few taxa from the trees could produce strongly supported pruned trees. In TNT, this approach can be implemented by saving the trees produced by resampling, and then calculating the rough recovery consensus; this requires that a tree (e.g. one of the most parsimonious trees) is defined as reference (Lemoine et al.'s 2018 implementation, available only for likelihood, differs in that it picks the tree defined by the majority of most similar groups automatically; Lutteropp et al. 2020 provide a faster implementation). It must be emphasized that such a summary of the resampled trees may be heuristically useful (e.g. to quickly figure out whether the group supports can be improved by pruning few rogue taxa, see Section 6.1.2.2), but it cannot be considered (contra Lemoine et al. 2018!) as strictly measuring group "supports".

8.4.4   SEARCH ALGORITHMS AND GROUP SUPPORTS

Sections 8.4.2 and 8.4.3 discuss details of the resampling and summary phases, assuming that all the *MPTs* for each of the pseudoreplicate datasets have been properly calculated. Obviously, exact searches are possible only for very small datasets. Only searches truly finding all possible *MPTs* are the "correct" values for the resampling method in question, but even those correctly calculated values may provide misleading evaluations of group supports. Therefore, the use of approximations to speed up calculations seems warranted. Although most prior discussions of resampling had assumed that one truly needs to find all the optimal trees for each pseudoreplicate, thus making resampling an extremely laborious procedure, the work of Farris et al. (1996) was instrumental in showing that faster—and more superficial—searches for each of the pseudoreplicates still provide useful approximations to group supports.

In terms of alternative searches for the analysis of pseudoreplicates, much the same considerations as for quick consensus estimations (Chapter 5, Section 5.8) can be made. Tree searches can be "intensive" in terms of two different, relatively independent aspects. One is the ability of the search to produce an optimal or near-optimal tree; in difficult datasets, some methods find trees that are more parsimonious. The other aspect in which a search can be thorough—or not—is in how it takes into account the ambiguity of the dataset, producing multiple trees of equal score if they exist, instead of just one or a few; this is the primary way to identify ambiguous groups.

As argued by Farris et al. (1996) even the most superficial search algorithms are unlikely to miss strongly supported groups. As searches fall shorter of maximum parsimony, they will miss more supported groups; the trees for different pseudoreplicates will then be less likely to agree in the presence of any given group. In other words, the variation in tree topologies produced by the superficial search algorithms is added to the intrinsic variation created by resampling. Therefore, more superficial searches tend to find *lower* supports for most groups. In the case of *GC* values, this correspondence between supports and search strength is less exact, because the contradicting groups themselves can also be found in lower frequencies, but the trend is still clear. This is shown in Figure 8.6c; the supports calculated with an intensive search (xmult=drift 5, keeping just the first tree found) are on the x-axis. Even a very quick algorithm, like a *RAS* Wagner tree, has a very good correlation with the intensively calculated values, although the values for most groups are lower. As the search is made more intensive (i.e. coming closer to maximum parsimony), the supports increase, approaching the correct values (i.e. the dots come closer to the diagonal indicating a perfect match). Therefore, the main drawback of shallow searches is that supports are generally underestimated. Note that the groups evaluated in Figure 8.6c are those in the strict consensus of *MPTs*; all the groups therefore should have positive *GC* values. Negative values in Figure 8.6c indicate cases where the *GC* method would misleadingly indicate contradiction for some groups.

Farris et al. (1996) also discussed the need to use randomized starting points for the analysis of each pseudoreplicate. The purpose of randomization is to eliminate as much determinism as possible. Some programs (e.g. MEGA) use Wagner trees and *TBR* for the analysis of every pseudoreplicate, but with a fixed addition sequence. This can repeatedly find wholly unsupported groups (an example is in Goloboff et al. 2021b, fig. 2). In PAUP*, the addition sequence can be optionally turned to "as is" (i.e. the sequence of the taxa in the matrix); for the dataset of Figure 8.7a, where two subtrees can slide within each other with no cost in parsimony, this produces 100% frequency (see Figure 8.7b) for groups with zero support (i.e. groups absent from some *MPTs*). Given the determinism introduced by Wagner trees with fixed addition sequences, Farris et al. (1996) emphasized the need to use randomized addition sequences.

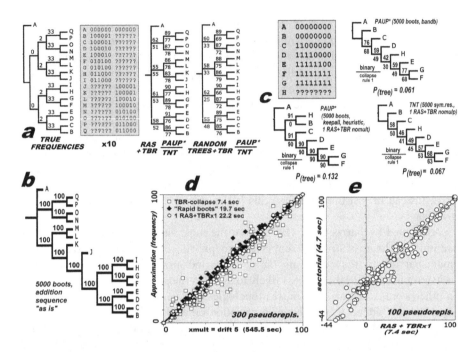

FIGURE 8.7 (a) A dataset showing the effect of bias in tree searches. The dataset contains two separate groups of taxa (B–I and J–Q), with no characters in common; it produces 1,027,839 most parsimonious binary trees. The frequencies of the groups in the most symmetric *MPT* are shown on the tree to the left of the dataset. Resampling with either PAUP* or TNT (using *TBR* but saving a single tree per replication), produces a substantial distortion of those expected frequencies (and regardless of whether each search starts from a *RAS* or a random tree), because some of the equally parsimonious trees are found much more frequently than others. (b) The results for the dataset in (a), with Wagner addition sequence "as is" (available in PAUP*, mandatory in MEGA). The determinism of the "as is" addition sequence produces 100% frequency for wholly unsupported groups. (c) A smaller dataset, where the bias in perfectly randomized search algorithms can be calculated analytically (see Tables 8.2 and 8.3). The dataset supports 11 binary *MPT*s (all possible locations of H). In the absence of bias, group frequencies if saving a single tree in a heuristic search should be as when using *FWR* and an exact search without collapsing zero-length branches (i.e. PAUP* with *bandb* option; note that collapsing criteria influence the group frequencies when using *FWR*, as they change the number of possible trees from 11 to 9). Heuristic searches saving a single tree in PAUP* or TNT produce a substantial distortion of those expected frequencies (whether or not zero-length branches are collapsed). Note that the tree found by PAUP* with a heuristic search is the most probable topology for a perfect randomization (Table 8.2), but it is found much more frequently than under perfect randomization (90% of the times, instead of 13%); PAUP* is further biased in favor of that tree. The frequencies in TNT are closer to those for perfect randomization in this particular case, but deviate more than those in PAUP* in other cases. (d) Comparison of resampling values obtained with different approximation methods (y-axis), against values obtained with a more exhaustive calculation (x-axis), for the dataset of Pei et al. (2020). (e) The values obtained for the dataset of Pei et al. (2020) with the approximation of sectorial resampling (y-axis) plotted against the values obtained with a *RAS+TBR* (x-axis). While not perfect, the values of sectorial resampling offer a rough approximation, and use much less time in very large datasets.

However, even with randomized addition sequences, some problems remain if a single tree is being saved for each pseudoreplicate dataset. Section 8.4.3.1 discussed *FWR*, which can attribute high support to groups that are present in a majority but not all of the near-optimal or optimal trees—which is to say, groups that are supported weakly or not at all. If the randomization of the starting point effectively precludes all deterministic effects (a big *if*; see Section 8.4.4.1), and there are *n* equally parsimonious trees for the pseudoreplicate, the search will be equally likely to find any of the *n* trees. Since every tree can be found with probability of *1 / n*, the results of saving a single tree are (in the absence of search bias) entirely equivalent to those produced by *FWR* (Goloboff and Farris 2001; Goloboff and Pol 2005; Simmons and Freudenstein 2011; Simmons and Goloboff 2013). The effect is of course intensified if a single binary tree per pseudoreplicate is saved (e.g. as in maximum likelihood programs; see Simmons and Kessenich 2019). To avoid this problem, it is necessary to eliminate groups that are not supported in a particular pseudoreplicate. To remove ambiguous groups, Farris et al. (1996) collapsed zero-length branches using "Rule 1" (the strictest optimization-based collapsing). In many cases, an optimization-based collapsing is insufficient, and to avoid attributing support to unsupported groups it is necessary to consider multiple trees. The difference between the effects of finding trees that are more parsimonious (i.e. improving the support values for truly supported groups) and considering multiple equally parsimonious trees (i.e. reducing the number of truly unsupported groups that spuriously appear supported) may not seem intuitive at first sight, but the logic of it becomes apparent on reflection. This difference has a clear parallel in methods for quick consensus estimation, discussed in Chapter 5 (Section 5.8, Figures 5.4a–b).

The number of equally optimal trees considered for each pseudoreplicate can be increased quickly and efficiently by using *TBR* collapsing (see Chapter 6, Section 6.4.3). Figure 8.6d shows the estimated supports (*GC* values) for the same dataset used to exemplify problems with *FWR* (i.e. Figure 8.5c), taking into account ambiguity with different levels of thoroughness. As the number of starting points (and the number of trees saved per starting point) used to analyze every pseudoreplicate increases, the spurious supports decrease. The most effective way to take into account ambiguity is by *TBR* collapsing, which produces supports of zero for all groups (the correct value for this dataset). Of course, in programs where *FWR* is mandatory (e.g. PAUP*, MPBoot), saving large numbers of trees per pseudoreplicate does not avoid this effect—tree search bias aside, the same results are obtained if saving one or many trees.

8.4.4.1   Search bias worsens the problems of saving a single tree

As discussed in the previous section, any program that saves a single tree is bound to reproduce, on average, the same results as *FWR*—if the tree search is unbiased. If the search algorithm is biased, then some of the equally optimal (or near-optimal) trees will be found with higher probability, introducing an additional distortion of the supports. Several authors (myself included: Goloboff and Farris 2001; but also Farris et al. 1996; Felsenstein 2004) had assumed that randomizing the addition (and insertion) sequence in Wagner trees would eliminate any systematic effects in the analysis of the pseudoreplicate datasets. When trying alternative locations for a taxon being added, most implementations of Wagner trees just pick the first best location, but it is quite obvious that the *insertion* sequence needs to be randomized as well if systematic effects are to be avoided. This is equivalent to randomly selecting all tied locations at each stage of the build of a Wagner tree; Takezaki (1998) thought that this (together with a random addition sequence) would eliminate all systematic effects.

However, Goloboff and Simmons (2014) examined the question more carefully and showed that *all* search algorithms are biased. This was discovered because of the most obvious symptom: group supports strongly deviate from the expected values. Consider the example of Figure 8.7a, a dataset where two subtrees can slide within each other at no cost in parsimony. Among *MPT*s for the original dataset, the frequency of the 2-taxon groups among all optimal trees is exactly 33%; the 4-taxon groups have a much lower frequency, 2%. Every group in one of the subtrees is supported by 10 uncontradicted characters, so most resampling pseudoreplicates will have the same structure (it is very unlikely that all 10 characters for a group are eliminated). An unbiased search thus should produce groups in approximately those frequencies, but resampling saving a single tree per pseudoreplicate with PAUP* (bootstrapping) or TNT (symmetric resampling, randomizing insertion sequence in the Wagner tree) produces very different values (see Figure 8.7a). This happens even if the starting point in the search for every pseudoreplicate is a random tree (instead of a Wagner tree) followed by *TBR* swapping. Those results are clearly artificial. Any computer program makes a number of design decisions that may remove it from perfect randomization. TNT certainly does; for example, trying reinsertions of clipped clades in specific sequences during *TBR* allows saving extra time. Therefore, the results of Figure 8.7a may well be just the result of imperfect randomizations in the particular implementations tested.

Goloboff and Simmons (2014) examined in more detail the behavior of Wagner and *TBR* algorithms under perfect randomization of every step of Wagner trees (i.e. addition and insertion sequences) and *TBR* (starting point, clipping and reinsertion sequences). That a Wagner tree with a specific addition and insertion sequence can be biased is obvious (see specific examples in Goloboff and Simmons 2014); for randomization to eliminate any bias, it would be necessary for the bias inherent to the different addition and insertion sequences to cancel out—which it does not. In the case of *SPR* or *TBR*, even starting from a wholly random tree still produces biased results: the neighborhood of different suboptimal trees contains different numbers of better alternatives (so that some trees are more likely to lead to optimal trees than others). Goloboff and Simmons (2014) provided a program to carry out the exact calculations for perfectly randomized Wagner and branch-swapping algorithms, although it can be used only on small datasets (up to 10 taxa). Figure 8.7c is an example where all search algorithms are biased; as there are 11 *MPT*s, an unbiased search would find each with $P = 1 / 11 = 0.09$, but no search algorithm produces those results. Tables 8.2 and 8.3 show the probabilities of finding every optimal tree, and different groups (i.e. the probability of group X,Y is the sum of probabilities of all the trees containing group X,Y). Some of the optimal trees are found up to 2.43 times more probably than others (between 0.6 to 1.46 times the probability expected if no bias were present). Note that, in the case of PAUP*, the majority rule tree for a *RAS+TBR* used to

TABLE 8.2
Probabilities of Finding Each Individual Optimal Tree

TREE	PROB	RATIO	BIAS
(A(B(C(D(E(H(FG)))))))	0.05448	0.59928	1.00000
(A(B(C(D(H(E(FG)))))))	0.06121	0.67331	1.12353
(A(B(C(H(D(E(FG)))))))	0.06656	0.73216	1.22173
(A(B(H(C(D(E(FG)))))))	0.07534	0.82874	1.38289
(A(B(C(D(E(G(FH)))))))	0.07802	0.85822	1.43209
(A(B(C(D(E(F(GH)))))))	0.07802	0.85822	1.43209
(A(B(C(D((EH)(FG))))))	0.09574	1.05314	1.75734
(A(B(C((DH)(E(FG))))))	0.10756	1.18316	1.97430
(A(B((CH)(D(E(FG))))))	0.11810	1.29910	2.16777
(A(H(B(C(D(E(FG)))))))	0.13249	1.45739	2.43190
(A((BH)(C(D(E(FG))))))	0.13249	1.45739	2.43190

Note: For the Dataset of Figure 8.7c, When Using Perfectly Randomized Addition and Insertion Sequences for the Wagner Tree Followed by Perfectly Randomized *TBR* Swapping (i.e. Equiprobably Choosing Any of the Better Trees in Each *TBR* Neighborhood). The table shows calculations for ideally randomized algorithms; in practice, computer programs (e.g. PAUP*, TNT) fall short of perfect randomization, because of design requirements. PROB is the absolute probability of each tree (as the dataset is simple, all sequences of *RAS+TBR* end up in an optimal tree, thus the sum of PROB equals one). RATIO equals PROB divided by the probability of finding each tree in the absence of bias (for 11 optimal trees, 1/11 = 0.9091); some trees are found with lower probability than expected in the absence of bias, others with higher. BIAS is the ratio between the probability of each tree, relative to the probability of the least probable tree (as the trees are arranged by increasing PROB, maximum BIAS is between last and first tree).

analyze each pseudoreplicate (Figure 8.7c) happens to be one of the most probable topologies, given the bias in Wagner trees and *TBR* swapping. TNT does not produce the same tree, but also produces results that are strongly influenced by bias.

The fact that all possible search algorithms find some optimal trees more probably than others is another argument against searching a single tree for each pseudoreplicate dataset. The problem of bias also exists under maximum likelihood, and thus it can also affect programs such as RAxML (Stamatakis 2014) or PhyML (Guindon and Gascuel 2003), designed to search a single tree per pseudoreplicate. In TNT, the search for each pseudoreplicate can be set to find multiple distinct *MPTs*, which eliminates or ameliorates the consequences of search bias. In the case of large and difficult datasets, this may be unfeasible and time constraints may force us to save only one or a few trees per pseudoreplicate—in that case, keeping in mind the possible distortions of support is important. Even if a single tree is to be saved per pseudoreplicate, it is always advisable to subject that tree to *TBR* collapsing, which can implicitly consider additional hundreds (or even thousands) of *MPTs* for each pseudoreplicate in very little additional time.

If an exact search is to be used for the analysis of each pseudoreplicate, you also need to remember that exact searches are usually not randomized, with alternative

TABLE 8.3
Group Probabilities, For the Dataset of Figure 8.7c

GROUP	EXPECT	REAL	RATIO
CDEFG	0.27273	0.34032	1.24783
DEFG	0.45455	0.52498	1.15494
EFG	0.63636	0.69375	1.09018
FG	0.81818	0.84397	1.03152
CDEFGH	0.81818	0.73502	0.89836
DEFGH	0.63636	0.54158	0.85106
EFGH	0.45455	0.36746	0.80840
FGH	0.27273	0.21051	0.77186
BH	0.09091	0.13249	1.45738
CH	0.09091	0.11810	1.29909
DH	0.09091	0.10756	1.18315
EH	0.09091	0.09574	1.05313
FH	0.09091	0.07802	0.85821
GH	0.09091	0.07802	0.85821

Note: When Using Perfectly Randomized Addition and Insertion Sequences for the Wagner Tree Followed by Perfectly Randomized *TBR* Swapping (i.e. Equiprobably Choosing Any of the Better Trees in Each *TBR* Neighborhood). EXPECT is the expected probability of finding a tree with the group (i.e. the number of optimal trees with the group, multiplied by the expected probability of each individual tree). REAL is the frequency with which trees containing the group are found, by a perfectly randomized *RAS+TBR*. RATIO is the ratio between REAL and EXPECT. For example, three optimal trees display group FGH, so that the expected probability of FGH is $0.09091 \times 3 = 0.27273$, but the group is found only in 0.21051, or 77% less frequently than expected without bias. Other groups are found more frequently than expected (e.g. BH, 46% more frequently than expected without bias).

paths to all possible trees followed in an orderly manner. If only one or a few trees are to be saved for every resampled dataset with an exact search, the situation is even worse than with randomized Wagner trees. In exact solution algorithms, the addition sequence for taxa is generally so that the most distinct taxa are chosen first. As this introduces little variation, the first tree found by an exact search for very ambiguous data will always be the same. Therefore, if one is going to use an exact search for each pseudoreplicate, it is important to make sure that there is enough space in the tree buffer to save large numbers of trees for each pseudoreplicate. Alternatively, you can use an exact search with reservoir sampling (Vitter 1986), which guarantees that the few trees found are randomly chosen from among all possible *MPTs* (see Goloboff and Simmons 2014 for details).

8.4.4.2 Approximations for further speedups

Given that resampling involves many tree searches, it naturally lends itself to various approximations. The use of those will depend on the size and difficulty of the dataset. Since identifying wildcard taxa may require an evaluation in two stages (first saving a single uncollapsed tree per pseudoreplicate to identify wildcards, see Section 8.5, then performing more thorough searches once the wildcards have been identified), some of these approximate methods may be particularly useful for that first stage.

One of the methods that have been proposed is the "rapid bootstrap" of Stamatakis et al. (2008), for speeding up support calculations in RAxML. The basic idea in the method is to avoid searching from scratch (i.e. from a *RAS*) for every pseudorep-licate, using instead the final tree for the previous pseudoreplicate as the starting point to perform branch-swapping for the present pseudoreplicate. After a certain number of pseudoreplicates using each previous tree, the tree is "reseeded" again, with a *RAS* Wagner tree; Stamatakis et al. (2008) reseeded the tree every tenth pseu-doreplication. By periodically reseeding the tree, the hope is that the co-dependence between pseudoreplicates is decreased or removed. Stamatakis et al. (2008) used this in conjunction with *SPR* swapping regrafting no more than a few nodes away from the pruning origin, and more superficial likelihood calculations, for further saving of time. In the parsimony context (where the use of the previous tree and periodical reseeding are easily handled with scripts) a *TBR* search is fast enough and (as men-tioned in Chapter 5, Section 5.3.2) provides a better tradeoff between time and results than more limited rearrangements (such as *SPR* or *TBR* with a reconnection limit). Figure 8.7d exemplifies the use of this "rapid bootstrap" in TNT, for the dataset of Pei et al. (2020). The method is only marginally faster than a full *RAS+TBR* per pseudoreplicate (19.7 vs 22.2 sec).

A faster method is the fast approximation to resampling (*FAR*) described by Torres et al. (2021) and Goloboff et al. (2021b). This is even easier to implement with TNT than the rapid bootstrap (see under Implementation), and it saves a more signif-icant amount of time; it is also a rougher approximation. Any well-supported group is likely to both be present in the *MPT* and show up in the pseudoreplicate; the main need is to eliminate those groups that the pseudoreplicate does not support. This can be done simply taking one of the *MPT*s from the original dataset as a guide for the pseudoreplicate, and removing groups that are not supported—done most efficiently with *TBR* collapsing. This saves all the search time for the pseudoreplicates. When a group X in the *MPT*s from the original dataset is actually contradicted by a group Y present in the *MPT*s for the pseudoreplicate dataset, this method will just find a polytomous tree with the group X collapsed, not with the contradicting group Y. Therefore, none of the groups in the *MPT* used as guide can be contradicted, just polytomized, and this method can only be used to estimate frequencies, not *GC* values. This is faster than the previous method, but also more imprecise and under-estimating the supports by a larger factor (see Figure 8.7d, white squares).

A third approximation is based on sectorial searches, and is useful for datasets with several hundreds or thousands of taxa (uncommon in morphological datasets). The scripts needed to implement this method are more complex than for the two previous ones. They are based on the `sectsch` command of TNT, which creates reduced datasets (by properly selecting terminals and *HTU*s based on the down-pass optimization, see Chapter 5, Sections 5.6.2 and 5.9.4). The support values can be calculated for the reduced dataset and then transferred to the equivalent node of the complete reference tree. The results are the same as if resampling with most of the tree constrained to a fixed topology, leaving only the sector in question free to vary, but obtained much faster. That the frequencies for some groups in the sector can be *over*estimated is obvious: as the limits of the sector are fixed, any rearrangement that places a taxon from the sector outside of it, or a taxon outside of the sector inside of it, will never be realized. But the frequencies for some groups in the sector can be *under*estimated. This is less obvious, but can also happen, because the resolution outside of the sector is fixed—choosing different possible resolutions (i.e. alternative *MPTs*) outside of the sector will change the reduced matrix; some of those extraneous resolutions will imply higher frequencies, others lower. Consider dataset 9:

```
              9
    01234 56 78
A   00000 00 00
B   00000 00 11
C   00000 11 11
D   00011 11 00
E   11111 11 00
F   11111 01 00
G   11111 00 00
H   11111 00 00
```

This dataset produces six distinct binary *MPTs*. D forms (because of characters 3 and 4) a group with E–H in one of those trees; then the relationships between E–H are well defined by reversals in characters 5 and 6, as (A(B(C(D(E(F(GH)))))). C can also form a group with B (because of the last two characters), and then the relationships between E–H can be resolved in five different ways, one of which has E–H resolved as in the first tree shown, (A((BC)(D(E(F(GH)))))). The correct frequencies for groups FGH and GH are about $F_{FGH} = F_{GH} = 0.20$ (i.e. calculated globally). If the first tree is used to analyze the sector E–H, with C fixed as sister to D–H, then $F_{FGH} = F_{GH} = 0.51$. But if the second tree, (A((BC)(D(E(F(GH)))))), is used to analyze the sector E–H, then the characters potentially resolving the subtree are no longer useful; $F_{FGH} = F_{GH} = 0$, instead of the correct 0.20.

Therefore, the sectorial approximation can either over or underestimate the frequency of any group; this can also cause *GC* values to be over or underestimated. The method is useful for saving time in large datasets. Figure 8.7e compares the *GC* values calculated with the sectorial (y-axis) and a global method (x-axis), for the dataset of Pei et al. (2020). The tree used to evaluate the support is a randomly selected *MPT*; the groups in that tree therefore can be either supported (in which case *GC* should be positive) or indifferent (in which case *GC* should be 0). Negative values in Figure 8.7e thus indicate cases where the *GC* method misleadingly indicates contradiction for some groups. Although the difference in times, relative to a single global *RAS+TBR* to analyze each pseudoreplicate (4.7 vs. 7.4 sec) is minor in this dataset (with only 164 taxa), it can be much more substantial in larger datasets.

8.4.4.3   Worse search methods cannot produce better results

Special consideration must be given to another proposal for saving time in bootstrap calculation, implemented in Hoang et al.'s (2018) program MPBoot. Hoang et al. (2018) proposed that their program both saves time in calculations and produces better estimates of "support", in the sense that a group with support X has a probability of being in the true tree that is closer to X than in standard bootstrapping. They evaluated the accuracy of the bootstrap estimates by means of simulations, following the approach of Hillis and Bull (1993) and Anisimova et al. (2011). Note that the conclusions by Hillis and Bull (1993) were contested by Felsenstein and Kishino (1993), and Efron et al. (1996). When generating datasets under the standard Poisson model for DNA sequences, such evaluations find that the bootstrap values generally underestimate the probability that a group is present in the model tree. Given that the explicit argument of Hoang et al. (2018) for defending their method is that the resulting bootstrap values underestimate the probability of a group being true with a lower margin of error, this section—unlike the rest of the sections in the present chapter—discusses group supports from that same perspective.

The idea in Hoang et al.'s (2018) MPBoot program is to avoid having to search the optimal trees for each pseudoreplicate, in a manner similar to the approach of Kishino et al. (1990) and the `jbremer` command of Nona (Goloboff 1993b). The program initially searches optimal trees for the original dataset, at the same time saving a substantial number of suboptimal trees. The number of steps of every character in every one of the trees in that sample is then calculated and stored in memory. As the dataset is resampled, the resulting length of each tree in the sample can be recalculated quickly, by simply multiplying (for each character) the number of steps by the new weights. This allows re-scoring all the trees in the sample without having to do a full optimization, and quickly selecting an estimate of the *MPT* for the resampled dataset without any actual tree searches.

The expediency of selecting from a sample of trees for each resampled dataset may save more or less time, depending on a number of factors. Prime among those is the type of characters. Character types that take more time to be optimized are candidates for the method to be more beneficial (Hoang et al. 2018 only examined step-matrix characters, where the method saved the most time, but the same would apply to other costly optimality criteria, such as self-weighted optimization, or landmark characters). Although this is not a serious problem for morphological datasets, Goloboff et al. (2021b) also find that, in MPBoot's implementation at least, run times increase more than lineally with numbers of characters, so that analysis of phylogenomic datasets with millions of characters becomes almost unfeasible. For datasets with fewer characters and no step-matrix characters with large numbers of states, run times of MPBoot for 100 bootstrap pseudoreplicates are comparable to those of TNT, so that not much is gained by using the method, timewise.

However, Hoang et al. (2018) claim that the results obtained when pre-sampling trees for the complete dataset are less biased than those where TNT or PAUP* are used to perform actual tree searches for each pseudoreplicate. If that were generally true, the approach of MPBoot would be preferable in terms of quality of results, not computing time. Hoang et al.'s (2018) approach to showing accuracy was simulating[6] datasets, and calculating the proportion of groups of support X that were true. If a proportion higher than X is true, then the evaluation of support is underestimating the correct probability that the group is true, $P_{(true)}$, and vice versa. On datasets simulated with the exponential model described in Goloboff et al. (2017), Figures 8.8a–b show the probability of a group being actually true (on the y-axis), for each value of support (x-axis) under symmetric resampling with a *RAS+TBR* for each pseudoreplicate, for both raw frequencies and *GC* values (Figure 8.8a is under equal weights, and Figure 8.8b implied weighting with *k = 10*). Figures 8.8a–b show that both raw frequency and *GC* underestimate $P_{(true)}$, as found by Hoang et al. (2018) and well known before; as *GC* equals the frequency of the group minus the frequency

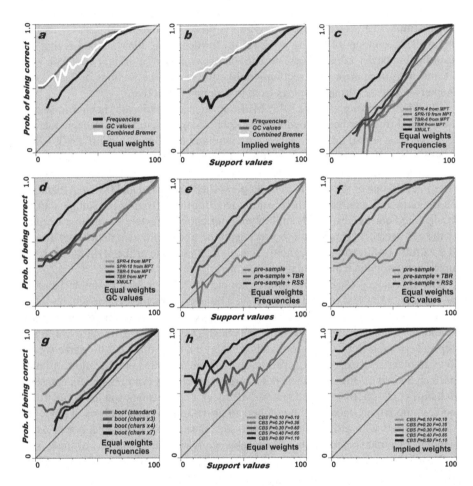

FIGURE 8.8 Summary of simulations under an exponential model, with λ a random number in the range 0.05–0.15. Cases (a)–(d) and (g)–(i) have 80 taxa and 120 characters, with (a) –(d) and (g) based on 3,000 simulated datasets, and (h)–(i) based on 10,000. Cases (e)–(f) have 40 taxa and 80 characters, based on 1,500 simulated datasets. The x-axis shows the support values obtained with different methods for resampling, summarizing, or searching trees. SPR-4 indicates an *SPR* search with reinsertion of clipped clades no more than 4 nodes away of original position; likewise for SPR-10 or TBR-4. The y-axis shows the proportion of groups for each value of support that were present in the model tree (i.e. an estimation of $P_{(true)}$). If support is intended to indicate $P_{(true)}$, then lines above the diagonal indicate underestimation, and vice versa. **(a)**–**(d)** Datasets resampled with symmetric resampling ($P_{(down)} = P_{(up)} = 0.33$); **(e)**–**(f)** Datasets resampled with bootstrapping; **(h)**–**(i)** Supports evaluated by recording cost of *TBR* moves on an *MPT*. See main text for discussion.

of the most frequent contradictory group, it must be less than or equal to the raw frequency of the group, therefore underestimating $P_{(true)}$ by a larger margin. The plot of $P_{(true)}$ as a function of *CBS* values is also shown, confirming that (as intended) the values of *CBS* tend to track resampling support values rather well. Note, however, that the magnitude of the deviation from perfect accuracy (i.e. the diagonal) also depends on the optimality criterion.

The results of Figures 8.8a–b are uncontroversial, but MPBoot's pre-sampling of trees produces plots of $P_{(true)}$ as a function of bootstrap frequency that are closer to the diagonal, and generally below (see Hoang et al. 2018, their fig. 4). Hoang et al. (2018) explain this on the grounds that resampling may create wildcard taxa and that selecting the pre-sample of trees based on the original dataset may therefore reduce the effect of wildcards. While undeniably imaginative, that explanation is also obviously incorrect (Goloboff et al. 2021b). Resampling by itself is unlikely to create wildcards, because wildcards are those *taxa* with more missing or ambiguous codings, and resampling modifies entire *characters* (i.e. removing or upweighting them for all taxa). The explanation is also self-disqualifying in a sense, because if the position of a taxon becomes ambiguous for a given pseudoreplicate, preferring the position of the taxon in the original dataset amounts to denying the utility of resampling itself—one might as well stick to the groups in the original dataset and assume 100% support for all of them.

Goloboff et al. (2021b) suggested instead a different reason for the better "accuracy" of MPBoot in Hoang et al.'s (2018) simulations. Among the pre-sampled trees, the best available tree for a pseudoreplicate can often be just the *MPT* for the original dataset, simply because the pre-sample failed to contain the tree that would indeed be optimal for the pseudoreplicate (Goloboff et al. 2021b provided a simple contrived example where this happens). In many simulations the *MPT* for the complete simulated dataset correctly recovers the true tree, and therefore this would *bias* results toward the correct tree.

The results of searching for each pseudoreplicate with one of the *MPT*s for the original dataset as the starting point of branch-swapping also indicate a dependence as suggested by Goloboff et al. (2021b), as in Figures 8.8c–d. More superficial swapping algorithms used to start searching from the *MPT* for the original dataset underestimate $P_{(true)}$ by a smaller factor. Freely searching the *MPT* for the resampled dataset (with `xmult`), much more likely to find shorter trees than just *TBR*, produces "worse" results according to the criterion of predicting $P_{(true)}$. Even more significant, Figures 8.8e–f reproduce the use of a sample of trees, using TNT scripts. For every pseudoreplicate, the script calculated the length of the trees under the new set of weights; as this was a simple script, it calculated the length by re-optimizing the trees, thus taking much longer than MPBoot to run (but producing similar results). Given the time required, this experiment simulated datasets with only 40 taxa and 80 characters (gathering a pre-sample of 10,000 trees, way more than what MPBoot samples by default). The results of just selecting the best available tree for each pseudoreplicate (Figures 8.8e–f, light gray line) match those presented by Hoang et al. (2018), but as the best tree is subsequently used as the starting point for *TBR* branch-swapping (`nomulpars`), the resulting values of support switch to underestimating $P_{(true)}$. For a given pseudoreplicate, if the tree produced by *TBR* differs from the best available tree it can only be because it is *more parsimonious*. The effect is even more pronounced if using a random sectorial search (default parameters) starting from the best available tree, which produces even shorter trees than *TBR*. That is, selecting the tree for each pseudoreplicate more carefully, leads to *worse* results, according to Hoang et al.'s (2018) criterion.

If the goal were indeed to produce measures of support that can interpreted as $P_{(true)}$, a more logical approach would be to modify the resampling procedure itself, rather than just expecting the inability of the tree-selection process for each pseudoreplicate to result in trees sufficiently similar to the optimal trees for the original dataset and magically straighten the accuracy of the method. Modifications of the resampling procedure would at least rely on a specific procedure and produce more predictable results, rather than rely on the failure to find better trees for the pseudoreplicates. An obvious alternative is to use a weaker resampling, introducing milder perturbations to the dataset. The case of bootstrapping is shown in Figure 8.8g; by resampling increasingly larger numbers of characters (darker lines), the resulting support values are higher, and better track $P_{(true)}$. The same would occur with a lower probability of change in jackknifing or symmetric resampling. As *CBS* can be calculated to approximate different values of $P_{(del)}$ in jackknifing, the resulting values of support can be made closer to values of $P_{(true)}$ by suitable adjustment, as shown in Figures 8.8h–i for both equal and implied weights.

As discussed in this chapter, measures of support are not meant to directly indicate probability, but instead intrinsic properties of the data. The correlation between support and probability can be made somewhat tighter under the conditions of a specific simulation, but real data—or different conditions for the simulation—are an entirely different question. Felsenstein and Kishino (1993) already noted that generating data under different conditions changes the extent to which bootstrap values exceed or fall short of $P_{(true)}$. Anisimova et al. (2011) also warned that their "conclusions should not be overgeneralized: the properties of the support measures investigated . . . may vary with divergence, tree shape, and model fit".

8.5   CONFIDENCE AND STABILITY ARE RELATED TO SUPPORT, BUT NOT THE SAME THING

Most of this chapter is focused on measures of support, intended only to indicate data structure. The hierarchical structure of a dataset is stronger to the extent that there are more characters supporting every group, and that the characters agree more with each other. But characters interact and conflict, and it is difficult to even define precisely the very property we are trying to measure, beyond a relatively vague statement of "more quantity, more coherence". All that makes exact measurements of data structure hard to design. All measures discussed in this chapter provide an approximation to the problem, but none seems perfect.

Scientists are often concerned with confidence, in the sense of ordinary, nonstatistical language: how strongly should we *believe* that a given result is correct? That problem is not addressed in this chapter in any direct way, but the methods and discussions in it still have relevance: we cannot strongly believe conclusions based on scarce and self-contradicting evidence. Only those conclusions based on substantial amounts of coherent evidence *can* be strongly believed—but this does not mean they *must*. Large amounts of coherent evidence could still be indicating incorrect conclusions, and so something *more* than just the degree of support is needed. For example, better agreement with geographical distribution or stratigraphy (as in, e.g. Sansom et al. 2018) can increase the confidence in a given result—those aspects are harder to quantify in a well-justified and unified framework, and go beyond simple measures of support. Any statistical interpretation of support measures (such as Felsenstein's or Anisimova et al.'s) that does not take into account such additional lines of non-character evidence must make strong assumptions.

Many authors have considered that resampling methods must be justified by reference to stability, not confidence (Hovenkamp 2009; Siddall 2002). Sometimes, Felsenstein himself referred to this notion: "[b]ootstrapping provides us with a confidence interval within which is contained not the true phylogeny, but the phylogeny that would be estimated on repeated sampling of many characters from the underlying pool of characters" (Felsenstein 1985b: 786). Again, support, stability, and confidence are related, but not the same (Giribet 2003). Weakly supported conclusions are less likely to be stable—they can be more easily overturned by subsequently found evidence. Conclusions that change with every new analysis cannot be strongly believed, no matter their degree of support.

But "stability" can be defined only by reference to specific contexts or circumstances. Bootstrapping, in the standard approach, addresses the question of what could happen if we forgot about all the characters we had collected to resolve a phylogeny, and had to gather again exactly the same number of characters for the same taxa. One may be interested in considering how stable the results can be under *other* circumstances. For example, when we add characters in the future—without forgetting any of the presently known characters. Goloboff et al. (2008a) emulated this situation by comparing the results obtained when removing some characters to those obtained for the whole dataset (with the `sibyl` option of the `resample` command of TNT); this calculates whether a group found for the reduced dataset is found in the complete one, instead of the other way around. It does not produce group supports, but overall evaluations of stability; if the complete dataset stands in relation to the reduced one as the future dataset stands in relation to the present one, it is the appropriate comparison. Of course, empirical evaluations of stability, based on comparison of successive updates of datasets (such as Murphy et al.'s 2021) would be preferable; those are much more laborious and rarely performed.

Therefore, the circumstances for which stability can be assessed are multiple. An obvious case is *sensitivity analysis* (Wheeler 1995; Giribet and Wheeler 2007). In the molecular realm, sensitivity analysis usually refers to transition/transversion ratios (*ti/tv*) and cost of gap opening and extension. The "true" values of those parameters are unknown, so the common approach is to explore a range of reasonable values, and evaluate whether the conclusion of a given group depends on (i.e. is "sensitive to") a specific value of *ti/tv* or gap opening/extension. The sensitivity analysis is sometimes used to maximize congruence between data partitions (i.e. choosing the parameters under which partitions produce most similar trees; e.g. Giribet and Wheeler 2002), which seems problematic (e.g. see discussion in Goloboff et al. 2008a). Clearly, not every value of *ti/tv* is reasonable; transitions are thought to be more frequent than transversions, and then *ti/tv = 0.001* makes no sense; there is no point in determining whether a group is absent from the results for *ti/tv = 0.001*. But the exact value of *ti/tv* is not known, and even if congruence is maximized by selecting a value of (say) *ti/tv = 1.674*, a group present under *ti/tv = 1.674* but absent from the results for *ti/tv = 1.675* cannot be seriously considered. A more conservative approach to summarize the results of a sensitivity analysis is then to retain only those groups present under the entirety (or the majority) of reasonable values for the parameters.

In the case of morphology, just as with molecules, "evidence" is never pristine, but always subject to interpretation. Characters can be considered additive, nonadditive or step-matrix; analyzed under equal weights, or implied weights—and in that case, under different concavities. A group can be supported when the evidence is considered from some perspective (e.g. "this character is additive"), but unsupported otherwise, even if the strings of 0's, 1's and 2's for each taxon are the same. Transformation costs can be based in part on empirical observations (see Chapter 2, Section 2.10, for discussion), but some doubts as to the best treatment can always remain. As in the case of sequences and *ti/tv* values, not every possible value is meaningful. Under implied weights, a value of $k = 0.01$ comes down to a compatibility analysis, with the implication that any character that shows even a single instance of homoplasy is almost entirely unreliable when compared to a perfect character; a value of $k = 1000$ implies that all the characters have (almost) the exact same weights,[7] regardless of homoplasy. None of those extreme values of k need to be considered, but even if (say) a range $6 < k < 15$ is to be checked, the exact value cannot be determined. Such uncertainty can—and should—be taken into account. For example, the supports for every group can be calculated under various plausible analytical conditions, and the most conservative decision is to use the lowest values.

Yet another possible evaluation of "stability"—not considered so far in this chapter—is the stability to the addition of new taxa. There have been only rare attempts to formalize evaluations of the influence of taxon selection. An example is Siddall's (1995) taxon jackknifing. One can explore the influence of a taxon on the relationships of the rest. A case where a terminal is crucial for the relationships between the other taxa was used in Chapter 5 (Section 5.3.1) to show how taxon addition sequences in Wagner trees can influence the results; inclusion or exclusion of taxon D in that matrix determines how A, B and C are related. Such changes in the relationships between taxon subsets can only happen in the presence of character conflict, but the results of removing taxa cannot be taken to indicate amounts of character support—and it is not clear exactly what they do indicate (a similar point is made by Wheeler 2012: 324–325).

8.6   IMPLEMENTATION IN TNT

TNT provides facilities for calculating both Bremer supports and resampling, with numerous options in both cases. The relevant commands and options are summarized in Table 8.4. Most evaluations of support require several steps in TNT; while this may seem inconvenient at first, it has the advantage that it makes calculations much more flexible, allowing the use of estimations of support with different search algorithms and criteria.

TABLE 8.4
TNT Commands for Calculating Group Supports

Command (option)	Minimum truncation	Action(s)
bsupport	bs	Calculate Bremer supports, from trees in memory
constrain	cons	Enable constraints (defined with force)
hold	ho	Set maximum number of trees to retain
majority	ma	Calculate majority rule from trees in memory
force	fo	Define constraints for mono- and non-monophyly
freqdifs	fr	Calculate *GC* values from trees in memory
resample	res	Resample matrix and search best tree(s) for every pseudoreplicate dataset (search commands within square brackets, [. . .])
boot	b	Resample using standard bootstrapping
from	fro	Plot supports on groups for specified tree
gc	g	Summarize using *GC* values
jak	j	Resample using jackknifing (independent character deletion)
poisson	p	Resample using Poisson bootstrapping (independent reweighting)
prob	pr	Define probability of deletion (or upweighting) for jak and sym
savet	sa	Save the trees from each pseudoreplicate for subsequent analysis
sym	sy	Resample with $p_{(down)}=p_{(up)}$
zerowt	z	Never eliminate characters, just downweight
rfreqs	rf	Calculate frequencies of rough recovery for groups in reference tree
subopt	su	Define value of suboptimality (needed for saving trees for bsupport, or for bsupport itself, if using the ! option)
ttags	tt	Handle overimposed branch labels (e.g. to plot different support measures on the same tree)
tvault	tv	Handle trees from the vault

8.6.1 CALCULATION OF BREMER SUPPORTS

The command to calculate Bremer supports is bsupport. Bsupport works in two alternative ways: one is by making comparisons between previously saved trees, the other is by subjecting existing tree(s) to *TBR* and recording the score differences for every move. Comparisons between previously saved trees require that several trees of different scores are present in the buffer; with a single tree (or several trees of the same score) held in memory no support calculations can be carried out by bsupport (see later).

To make comparisons between previously saved trees, it is entirely up to the user which (and how many) trees to include in the buffer; for correct calculations, the tree buffer must include trees of optimal score, and a representative set of suboptimal trees. Ideally, for every group present in the best trees, there should be at least one tree where the group is absent. If a group is present in each and every one of the trees in the set, the support is reported simply as "M?" for *BS*, where M is the score difference between the best and worst trees in the set (i.e. the only possible estimation of support: M or more; if running `bsupport` on a single tree or a set of trees of the same score, "0?"). In a similar case for *RBS* or *CBS* calculations, the support is reported as "100?" (i.e. apparently not contradicted by any character). Whenever there is a group with its support indicated with a question mark, it means that the set of trees in the tree buffer is insufficient for estimating the support for the group. The suboptimal trees can be calculated by searching for suboptimal trees (the `subopt` and `hold` commands determine the acceptable degree of suboptimality, and the maximum number of trees to save), or by searching under reverse constraints (constraints are defined with the `force` command, and enforced with `constrain=`).

8.6.1.1 Searching suboptimal trees

Remember that saving increasingly larger sets of trees, with suboptimality increasing gradually, is the best approach to search suboptimal trees. In this case, the `hold+N` option will increase the size of the tree buffer to be able to hold an additional N trees. The `fillonly` option of the `bbreak` command is also very useful to make sure that branch-swapping for suboptimal trees stops as soon as the memory is filled (otherwise, swapping needlessly continues through the thousands of trees saved). Thus,

```
tnt*>sub 2; hold +10000; bb = fill;
. . . program output here. . .
tnt*>sub 4; hold +10000; bb = fill;
. . . program output here. . .
tnt*>sub 6; hold +20000; bb = fill;
. . . program output here. . .
tnt*>bsupport;
```

although this depends on the numbers of taxa and structure of the dataset, every step generally takes only a few seconds (interrupting the swapping as soon as the specified number of trees is found). A general criterion to decide whether a reasonable tradeoff suboptimality/number of trees has been set at every stage is to see how many trees are swapped to completion before filling the tree buffer. For the dataset example.tnt (the spider dataset from Goloboff 1995a, distributed with the TNT package), assume that the 72 trees distinct under collapsing Rule 1 are initially held in memory. Consider the following commands and their output,

```
tnt*> sub 3 ; ho 1000 ; bb = fill ;
Suboptimal 3.000 x 0.000
Space for 1000 trees in memory
Start swapping from 72 trees (score 382-385). . .
Repl. Algor. Tree      Score    Best Score  Time     Rearrangs.
---    TBR   1 of 1000 ------   382         0:00:00  321,430
Completed TBR branch-swapping.
Total rearrangements examined: 321,430.
Maxtrees was reached when swapping tree 1. Didn't swap tree(s)
2-999.
Note: some trees of different length may become identical if
collapsed.
Best score (TBR): 382-385. 1000 trees found (overflow).
```

The message specifies that maxtrees was reached when swapping tree 1, suggesting that 1,000 trees was too few to consider the trees within 3 steps of optimal: just the first few rearrangements filled the memory buffer. The 1,000 trees are found very quickly, but that is probably a bad set of trees to calculate Bremer supports. In the other extreme, for the same dataset, consider saving up to 15,000 trees one step longer than the optimal:

```
tnt*>sub 1 ; ho 15000 ; bb = fill ;
Suboptimal 1.000 x 0.000
Space for 15000 trees in memory
Start swapping from 72 trees (score 382-383). . .
Repl. Algor.    Tree       Score  Best Score Time   Rearrangs.
---    TBR    11536 of 15000  ------  382          0:00:26 1,964,504,455
Completed TBR branch-swapping.
Total rearrangements examined: 1,964,504,455.
Maxtrees was reached when swapping tree 11536. Didn't swap
tree(s) 11537-14999
Note: some trees of different length may become identical if
collapsed.
Best score (TBR): 382-383. 15000 trees found (overflow).
```

This, instead, did not fill the tree buffer until after swapping 11,536 trees, taking 26 sec instead of 0 sec as the previous routine. The estimations of Bremer supports for groups within one step will be quite precise, but—depending on your time availability—such large numbers of trees to save for that value of suboptimality may be excessive. The precision is of course decided by the user, but a useful number to decide on a specific strategy is the number of trees swapped to completion before the tree buffer is filled (this number can also be accessed via the scripting expression cutswap, so that the process can be automated in scripts, see Chapter 10).

If you wish to consider different starting points as well (rather than just swapping from some of the optimal trees), you can also use a mult command after searching suboptimal trees, saving previous trees to the tree vault for subsequent retrieval. Thus, the following sequence of commands,

```
tnt*>hold +100/+0; tvault >.; keep0 ;
tnt*>sub 0; mult = rep 100 keepall hold 1;
tnt*>tvault <.; bsupport ;
```

places in a large vault the suboptimal trees (note: setting such a large vault may require that you properly set mxram before reading your dataset), searches with mult (note: it is *very* important to remember to reset subopt to zero before doing any subsequent tree searches; otherwise, the quality of the searches can be seriously compromised, particularly if doing multiple random addition sequences with a few trees per replication), brings back the suboptimal trees from the vault,[8] and calculates the supports. If the dataset is so well structured that the default *TBR* swapper always finds optimal trees, you can switch to a more superficial swapping algorithm (e.g. mult = rep 100 keepall hold 1 spr). Adding the suboptimal trees from independent starting points can provide more accurate estimations of Bremer supports. Once the suboptimal trees are in memory, together with the optimal trees, the bsupport command checks the trees and displays the values on the tree branches.

Except for the vault operations (not available from the menus), the same sequence of commands can be done by appropriate repetition of *Analyze > Suboptimal*, *Settings > Memory*, and *Analyze > Traditional Search*. Rather than protecting the trees in a vault, you can just save them to a tree file (with *File > Tree Save File*, then closing file with the same option), for subsequent re-reading (with *File > Open Input File*, or *File > Read Compact Tree File*, depending on the format you chose to save the trees).

8.6.1.2   Searching with reverse constraints

The reverse constraints are defined with the force- command, and applied with constrain=. You may save to the tree vault the constrained trees, then subsequently retrieve them for evaluation with bsupport. For listing the groups to constrain for non-monophyly, the easiest way is by referring to nodes of a specific tree, for example node 47 of tree 0:

```
tnt*>hold 100 / 100;
tnt*>force - [@0 47] ; constrain=;
tnt*>xmult; tvault > 0;
```

Evaluations of support via reverse constraints for all groups are easily done with scripts (see Chapter 10). For the Windows version, the reverse constraints can be defined with *Data > Define Constraints*, selecting "*for non-monophyly*" and "*use selected groups*", then clicking on the desired group when the tree image shows up. Note that this requires that a tree with the group to be constrained for non-monophyly exists in memory. The trees where every possible group of interest has been made non-monophyletic must be present in memory, together with optimal trees, and then the bsupport command is applied as in the previous case (the bsupport command does not need to know where the trees came from).

8.6.1.3   Estimation of Bremer supports via TBR

The estimation with *TBR* operates from the tree(s) in memory; it is invoked with the ! option of the bsupport command. The ! symbol must be followed by the number of tree used as reference. With a single !, it collapses the groups detected to have zero support; with two (!!), it keeps all the groups in reference tree (and if finding a shorter tree, reports negative values for the groups of reference tree that are absent from the shorter tree).

Rather than accepting all possible moves, *TBR* accepts the moves that are within a pre-specified value of suboptimality (considering both absolute and relative differences in score). This value of suboptimality must be set prior to branch-swapping, with the subopt command. Alternatively, a value of suboptimality that will guarantee finding that at least one acceptable rearrangement will violate the monophyly of every group can be set automatically, with the : (colon) option of the bsupport command; this option also exists for the subopt command. The : symbol is followed by a tree number, a slash, and the proportion of groups with branch lengths larger than the resulting suboptimal value; subopt: 0/0 will set suboptimal to the length of the longest branch in tree 0 (or, in the case of implied weights, the cost of collapsing the branch that costs the most to collapse). That is,

```
tnt*>bsup:0/0 !0;
```

If the suboptimality is manually defined with the subopt command, and no tree lacking a given group is found that is within the defined level of suboptimality, the branch displays the value of suboptimality followed by a question mark. Also keep in mind that all the commands that effect searches or branch-swapping (this one included) obey currently enforced constraints. If you have defined and enforced constraints, the swaps that violate the constraints will *not* be considered for calculation of Bremer support. Thus, unless you have a specific kind of evaluation or test in mind, you have to make sure that no constraints are enabled (they can be disabled with constrain-) before estimating Bremer supports via *TBR*.

8.6.1.4 Variants of Bremer support

The three types of Bremer support implemented in TNT are the absolute (*BS*, the default), relative (*RBS*), and combined (*CBS*) supports. The *RBS* is invoked with bsupport], which only compares with trees within *BS* value for every group. To compare against any of the trees present (which can produce values below 100% for some groups in the absence of contradicting characters), use bsupport[. Remember that if searching suboptimal trees for calculation of *RBS* or *CBS*, using a margin of relative suboptimality (as discussed in Section 8.3.2) produces more accurate estimations. The relative suboptimality is set as a second fractional number after the value of absolute suboptimality, preceded by an "x": subopt A x R, where A is the value of absolute suboptimality (unbounded), and R the value of relative suboptimality ($0 \leq R \leq 1$).

The *CBS* is calculated with bsupport&. The & symbol can be followed by a fractional number P ($0 \leq P \leq 1$), the jackknife deletion frequency that bsupport& will emulate (i.e. values of P closer to 1 produce lower supports). If implied weighting is on, the exponent for calculation of *CBS* is automatically multiplied by the (absolute) cost of the first step of homoplasy, to obtain an equivalence with analyses under prior weights. Optionally, you can also multiply the exponent[9] by a factor F; as the factor F increases (unbounded), the scale along which supports are measured is shifted toward zero (this can be used to spread support values in phylogenomic datasets with millions of characters, where all *CBS* values—just like standard resampling—tend to unity; see Torres et al. 2021). *CBS* cannot be applied to landmark data: how to rescale the exponent of the expression for *CBS* is not obvious in that case. If you think you can find a meaningful rescaling—for example to make absolute values for all landmarks comparable—then you can always calculate *CBS* manually from the *BS* and *RBS*. An alternative combination of *BS* and *RBS* (producing an even closer correspondence with jackknifing in the case of groups supported by uncontradicted characters), is obtained with bsupport|; this uses as supports $CBS' = 1 - (1 - (RBS \times R_{(del)}))^{BS/F}$.

After determining type of support, the bsupport command can read a list of trees to use for calculating the supports, and (preceded by a slash), the list of taxa whose position is to be ignored when calculating supports. This can be used to prevent wildcard taxa from decreasing the supports; the wildcard taxa must have been identified already—the `bsupport` command does not find wildcard taxa by itself. See later (Section 8.6.4) for advice on how to identify taxa that decrease group supports.

8.6.2 Resampling

Resampling is done in TNT with just one command, `resample`. Unlike Bremer support calculations, `resample` does not require setting values of suboptimal trees or doing prior searches, but the command has more options than the `bsupport` command. Some options determine the way in which resampling itself is done, others how the results are summarized. For every pseudoreplicate dataset, the search is performed with the commands indicated within square brackets, retaining the strict consensus of the trees found by the search (collapsed according to currently enforced rules, set with the `collapse` command). Resample with no arguments will just run with default options (symmetric resampling with *P(down)* = *P(up)* = *0.33*, *GC* values with cut 0, and a `mult` search with 5 *RAS* followed by *TBR* saving up to 5 trees per pseudoreplicate), or the same options used in the last run of `resample`. The options can be given in any order (except the option to exclude some taxa, which *must* be the last one).

In the Windows versions, most of the options for resampling are accessed through *Analyze > Resampling*. It may also be necessary to change some settings, such as the maximum number of trees to hold in memory (with *Settings > Memory*) or the way in which ambiguity is taken into account to consense the trees for each pseudoreplicate (*Settings > Collapsing Rules*).

8.6.2.1   Options to determine how resampling is done

These options determine both the type of resampling, and the strength of the perturbation. The types of resampling can be `boot` (standard bootstrapping), `poisson` (Poisson bootstrapping), `jak` (jackknifing), `sym` (symmetric resampling); these are all mutually exclusive; if you specify more than one among the arguments to `resample`, only the last one will take effect. Two less commonly used options are `nozerowt` and `sibyl`. With `nozerowt` the weights never go to zero; this is used only with `sym` (use `zerowt` to return to allow full elimination). The `sibyl` option deletes characters, infers trees for the resampled datasets, and compares them to optimal trees in memory.

The strength of the resampling can be changed in all cases. For `jak`, `sym`, `nozerowt`, or `sibyl`, the strength is set with `prob`. For `boot` and `poisson`, the strength is set with `softboot`, an integer between 0–100 (100 is the weakest perturbation), which determines the number of characters that are resampled (in `boot`), or the mean of the Poisson distribution used to draw resampled character weights (for `poisson`).

8.6.2.2   Options to determine how results are summarized

The trees found by the search for each pseudoreplicate are used automatically to produce a strict consensus, according to current settings (ideally, *TBR* collapsing, or different rules for elimination of zero-length branches) and taxa to be included. Those trees (one per pseudoreplicate) are then used to produce the final summary, which can be done using *GC* values (gc, default), raw frequencies (freq), or frequency slopes (slope). The summary may either use the tree that displays all the groups with *GC* or frequency above a certain value (cut, with default 0% for gc, and 50% for freq), or show the values on a tree given by the user as reference (from N, where N is the number of reference tree). The use of a reference tree is mandatory in the case of slope. The exclusion of some taxa from the trees is indicated by using, as last option for resample, a list of taxa (preceded by a slash, /).

8.6.2.3 Tree searches

The commands for searching each pseudoreplicate dataset are given within square brackets. The search for the pseudoreplicates can be done with, basically, any search command. The random seed (rseed command) needs to be different for each pseudoreplicate (to reduce systematic effects; see Section 8.4.4); TNT takes care of this automatically, so that you do not need to include commands to change random seeds as part of the instructions. For example:

```
tnt*>resample [mult1 = hold 1 tbr;];
```
or
```
tnt*>resample [xmult;];
```

The reporting and writing to screen are automatically disconnected for the searches; you can reconnect them as part of the instructions (which produces a lot of probably unnecessary output):

```
tnt*>resample [sil-all; rep=; mult;] allow;
```

For search instructions too large and complex to fit in a few lines, you can just place the instructions in a file (say, how_to_search.txt) and invoke the file within square brackets:

```
tnt*>resample [p how_to_search.txt];
```

While that is generally not advisable, it is possible to set a timeout for searches (see Chapter 5, Section 5.9.1). Note that a timeout set before the resample command itself will affect the overall time used to resample; a timeout as part of the instructions for the individual searches (i.e. within the square brackets) will affect the time used to search for every pseudoreplicate.

Some commands are disabled for the search, because they can interfere with the resampling. An example is the report command, in the example earlier; another example is reactivating or reweighting a character with the ccode command during a pseudoreplicate. These forbidden commands can be enabled (as done in the previous example) with the allow option of the resample command (which, obviously, needs to be used with caution and only if you know what you are doing!).

The savetrees option of the resample command saves to the memory buffer the consensus of the trees found for each pseudoreplicate. Alternatively,

you can always open a disk tree-file, and save the trees for each pseudoreplicate
to the file:

```
tnt*>tsave * sampled_trees.tre;
tnt*>resample [mult1 = hold 1 tbr; save;];
tnt*>tsave/;
```

The difference with the savetrees option is that this allows saving the individual trees (the savetrees option only retains the strict consensus at the end of
every pseudoreplicate). Yet further possibilities exist—for example, the trees for each
pseudoreplicate can be placed in the tree vault (with the tvault command), then
retrieved at the end. Just use your imagination.

8.6.3   SUPERPOSING LABELS ON TREE BRANCHES

As not all measures of support perform equally well under all circumstances, and
they can measure different aspects of the support (e.g. *BS* vs. *RBS*), it is often advisable to display the values of several measures on the branches of the tree. The same
can be useful for sensitivity analyses, to display the variation of the same measure
of support under, for example, different values of concavity when doing implied
weighting.

   The command ttag (already discussed in connection with graphic trees,
Chapter 1, Section 1.12.9) allows handling multiple labels for tree branches. The
ttag[option separates every new value in a new line. Thus, if you already have
suboptimal trees in your tree buffer (and assuming that the first tree is optimal; you
can change this), a simple series of commands like,

```
tnt*>ttag-;ttag=;bsup;ttag[;
tnt*>resample [mul1=hol] from 0;
```

will plot *BS* and *GC* values on the branches (the ttag command itself takes care of
always placing the labels on the corresponding branch; if a group in the tree to be
plotted does not exist in the tag-tree, its label is ignored). Note that the ttag- command discards all preexisting labels—this may or may not be what you want to do.
Thus, the commands,

```
tnt*>ttag-; ttag=; bsup; bsupp &;
tnt*>ttag[; resample gc [mul1=hol] from 0;
tnt*>ttag]; resample nogc freq [mul1=hol] from 0;
```

will superpose *four* labels on tree branches, with absolute and combined Bremer
supports above branches, and *GC* values and raw frequencies below. Note that new
lines for every label are activated prior to calculating *GC* values, but then *deactivated*
prior to calculating absolute frequencies, with ttag]. Otherwise *GC* and frequencies would be in two different lines (three lines total, instead of a total two). To

display (as text output) the tags in multiple lines you need to reconnect the multiple lines, with ttag[again, and display with ttag followed by no arguments:

```
tnt*>ttag[; ttag;
```

The terminal branches in this case will display their tags as only slashes (they never have any value of branch support); you can erase all tags for the terminal branches with ttag <<. In addition to displaying the multiple labels as text output, you can also save them in graphic form (see Chapters 1 and 10).

8.6.4 Wildcard Taxa and Supports

Just like some taxa can switch between alternative equally parsimonious positions in the tree and reduce the resolution of the strict consensus tree, some taxa can also switch between positions that are only slightly suboptimal, thus reducing the supports for the relationships between the rest of the taxa. Both the bsupport and the resample commands include the option to ignore the positions of some taxa, but neither command identifies wildcards by itself. They have to be identified separately, prior to the final calculation of supports. One of the commands that may help identify taxa that decrease group supports is chkmoves (see Chapter 6, Section 6.3.5), using a value of acceptable suboptimality instead of just equally optimal moves. To use other commands that improve consensuses, the mechanics may be different, depending on whether the supports are to be measured with Bremer supports or resampling.

Bremer supports are, basically, strict consensuses of trees of decreasing optimality. The commands to improve strict consensus trees (e.g. prunnelsen, pcrprune) can then be applied to a subset of the suboptimal trees to be used for bsupport. These commands allow saving groups of taxa to taxon groups. With the tgroup command, it is easy to create groups with specific lengths. For example, if optimal trees are 382 steps, then

```
tnt*>tgroup =0 len<384 ;
tnt*>prunnelsen >4 {0} / > 0 ;
tnt*>bsupport / {0} ;
```

will place in tree-group 0 all the trees within one step of the best, and then find out which taxa to cut to improve at least 4 nodes of their strict consensus (placing those taxa in taxon group 0). The taxon group is then used to tell the bsupport command which taxa to exclude. Alternatively, one can use the sort command to re-order all the trees so that the best trees are the first in file. Applying the consensus-improving commands to successively larger sets of trees will then find the wildcards that decrease the supports of groups.

The commands that help find taxa decreasing resolution of majority or frequency difference consensus can be used on the trees produced by resampling. A problem arises here, as the tree saved to memory under the savetrees option is the strict consensus tree for each pseudoreplicate, and this may in turn be influenced by taxa that wander around the tree *within* the pseudoreplicate—remember that first consensing and then removing wildcards is not the same thing as first removing wildcards and then consensing (Section 6.8.2). The `resample` command, if run with search algorithms that find a single tree, will store it as uncollapsed if using `collapse none`. The commands that help improve liberal consensuses for given trees (i.e. `pcrprune`, `prunmajor`, `prupdn`, and `rfreqs`) can then be used on that set of trees. But, ideally, the trees themselves should be collapsed after having identified wildcards, eliminating zero-length branches or groups that disappear on *TBR* rearrangements producing trees of equal score. That elimination of unsupported groups *cannot* be done on the original dataset—what needs to be checked is whether the resampled dataset itself supported the group, not the original dataset. This requires that resampling is repeated, generating the pseudoreplicate datasets again. If `resample` is run with the same random seed (`rseed` command), identical pseudoreplicate datasets will be generated in the same sequence as before. This can be used, in the special case of time-consuming evaluations of support, to avoid having to search again for every pseudoreplicate, simply storing the trees in a tree-file or in the tree vault, including instead of search instructions for every pseudoreplicate, instructions to retrieve the corresponding tree from the file or the vault. This needs to be done with scripts (enabled with the `macro=` command), but they can be relatively simple:

```
tnt*>rseed 1; collap none; resample savetrees [mul=hol];
tnt*>hold /+0; tvault > 1.;
tnt*>... identify wildcards here, place in taxon group 0...
tnt*>macro=; var: i; set i 0; collap tbr;
tnt*>resample savetrees [tvault < 'i'; set i ++;]/{0};
```

The `hold/+0` command makes a vault to store as many trees as are now in memory (i.e. +0); the first `tvault` command stores in the vault trees from 1 on (first tree is always the summary tree, unless both *GC* and raw frequencies are being used for summary, which produces *two* trees, followed by individual trees; to prevent that, make sure you also include `nofreq` as an option to the first resample). The final resample, instead of searching, just brings the tree from the vault and (as *TBR* collapsing is now connected) automatically eliminates all unsupported groups from each pseudoreplicate (with taxa from group 0 ignored).

NOTES

1   These branch "lengths" can also be calculated under implied weights, with the `blength/` option. This command takes into account the step increase, but also the number of steps elsewhere in the tree.

2   Repeated many times. In their 2008b paper, the roots "objecti—" and "subjecti—" occur a whopping 53 times (39 and 14 times, respectively). Farris and Goloboff (2008) noted that since Grant and Kluge (2008b) claimed that $BS = (F—C)$ is "objective" and $RFD = (F—C)/F$ is "subjective", Grant and Kluge must have considered that it is the division by F that introduces "subjectivity". Perhaps having learnt that "division" and "subjectivity" are not the same thing, in (2010) Grant and Kluge no longer discussed support measures in terms of "objectivity".

3 Minh et al. (2020) proposed using only a *sample* of the possible quartets, to save time. As the examples shown here are small, I have calculated *sCF* values by exhaustive enumeration of quartets (using the ***sconcfact.pic*** script for TNT).

4 It is interesting to note that Bui Quang Minh did his Ph.D. and postdoc in the lab of Arndt von Haeseler, of PUZZLE fame (a quartet-puzzling program; Strimmer and von Haeseler 1996). "When all you have is a hammer . . ."

5 When the results for every pseudoreplicate are summarized by means of a strict consensus tree.

6 In their simulation, Hoang et al. (2018) used model trees with different branch lengths (p. 4), which can easily make parsimony inconsistent. Given the similarity in results reported here (where data are simulated under a model where parsimony is consistent), that seems not to have affected their results.

7 Unless the number of taxa is very large, see Chapter 7, Section 7.5.

8 Note that the alternative sequence of commands (first searching with `mult` and `keepall`, storing those trees in the vault, next searching for successively larger numbers of longer trees, and then retrieving the `mult` trees from the vault) will not work, because when you increase the tree-buffer size with `hold` the vault is reset (discarding the trees in the vault).

9 The exponent itself is a division; the `help` command of TNT refers to the factor `F` as dividing the denominator of the exponent, which is the same thing.

9 Morphometric characters

Most datasets include only discrete characters, where the states can be defined qualitatively. But qualitative definition is hampered in many cases (e.g. fine intergradations, shape characters), and then the different conditions observed in the terminal taxa are best represented quantitatively. An obvious example is the differences in proportional length of an otherwise identical body part. The distinction between qualitative and quantitative characters itself is to some extent a matter of degree more than a matter of kind. If the distribution of lengths is clearly bimodal or multimodal (e.g. some species have long legs and some short legs, with no intermediates), then the character is often treated as discrete. As the taxa under study intergrade more for the variable in question, the best way to represent the differences among taxa is as a *continuous* character. The same applies to *meristic* characters (representing counts, e.g. numbers of teeth, scales, or hairs that can vary between tens or hundreds). In those cases, the use of real values (instead of a few distinct state categories) is required to best represent the differences among taxa.

Another case where the differences among taxa cannot be easily represented by means of qualitative characters is in the case of shapes. This is often observed in characters that are highly variable between species, yet rather conservative within. A typical example is in arthropod genitalic characters, which often allow recognizing most species within a genus. If the variability in those characters can be dissected into separate components (i.e. tip wide or narrow, constriction present or absent, etc.), then the observed variation can be decomposed into a series of discrete characters. When that is not the case, simply assigning a unique state—its own "shape"—to every species will produce characters with very low information content: the shapes usually cannot be arranged in linear ordered series, but making the character nonadditive leads to all trees being of the same length for the character (i.e. the character is useless for selecting trees). The methods of *geometric morphometrics* focus on mathematical characterization of shapes and their differences. The most widely used approach consists of tracking variation in Cartesian coordinates of specific points—or *landmarks*—of the structure in question. These methods allow comparing shapes and measuring their degrees of similarity, and nothing precludes adapting them for phylogenetic studies—a phylogeny being a scheme which allows explaining similarities by reference to common ancestry.

Both continuous and landmark data constitute morphometric evidence. There are two main questions to be considered in relation to the analysis of such types of data. The first is whether it is possible to analyze such characters by means of the same principles and criteria used to analyze standard (discrete) characters. Although several authors argue against this possibility (e.g. Pimentel and Riggins 1987; Cranston

DOI: 10.1201/9780367823412-9

and Humphries 1988; Bookstein 1994), most of those criticisms stem from an incomplete understanding of phylogenetic goals and methods. This chapter shows that continuous characters and landmarks can be analyzed under the same principles as other types of characters.

The second question is whether there is a point in using morphometric characters in phylogenetic inference. A number of authors (e.g. Felsenstein 1988; Hendrixson and Bond 2009; Klingenberg and Gidaszewski 2010) have argued against morphometric characters, not so much from first principles, but instead from the idea that they produce unreliable results. The main argument, however, boils down to the idea that those characters can retrieve phylogenies that are less than perfect. Of course, the use of morphometric (or any other kind of) data does not presuppose absolute infallibility. All that matters is whether the differences recorded are worthy of phylogenetic explanation—and if they have a heritable component, then they are. A simple variable such as total weight is more phenotypically labile than the number of leg segments, but this can hardly mean that there is no heritable component whatsoever in the weight—you can feed a mouse as much as you like, but it will never weight as much as a whale. Therefore, morphometric characters also provide evidence bearing on phylogenetic relationships, and is hard to see any defensible reason to proscribe their use in general, particularly at lower taxonomic levels. Even if perhaps weaker evidence than discrete characters, morphometric characters may be important in those cases where other types of evidence are scant (as in many fossil groups).

9.1 CONTINUOUS CHARACTERS

Continuous characters must be treated as additive, on the basis of the same criterion proposed for discrete additive characters: relative degrees of primary homology among states (see Chapter 2, Section 2.10.1). Pimentel and Riggins (1987) believed that none of the standard tests for homology are applicable to quantitative variables (a conclusion many authors have repeated since). That is not strictly true: the size, length or count is one of the components of the similarity. A leg of 1.14 mm is indeed more *similar* to a leg of 1.15 mm than it is to a leg of 2.37 mm. The "test" in this case may be weaker than more complex or subtle details, as used for homologizing discrete characters or states. However, one needs to recall that the goal of a cladistic analysis is to find the tree that attributes to common ancestry as much observed similarity as possible. Therefore, the relative degrees of similarity between the observed conditions should be used to establish transformation costs, and this varies in direct proportion to their size, length, or count. The relative differences between states therefore must depend on their numerical difference. Legs of 1.15 mm originating twice from ancestors with 1.14 leave unexplained a minor component of similarity, while legs of 2.37 originating twice from ancestors with 1.14 represent cases of independent originations from much more distinct conditions. The first case would have a cost $(1.15–1.14) \times 2 = 0.02$, while the second would have $(2.37–1.14) \times 2 = 2.46$; the difference in costs appropriately reflects the difference between the two cases.

Continuous characters therefore are just like additive characters with a very large number of possible states, and they can be optimized with Farris' (1970) algorithms (see Chapter 3). However, computer programs for phylogenetic analysis did not include, for several decades, any actual implementation of continuous characters. Some authors (e.g. MacLeod 2002; Lockwood et al. 2004; Adams et al. 2013) went as far as confusing this lack of implementation with a basic incompatibility of cladistic principles and continuous characters. Finally, Goloboff et al. (2006) described a TNT implementation handling continuous characters with Farris' (1970) algorithms, and papers using continuous characters are no longer a rarity.

9.1.1 Ancestral States, Explanation, and Homology

As many phylogeneticists used to think of character-state reconstruction as mandating discrete states, a number of "coding" methods were proposed (see Section 9.1.3), with the intent of partitioning the continuous variation into a small number of distinct conditions. But parsimony consists of evaluating the best possible set of ancestral states, whether they are qualitative conditions, or real values. One of the critical properties of parsimony (unlike e.g. distances, see Section 3.12) is that all ancestors are realizable—it is always possible for the ancestors to have had the exact values reconstructed via optimization, and these would indeed produce the tree scores calculated by parsimony. That applies to continuous characters as much as it does to discrete characters. An example is in Figure 9.1a, where specific values for the terminals are observed, and values for the internal nodes must be assigned so that the ancestor-descendant differences along all the branches of the tree sum up to a minimum. Figure 9.1a shows an optimal assignment of ancestral values, producing differences that sum up to 10.552; no other assignment of ancestral values can do better than the one shown in Figure 9.1a. Those assignments can be found with Farris' (1970) algorithms (or Goloboff's 1993d modifications). In Figure 9.1a, a single possible value at every node minimizes sum of changes for the given tree, but other cases may be compatible with multiple assignments. An example is in Figure 9.1b, a tree with four taxa scored for the length of an organ, where every one of the species B, F, H or I, has some variability in length and is then scored with a range. For those data, each ancestor can be scored with any value within the range shown in Figure 9.1b. Algorithms for optimization of additive characters find the correct ranges in that case, and any one of the possible intermediate conditions within the range is a valid value. In Chapter 3, the description of Farris' (1970) algorithms glossed over whether the intermediate values of an ancestral range could be any value or only possible discrete conditions,[1] but any real value will do. As discussed in Chapter 3, Section 3.7, individual reconstructions can select only some combinations of values from the ranges for each node; Figure 9.1b shows *ranges* (i.e. the equivalent of *MP*-sets in discrete characters), not *reconstructions*.

In Figure 9.1b, the common node of HI in the tree (B(F(HI))) can be reconstructed with any value between 5.75 and 6.25. Remember that explanations (genealogical or otherwise) consist of why/because pairs. Given the tree of Figure 9.1b, the question of "*why* taxa H and I have values of 5.75 or more?" can be answered with "*because* they are descended from ancestors with at least 5.75". In other words, according to that tree, the length over 5.75 shared by H and I corresponds to a (secondary)

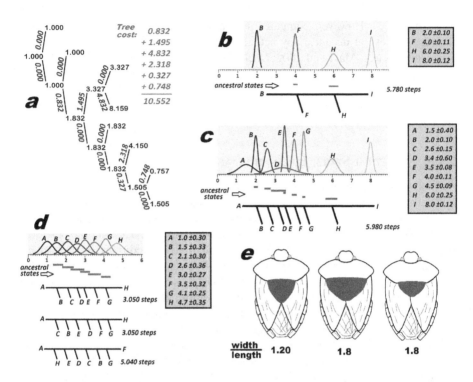

FIGURE 9.1 (a) In the optimization of continuous characters, the values of terminal taxa are observed, and ancestral values are calculated so as to minimize the differences along all the branches of the tree, just as for discrete characters. The ancestral assignment shown produces a total difference of 10.552; no other assignment can produce a lower value. (b) In a continuous character, the variability needs to be considered. The diagram shows the probability distribution of a variable in four taxa. Even considering the variability, there is still no overlap between the values of the terminal taxa. Although the optimal ancestral states are single values for each node, several alternative assignments optimize score, resulting in ranges of values for the ancestral assignments. The tree at the bottom shows the range of optimal ancestral assignments for each internal node. (c) When adding some taxa to the problem in (b), some of the new taxa (e.g. D) do not significantly differ from two distinct taxa (E, F, significantly different). This situation cannot be properly handled by any single-state scoring: D should have the same state as E and F, but E and F should have different states. The significant and non-significant differences can be properly handled only by using ranges. (d) Even when any two adjacent taxa have no significant differences, the character continues providing *some* information on grouping, as some of the trees are less parsimonious than others. (e) When calculating ratios (e.g. scutellum width/length), it may be necessary to distinguish the situation where (relative to the case in the left, ratio 1.20) the numerator grows (middle) from that where the denominator decreases (right, same ratio but for different reasons). Such cases may be best treated with two ratios rather than one.

homology—a similarity inherited from a common ancestor. There is no substantial difference in the way in which trees and hypotheses of common ancestry allow explaining similarities in discrete or continuous characters; the logic is the same.

A peculiarity of the minimization of lineal differences in a continuous variable is that the limits of the intervals at internal nodes always agree with the limits of the intervals observed in the descendants. This holds during either down or up-pass optimizations, where only the intervals for the descendants (and the ancestor, in the up-pass) need to be considered as possible limits. Goloboff et al. (2006) used this property to facilitate reconstruction of internal polytomous nodes (where Farris' 1970 or Goloboff's 1993d algorithms cannot be used).

9.1.2 HERITABILITY AND THE PHYLOGENETIC MEANING OF DESCRIPTIVE STATISTICS

When a character refers to anything that can be measured exactly, no two individuals—conspecific or not—will have the same value, if measured with enough precision. Thus, quantitative characters are intrinsically variable, and descriptive statistics are needed to cope with that variability. This has formed the basis of one of the common arguments against phylogenetic analysis of continuous characters (Pimentel and Riggins 1987; Cranston and Humphries 1988): that descriptive statistics (such as mean or standard deviation) have no cladistic meaning. Depending on what the descriptive statistics can be assumed to represent, the argument can be correct or not. In the case of higher groups, those descriptive statistics are indeed devoid of meaning. What is the point of calculating the average weight of a mammal? Why pooling mice and whales in the same measure? Should the calculation of the average weight of a mammal use every rodent and whale species? Every individual?

The mammalian ancestor must have had some weight, and then in a phylogenetic analysis the weight of a taxon "mammalia" should be that of its most recent common ancestor, if this could be observed (or reconstructed). This is exactly what is done with discrete characters and higher groups. Wings are present in some hymenopterans and absent in others, but a phylogenetic analysis including "Hymenoptera" as a taxon would not use the proportion of hymenopteran species with and without wings. Rather, the ancestral state for wings in Hymenoptera should ideally be used. The same applies to continuous variables; the weight of the ancestral mammal did not change when—millions of years later—some of its cetartiodactylan descendants re-invaded water and increased their size.

The previous examples illustrate the cases where average lengths, sizes, or counts cannot be used in a phylogenetic analysis: when the intra-taxon differences correspond to heritable differences, differences in the genetic makeup of the members of a higher taxon. But the other side of the same coin is that the descriptive statistics become meaningful when the differences in lengths, sizes, or counts correspond to individuals that can be assumed to have the same genetic constitution. That is, when the differences can be assumed to represent a different phenotypical expression, a consequence of the environment only. Therefore, continuous characters and the associate statistics are meaningful when the terminal taxa represent the lowest taxonomic levels.

9.1.3 SIGNIFICANT DIFFERENCES AND METHODS FOR DISCRETIZATION

As noted by several authors, the variability intrinsic to morphometric characters makes it necessary to consider whether the differences observed among terminals are significant (Simon 1983; Chappill 1989). One of the ways in which several authors tried to cope with this problem is by means of discretization methods that would assign the same state to taxa with no significant differences, and different states to taxa with significant differences. Discretization methods initially stemmed from the notion that cladistic analysis requires discrete states, but the problem of significance also played a substantial role in the design of some of them.

Ironically, none of the discretization methods proposed really solves the problem of significance (Farris 1990, 2007)—no method using a single state for each terminal can fulfill at the same time the condition of taxa with significant differences having different states and taxa with non-significant differences having the same state. As example, consider the case shown in Figure 9.1c (the same example of Figure 9.1b with an additional five species). The species D is highly variable, and overlaps with species E and F; for example, an individual with value 3.5 might equally well belong to population D or E, and an individual with value 4.0 to populations D or F.[2] This would suggest that the *same score* needs to be assigned to D–F. However, the populations E and F are themselves significantly different—it is extremely unlikely that an individual with value 3.5 belongs to population F, or an individual with value 4.0 belongs to population E. Therefore, this would suggest that taxa E and F must be assigned *different scores*. With single discrete states, any assignment satisfying both requirements at the same time is impossible. Satisfying the requirement is even more unlikely in methods (such as those proposed by Archie 1985; Thiele 1993; Schols et al. 2004) where the final code depends only on whether the mean falls at a given interval of the whole range of variation (Archie's), or where the intervals are equally sized (Thiele's, Schols et al.'s). And if the intervals are not equally sized (as in Archie's method), the distance between (say) taxon E or F to taxon I will not correspond to the magnitude of their observed difference (as required for appropriate representation of differences, see Section 9.1.1).

The only way to satisfy both requirements (significance and non-significance) at the same time is by assigning D, E and F different *ranges* (Goloboff et al. 2006), so that D overlaps with both E and F, but E and F do not. When the variation within each taxon (=population) approximately follows a normal distribution,[3] using a range from the mean minus one standard deviation to the mean plus one standard deviation will automatically take care of significance (using more than one standard deviation will of course increase the statistical level). These codes are shown to the right of Figure 9.1c. With such coding, taxon D (2.80–4.00) overlaps with E (3.42–3.58) and F (3.89–4.11), but E and F themselves do not overlap. In phylogenetic terms, the width of the range of taxon D means ambiguity in the placement of D, without sacrificing the ability to express differences between the *other* taxa. A tree placing D next to taxa E or F will require no extra steps (D overlaps with E–F), but a tree placing D together with H or I instead will require extra steps (as there is no overlap between D and H or I). A proper representation of the significant and non-significant differences cannot be achieved by using unique states for the terminal taxa (Goloboff et al. 2006).

It is important to keep in mind that the high variability of taxon D, in Figure 9.1c, implies ambiguity in the placement of D, but this does not invalidate the whole character. The overlap between some of the taxa may depend on the high variability of those particular taxa, but that does not detract from the fact that the *other* taxa can be perfectly well distinguished on the basis of the size, length or count (Farris 1990). The possibility that some taxa overlap has led certain authors to propose that, in such cases, the character should be abandoned altogether:

> if no discontinuities in the distribution of observations along a quantitative variable axis are found ... the quantitative variable must be regarded as unsuitable for inclusion in a cladistic investigation. Under these circumstances the variable/character cannot be used for diagnosing a group.

(Humphries 2002: 15)

Consider the example of Figure 9.1d, where the values of adjacent terminals fully overlap; no two adjacent taxa can be distinguished on the basis of this variable. But non-adjacent taxa are significantly different. Therefore, not all trees fit the data equally well. A pectinate tree where taxa split off in the sequence of increasing means (i.e. the top tree in Figure 9.1d, for which the ancestral reconstructions are shown) has the best possible fit to the data, 3.05 steps. Transposing adjacent taxa (e.g. B with C, and D with E, as in the middle tree of Figure 9.1d) still produces the same fit—there are no significant differences between either pair of terminals. But more drastic rearrangements produce an inferior fit, if they imply separate occurrences of significantly different values, as in the bottom tree of Figure 9.1d (with 5.04 steps instead of 3.05). It is clear, then, that the data are informative, and indicate a preference for either the top or middle trees over the bottom tree of Figure 9.1d.

9.1.4 SCALING AND RATIOS

Morphometric characters are not used in a void—real datasets will usually combine them with other characters. This begs the question of the relative costs of transformations between conditions in the different characters, a problem known as *scaling* (often called also *standardization*). This problem arises regardless of whether the other characters are discrete or morphometric. For many quantitative variables, there is no natural unit of measurement; if the character is expressed in millimeters an increase in femur length will cost ten times more than if expressed in centimeters. Meristic characters have natural units of measurement, but they can vary among large numbers—having 95 serrations instead of 90 is a minor change, yet if this difference is expressed directly it would take six individual discrete characters to overcome a group potentially defined by such a subtle difference in number of serrations.

It seems that the problem of scaling does not admit of absolute solutions (Goloboff et al. 2006); even if one were to use specific models of evolution to inform the probabilities of change among different conditions, nothing can guarantee that whatever mechanism determining such probabilities remains constant during evolution. It seems best to view the problem of establishing scale under the same light as other problems discussed in this book: how to best describe observed variation and avoid making some characters much more influential than others *a priori*. Some general guidelines follow. In Section 7.1, the idea of making weights inversely proportional to the number of states in discrete characters was discussed and rejected—a character with more steps does not necessarily exert a stronger influence on the results if those steps result from transitions among different conditions; what determines the influence is the cost of every step. The situation is different with continuous characters. First, depending on the scale used, transforming between the conditions in some of the taxa may cost tens or hundreds of times more—or less—than in discrete characters. A rather natural way to rescale a morphometric character is then to make the maximum difference between any two terminals equal to one step. The rescaling is usually done shifting the values so that the smallest observation equals 0 (this affects none of the results), and then multiplying by an appropriate factor. This, obviously, will make the smaller differences between terminals cost less than in discrete characters—for example, a character representing values of 1.20, 2.50, and 5.00 rescaled to unity (and shifted toward 0) will have states 0.000, 0.342, and 1.000. The difference between 0.000 and 0.342 is much smaller than the difference between two states in a discrete character. This effect is alleviated by implied weighting (Goloboff et al. 2006), because the shape of the weighting curve (see Chapter 7) decreases the costs of additional steps faster if additional steps of homoplasy cost more. Consider the cost of an extra derivation of 0.342 from 0.000 in the absence of homoplasy (and using $k = 6$); this will be $0.342 / (0.342 + 6) = 0.054$, while the difference between no homoplasy and the first step of homoplasy in a discrete character is $1 / (1 + 6) = 0.143$. The ratio of costs under implied weights are $0.143/0.054 = 2.648$, instead of the ratio under equal prior weights $1 / 0.342 = 2.94$. The effect is even more pronounced for successive independent derivations; the fifth independent derivation of 0.342 from 0.000 under implied weights will cost $(4 \times 0.342) / (4 \times 0.342 + 6) - (3 \times 0.342) / (3 \times 0.342 + 6) = 0.040$, while the fifth independent derivation in a discrete character will cost $4 / (4 + 6) - 3 / (3 + 6) = 0.066$. Ratios under prior weights remain the same (2.94), but under implied weights they are now only 1.67. If the dataset includes a substantial number of taxa (thus giving character unreliability the opportunity to manifest as homoplasy), then implied weights will reduce the problem of differences in scale (Goloboff et al. 2006). Mongiardino et al. (2015a), in an empirical analysis, also found that implied weights reduces the problem of differences in scale, although without eliminating it completely—appropriately scaling morphometric characters is still advised.

9.1.4.1 Shifting scale using logarithms

In morphometric characters, a common problem is that some changes of the same absolute magnitude may represent changes of a different *proportional* magnitude. Assume a character describing the number of serrations in an edge varies between 1 and 300. A change from 5 to 10, if measured on a lineal scale, will cost half as much as a change from 290 to 300. However, $5 \rightarrow 10$ amounts to duplicating the number of teeth, and $290 \rightarrow 300$ only increases them by about 3.4%. Rescaling the original variable so that the maximum difference costs one step does not solve this problem; a change $5 \rightarrow 10$ becomes $0.013 \rightarrow 0.030$ ($= 0.017$), and a change $290 \rightarrow 300$ becomes

$0.966 \rightarrow 1.000$ (= 0.034). The problem can be alleviated by applying logarithms to the original values; the original values then become

	Original Values	Natural log		log base 10	
		absol	rescal	absol	rescal
Bass	1	0.000	0.000	0.000	0.000
Tuna	5	1.609	0.282	0.699	0.282
Catfish	10	2.303	0.403	1.000	0.403
Flounder	290	5.669	0.993	2.462	0.993
Mackerel	300	5.704	1.000	2.477	1.000

The absolute values (absol) depend on the base of the logarithm, but if subsequently rescaled (second column for each type of base, rescal) so that maximum difference in the logarithmic rescaled values is the same (unity in this case), the values are realigned and become identical for both bases. Thus, applying logarithms first and then rescaling effectively makes the conversion independent of the base chosen for the logarithm (when smallest original value is 1 and the largest value is $Vmax$, converting to logarithms followed by unit rescaling is exactly the same as converting the original values to $\log_{Vmax}$, effectively providing a non-arbitrary base for the logarithm). For the logarithmic rescaled values, the difference between $5 \rightarrow 10$ becomes much more substantial than the difference between $290 \rightarrow 300$ (0.403–0.282 = 0.121, as opposed to 1.000–0.993 = 0.007). An obvious problem with any logarithmic approach is that it cannot handle values of zero; that is not a problem for actual measurements, but it can affect counts; shifting the original scale in that case will eliminate the zero values.

9.1.4.2 Ratios
Several of the preceding examples referred to the "length" or "size". Evidently, the absolute size or length of many body parts will be influenced by the overall size of the organism. Not that the overall size is an irrelevant variable, but using absolute magnitudes of numerous body parts will introduce overall size as a factor repeated many times in the dataset. To prevent this, lengths of body parts should always be rescaled by absolute size, or whatever part appropriately represents the size; for example, in spiders, the cephalothorax length is generally used as a proxy for "size" (because the length including the abdomen changes with feeding and reproductive stage).

Ratios are also often used to describe simple variations in shape, such as the relation length/width. Two caveats apply here. The first is that, as noted by Mongiardino et al. (2015b), the ratios—and sometimes the results—depend on the (arbitrary) choice of denominator; it may well be that representing some shape as length/width produces results different from width/length. The use of logarithms decreases this problem (if some original values are smaller than 1, the resulting negative logarithms obviously need to be all shifted). The second caveat is that the same ratios may result from either increasing the numerator or decreasing the denominator. As an example, consider Figure 9.1e, showing the shape of the scutum in three insects. The width/length ratio in most species is 1.2; other species have a width/length ratio of 1.8, but this can result either from widening (middle) or shortening (right) the scutum. It seems obvious in this case that two characters are involved, and perhaps it is better to express the observed differences as two ratios, scutum width/total size and scutum length/total size.

9.1.5 SQUARED CHANGES "PARSIMONY" AND OTHER MODELS FOR CONTINUOUS CHARACTERS

Besides minimization of linear differences, other criteria for reconstructing ancestral states in quantitative characters exist. Those criteria have been used almost exclusively to reconstruct ancestral states on given trees. The most common model adduced for a likelihood treatment of continuous characters is Brownian motion (the model that describes e.g. movement of gas particles). It is the only explicit model for continuous characters that has been used to estimate trees on a handful of papers or programs (e.g. the contML program in the Phylip package, Felsenstein 2005; RevBayes, Höhna et al. 2016). In this model, the characters follow separate trajectories after each split in the tree, and randomly increase or decrease their values at any point in time. The time (length of the branch) affects the probabilities, and this model also uses common branch lengths (of dubious application in the case of discrete characters, see Chapter 4; to my knowledge, this assumption was never tested empirically in continuous characters). Felsenstein (2004: 391–410) provides an excellent treatment of the mathematical intricacies of phylogenetic models based on Brownian motion.

Maddison (1991) showed that the state assignments of maximum likelihood under Brownian motion coincide with those calculated by *squared changes parsimony*. The idea in squared changes is to find the ancestral state assignments such that the sum of squared differences between all ancestor/descendant values is minimum. The minimization of squared differences has the effect of spreading change (Maddison 1991; Hormiga et al. 2000) over several branches of the tree; this is so because the sum of several small changes squared is less than a single big change squared. Figure 9.2a shows the behavior that is typical of minimization of squared changes (optimized here with Mesquite, Maddison and Maddison 2008). Taxa A–E have state 0, and taxa F–J have state 5; it seems plain from looking at the data that the best conclusion, given that tree, is that the common ancestor of taxa F–J had an increase $0 \rightarrow 5$, and there is no need to posit change along any other branch of the tree. Such set of assignments implies that every observed similarity can be accounted for by common ancestry, without exceptions. With squared changes parsimony, however, all internal branches have some change (see Figure 9.2a), with the character starting at value 0.041 at the root (instead of 0.000), gradually increasing toward the tip of the tree (but never quite reaching value 5.000 at any internal node). The sum of squared changes calculated by Mesquite is 11.182 (while the sum of squared changes of the reconstruction found by linear parsimony is $5^2 = 25$). According to the squared changes reconstruction, there is some change in every single branch of the tree, even in those branches interconnecting identical taxa. Squared changes may be justified if one assumes a Brownian motion, but it is "parsimony" just in name; only linear parsimony—that is, Farris optimization—is properly called parsimony (Catalano et al. 2010).

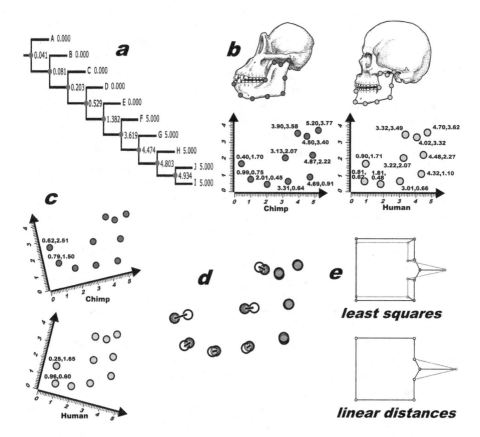

FIGURE 9.2 (a) Squared changes "parsimony" (which minimizes squared differences along branches) produces ancestral values that cannot be justified on the grounds of parsimony. The taxa in clade F–I have value 5.000, and all the other taxa have value 0.000; assigning value 5.000 to all the nodes interconnecting F–I, and 0.000 to all the nodes interconnecting A–E, implies that *all* the similarities in the character are due to common ancestry. Yet, squared changes finds ancestral values where no two taxa are identical by virtue of common ancestry. (b) In geometric morphometrics, comparison between forms (in the example, chimp and human jaws) can be achieved by comparing coordinates of landmarks (well-defined points). (c) Any system of coordinates is arbitrary, yet the "shapes" remain the same. The shapes result from the position of every point, relative to each other point, and this is independent of the coordinate system. The "shape" is preserved when the system of coordinates is translated, rotated, and scaled. (d) Proper comparison between shapes requires that they are translated, rotated, and sized (in their respective coordinate systems), so that the distances between points are minimized. This is known as *superimposition*. (e) Minimization of squared differences between points (instead of linear differences) results in superimpositions with fewer large displacements, and many smaller ones, spreading differences among many landmarks. This is known as the "Pinocchio effect".

The values in example of Figure 9.2a were reconstructed with all the tree branches identical. In squared changes parsimony, or in Brownian motion, changes are more probable (i.e. cost less; see Maddison 1991; Felsenstein 2004) along long branches. The values in the example of Figure 9.2a could have been calculated by making all the branches of the tree of near-zero length, except the branch splitting off group F–J, very long. This would have assigned identical values to nodes interconnecting similar taxa, producing the same ancestral values as linear parsimony. In any real case, however, this begs the question of where the branch lengths are supposed to come from. If branch lengths are optimized for the character alone, they will be optimally very short among identical taxa, and very long along branches where parsimony reconstructs changes—just as it happens with discrete characters when optimized by themselves (Tuffley and Steel 1997; see Chapter 4, Section 4.10.4). Using branch lengths optimized for a molecular dataset on the same tree seems inappropriate, but so is using branch lengths from discrete morphological characters. In empirical datasets, morphological characters do not to share common branch lengths for subgroups of characters (Goloboff et al. 2018a); thus, it seems unreasonable to expect that common branch lengths will be shared by discrete and continuous characters (or even different groups of continuous characters, probably). Therefore, when applying squared changes to reconstruct ancestral values, it is common to use uniform branch lengths (as done in Figure 9.2a).

A modification of the Brownian motion model that has become fashionable in recent years is the Ornstein-Uhlenbeck (OU) model (Martins 1994; O'Meara et al. 2006; O'Meara and Beaulieu 2014). In a standard Brownian model, the mean value remains the same, and the variable can randomly drift far away from that mean (with variance increasing over time). In the OU model, the variable moves around a mean that is the optimal value; as it shifts away from the mean, then it becomes "attracted" back to that value, so that it never drifts too far. This model is only used to map characters on given trees; I am not aware of any paper using it to infer trees. The OU model seems more reasonable, from the biological point of view, than the unrestricted Brownian model—there are of course limits to the magnitudes that quantitative characters can achieve. However, the OU needs to estimate many more parameters, possibly making it inappropriate for proper statistical estimations (Cooper et al. 2016). And if realism is what is at stake, although trait values are expected to tend to their optima, the optima will obviously be different for different taxa. Optimal length neck of an animal with the general morphology and habits of a giraffe is obviously different from optimal length neck of an animal with the general morphology and habits of a rhinoceros. In other words, OU optima must change during evolution. Thus, optima for the individual branches of the tree would need to be estimated, and this would have to be done for each of the quantitative traits included in the dataset, increasing the number of parameters that need to be estimated (easily leading to overparameterization). It is then difficult to see how the OU model could be reasonably used to select among trees.

An interesting modification of Farris optimization that is not based on models is the asymmetric-linear-parsimonious (ALP) ancestral state reconstruction (Csűrös 2008, Didier 2017). For any given assignment of ancestral states, the cost of increases and decreases is factored differently; the optimal assignment is the one minimizing cost, and it obviously depends on the asymmetry parameter. Csürös and Didier being mathematicians, the original papers are complex and do not provide much biological justification or discussion in terms of first principles. It is then unclear whether *ALP* produces reconstructions that can be called "parsimonious" in the same sense as standard linear parsimony optimization. The only application of *ALP* of which we are aware is the paper by Didier et al. (2019). Like the other methods discussed in this section, *ALP* seems applicable only for reconstructing ancestral states on given trees, not to tree inference.

9.2 GEOMETRIC MORPHOMETRICS

Farris optimization is applicable to the evaluation of any character that can be defined with just one magnitude. These variables can be used to capture many aspects of the taxonomic variation, but are less apt for expressing differences in shapes: shapes really vary along two or three dimensions. The best way to quantify subtle differences in shapes is provided by *geometric morphometrics* (*GM*). *GM* can be traced back to D'Arcy Thompson (1917), and took a tremendous impetus in the 1970s and 1980s with the work of F. Bookstein, D. Kendall, and J. Rohlf (e.g. Bookstein 1978, 1991; Kendall 1984; Rohlf and Bookstein 1990). The field originally developed separately from advances in phylogenetic theory, and was concerned mostly with the abstract study of form. Although *GM* is often associated with phenetics, nothing precludes incorporation of shape descriptors in a phylogenetic context. After several controversial attempts to combine *GM* and phylogenetic analysis starting in the 1990s (e.g. Zelditch et al. 1995; González-José et al. 2008), Catalano et al. (2010) finally showed how the parsimony criterion can be applied to shape descriptors.

Below, I provide first a brief introduction to the general ideas of *GM*, then followed by more detailed discussion of the use of parsimony for analysis of landmark configurations. Particularly in terms of phylogenetic applications and *GM*, there have been in recent decades so many publications and conflictive positions (including doubts on parsimony itself) that the amount of background noise is almost deafening; although I try to provide a balanced summary, this section inevitably pays more attention to some voices over others. For a more complete introduction to the general ideas of *GM*, a very good and detailed treatment for biologists is given by Zelditch et al. (2004).

9.2.1 Geometric Morphometrics in a Nutshell

The most common way to enable the mathematical treatment of shape is with the use of *landmarks*. A landmark is a point that can be defined in both of the shapes to be compared. The amount of difference in the two shapes to be compared results from the sum of the differences in positions of the comparable landmarks. Ideally, the landmark should be defined on the basis of anatomical details (e.g. nerve insertions, bone sutures, condyles, etc.). In *GM*, corresponding landmarks are usually termed homologous, but that "homology" is clearly different from "homology" as used in cladistics (and throughout this book); the term *comparable* will be used here. Sometimes, to better describe shapes without discontinuities, multiple points can be used to track contours, edges, or curves between two identifiable anatomical features; these points are called *semilandmarks*. Note that semilandmarks are associated with recognizable traits only indirectly, and their definition is then more arbitrary (or conventional) than for landmarks. The set of landmarks (or semilandmarks) that define a shape is known as a *configuration*; note that at least three landmarks are required to define a flat shape, and four to define a shape in tridimensional space.

Figure 9.2b illustrates the basic comparison of two shapes, for chimp and human lower jaws. The landmarks are positioned on the condyles, processes, and so on. There is thus a one-to-one correspondence: there is a landmark for the condyle, one for the coronoid process, and so on. The relative positions of each landmark are then recorded, on a system of coordinates. The coordinates shown in Figure 9.2b are in two dimensions (2D); exactly the same reasoning can be used to compare tridimensional (3D) shapes, by recording x, y, and z coordinates. The differences between the two shapes can be measured by comparing the positions of each landmark in chimp and human. A problem arises here: the system of coordinates is arbitrary; the axes could be positioned in any way. For example, Figure 9.2c shows the same original shapes, mapped on two different sets of axes. The "shapes" are exactly the same as before, but all the coordinates have changed (Figure 9.2c shows only the coordinates of the insertion of the first incisor and the chin). As the difference in shapes is to be measured by the differences in the landmark positions, the arbitrary element introduced by choice of axes needs to be eliminated. This is achieved by translating and rotating the shapes, until the differences in landmark positions (Figure 9.2d) become minimum. The magnitude of the difference in the shapes is then given by the sum of the differences in position of the individual landmarks used to represent each of the shapes. In Figure 9.2b, both skulls are drawn to approximately the same size, but this need not to be so; when adjusting the two shapes so as to minimize their difference, the sizes must be adjusted as well. The process of adjusting the shape coordinates to eliminate differences in size, rotation, and translation, is known as *superimposition* (Kendall 1984) or *alignment*.

9.2.1.1 Superimposition and criteria for measuring shape differences

The process of translating, rotating, and resizing (one of) the shapes must be done so as to minimize the differences. The optimal superimposition of the shapes will depend on how the differences in landmark positions are evaluated. The two main options are minimizing the sum of squared landmark distances, and the sum of lineal distances.

In standard *GM*, the most commonly used criterion is the sum of squared distances between landmarks, already proposed in early work by Sneath (1967). When multiple shapes must be compared, they can all be superimposed on the same reference shape. The reference can be an observed shape (*ordinary superimposition*), or a summary shape representing the "average" (or *consensus*) of all shapes (*Procrustes generalized least squares* or *GLS*). Gower (1975) and Rohlf and Slice (1990) provided mathematical treatments of *GLS*, showing that the resizing, rotation, and translation needed to minimize the sum of squared distances between landmarks can be calculated analytically. Minimization of squared differences is generally useful when the differences in landmark positions are expected to be normally distributed. In the case of superimpositions, it is well known that *GLS* produces an effect akin to squared changes parsimony, by spreading changes over many landmarks when there is one or a few landmarks with a big displacement—the sum of many smaller distances squared is less than the sum of a few large distances squared. This is widely known as the "Pinocchio effect" (Figure 9.2e, top superimposition). This spreading of changes may or may not be problematic, depending on how the differences between landmark positions can be interpreted. For example, if they can be assimilated to errors, then the spreading is well justified (see Section 9.2.9 for further discussion of minimization of squared changes in phylogenetic approximations).

As further discussed later (Section 9.2.9), the minimization of linear distances is more appropriate for phylogenetic treatment of landmarks. The minimization of linear distances can be approximated by a method called resistant fit theta-rho analysis, *RFTRA* (Siegel and Benson 1982; Rohlf and Slice 1990; Torcida et al. 2014 extend it into 3D). The goal of *RFTRA* is to provide superimpositions that are less sensitive to a large displacement of a few points; this is achieved in practice by using medians instead of means, and comes close to minimizing linear displacements. In a phylogenetic context, it seems preferable to use an explicit minimization of linear distances as criterion for superimposition; this agrees with the general approach outlined in Section 9.2.3. In TNT (Goloboff and Catalano 2016), the linear distances between landmarks are minimized with a heuristic (by translating, rotating, and optionally resizing in small steps, until the optimum is found); this does not guarantee optimal results, but produces very good results in practice. Like *RFTRA*, the alignment minimizing linear distances is insensitive to (i.e. "resists") larger displacements of one or a few landmarks (Figure 9.2e, bottom superimposition).

Another alternative is known as *two-point registration* (sometimes called "Bookstein coordinates", after Bookstein 1986), where the two shapes are aligned so as to get an exact match between two landmarks in particular. This is rarely used, but can be useful when the two particular landmarks represent an important anatomical feature under which the features tracked by the other landmarks are subordinate. In this case, the exact matching of the two points defines a unique position, angle, and size for the shape to be matched to a reference. Once the shape in question is adjusted to the reference, the difference between the two shapes can be quantified either with linear or squared landmark distances. Note that the two-point registration is possible only for 2D points; under 3D, two points do not determine a unique alignment. It is also possible for only two points: an exact matching of three points will rarely be possible.

9.2.1.2 Symmetries

A potential problem when aligning configurations that represent just one side of symmetric specimens is that the rotation can change the axis of symmetry. An option to remedy that problem in 2D space is to align first using a two-point registration (on two landmarks that represent the axis of symmetry), then using a heuristic alignment that does not rotate the shapes. In the case of 3D specimens, the desired rotation is usually forced by doubling and symmetrizing the points on each of side of the plane of symmetry, and then aligning.

9.2.2 PROBLEMATIC PROPOSALS TO EXTRACT CHARACTERS FROM LANDMARKS

Early attempts to use *GM* data in cladistic analysis worked under the constraints of the phylogenetic implementations then available. Therefore, those attempts focused on possible ways to extract qualitative characters (Zelditch et al. 1995) from partial warps, or creating continuous characters (González-José et al. 2008) from scores along main axes of a principal component analysis (*PCA*). Other attempts departed even more from cladistic analysis as outlined in this book, directly using the distances between configurations (e.g. Lockwood et al. 2004) as data.

The partial warps are based on thin-plate splines (Bookstein 1989). The idea in thin-plate splines is to treat the shapes as overlaid on a thin metal grid. The partial warps are calculated as the energy required to deform the grid when converting one shape into the other; they have directions and magnitudes. Zelditch et al. (1995) attempted to convert these values into discrete characters; they received vigorous criticisms (e.g. Rohlf 1998; Monteiro 2000). Although they initially tried to hold their ground (Zelditch et al. 1998), they subsequently (and much to their credit) recanted their position explicitly (Zelditch et al. 2004: 365). Zelditch et al. (2004) then suggested the use of *PCA* scores instead of partial warps. González-José et al. (2008) combined the proposal of using *PCA* scores with continuous characters (which had just been implemented in TNT; Goloboff et al. 2006), and the same did Smith and Hendricks (2013, calculated in this case from Eigenshape analysis, MacLeod 1999, instead of landmark data). *PCA* finds the main axes of variation by projecting a set of points onto a lower number of dimensions; the main axis thus reduces a multidimensional space into a single unidimensional value. The main problem with both partial warps and *PCA* scores is that they imply a reduction (of a 2D or 3D shape) into a single dimension—that is, a "projection". Therefore, using partial warps or *PCA* scores for a phylogenetic analysis, it is not possible to reconstruct the ancestral shapes themselves, only the points at which they would have projected. The worst problem is that two different shapes can project onto the same point, and using partial warps or *PCA* scores for only the main axes of variation cannot distinguish between those two (Monteiro 2000; Catalano et al. 2010). Conversely, it is not possible to reconstruct an ancestral shape from the reconstructed variables; the analysis is no longer a shape-based analysis, but instead one based on one-dimensional variables with no obvious implications in terms of actual shapes.

Although both partial warps and *PCA* scores fail to capture the complexities of the problem in a phylogenetic context, the use of distances (Lockwood 2004) is probably even worse. This will suffer from all the drawbacks pointed out for distance analysis in Vol. 1 (Sections 1.5 and 3.12), while also making it impossible to reconstruct actual ancestral shapes. At the most, one could calculate the distances from observed taxa at which the ancestral shapes would have been located. However, just as in the case of discrete characters, it may well be that the "distances" so calculated cannot result from any actual physical shape—thus leading to a paradox. Catalano et al. (2010: 544) provided a simple example where this happens. On the practical side, Wheeler (2021a) recently proposed that using distance-based trees as starting points for further searches to improve trees with a more appropriate optimality criterion could perhaps save computing time in phylogenetic morphometric analyses. Note, however, that Wheeler (2021a) only refers to a time-saving shortcut; Wheeler

(2021a) still agrees that methods reconstructing actual ancestral conditions (e.g. parsimony) are preferable to distance analyses, and that final results always need to be evaluated from that perspective.

9.2.3 Application of the Parsimony Criterion: Phylogenetic Morphometrics

Catalano et al. (2010) proposed a solution to the problem of how to use landmark data in a phylogenetic analysis, a solution that is based directly on the criterion of parsimony and the idea of selecting trees on the basis of how well they allow explaining similarities in terms of common ancestry. They called this approach *phylogenetic morphometrics*. Most of Catalano et al.'s (2010) ideas have been reviewed in depth (and supported) by Palci and Lee (2019).

In the case of traditional characters, the states in the terminal taxa are observed, and the states in the ancestors are reconstructed so as to maximize the degree to which observed similarities correspond to common ancestry. The possible states that the ancestors can take in traditional characters are limited to just a few possibilities; in the case of continuous characters, there may be a large number of possible values, but the rationale is identical.

The same reasoning can be applied when the terminal taxa have been observed to have different shapes, with each shape represented via their landmarks and the degree of difference between shapes measured by the sum of landmark distances. Consider the example in Figure 9.3, where fishes with two different shapes are arranged in a tree. Given the tree in Figures 9.3a–c, which places the taxa with one shape (C=D, narrower) within a clade and the taxa with the other shape (A=B, wider) outside, the ancestral nodes can be reconstructed to have different shapes. The shapes that must be reconstructed at internal nodes are those that minimize the differences when interconnecting the observed shapes.

In Figure 9.3a all the ancestral nodes are reconstructed as the wider fishes. The implication of such reconstruction is that taxa C and D have not inherited their similar shape from a common ancestor—according to that reconstruction, the shape of C and D became identical by *independent* transformations from an ancestor of shape A=B. The degree of difference between a terminal and its ancestor is measured by the sum of distances between landmarks (i.e. the thin gray lines connecting a gray with a black landmark)—those distances measure the amount of similarity that the reconstruction fails to explain.

Each of the ancestors reconstructed in Figure 9.3a is identical to one of the observed taxa, but this need not be so. That is, the shapes to be considered for ancestral assignments are *all possible shapes*, whether observed or not. For example, in Figure 9.3b, a shape (marked with stars) different from each of the observed taxa has been placed at the common node of taxa C and D. Such a shape implies a poorer fit to the observations—the summed differences between the starred shape and its ancestors, and the starred shape and its descendants, are larger than before. If the landmark coordinates are rescaled so that the individual pairwise comparison of shapes A=B versus C=D has a cost of 1.000, then the summed differences in the reconstruction of Figure 9.3b is 2.485.

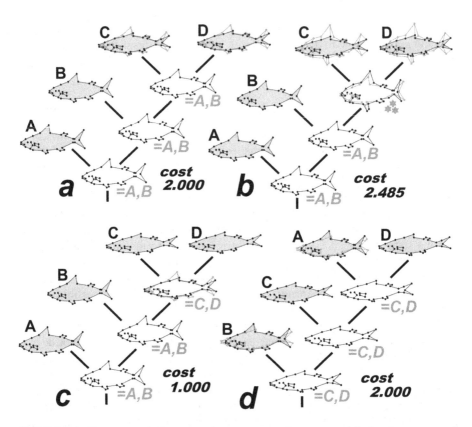

FIGURE 9.3 The ancestral shapes can be reconstructed so as to minimize the amount of differences (i.e. landmark distances) between all ancestors and descendants. In the example, only two distinct shapes (A=B, and C=D) have been observed. In the example, the landmarks are rescaled so that the total difference between these two forms costs unity. Trees (a)–(c) are identical, tree (d) groups the taxa differently. **(a)** Assigning the observed shape A=B to all the ancestral nodes implies that taxa C, D obtained their identical shape independently; this has a cost of 2.000. **(b)** On the same tree, the node common to C, D can be reconstructed as a hypothetical shape; this implies even more differences than the assignment in (a), and a worse cost. **(c)** Reconstructing the ancestors so as to minimize change leads to connect identical shapes with similar ancestors, in a proper application of the parsimony criterion. **(d)** On a different tree, even an optimal ancestral reconstruction (the figure shows one, there can be others with equal cost) implies a higher cost than the best reconstruction possible on a tree that groups similar taxa together [i.e. reconstruction (c)].

In Figure 9.3c the common ancestor of the similar C and D has been reconstructed as identical to its descendants. The only difference in shape, in that reconstruction, is in the branch where a change from the A=D shape into the C=D shape occurs; if size is rescaled as pointed out earlier, the cost of this reconstruction is 1.000. That is the best possible reconstruction, given the tree (A(B(CD))); it is the reconstruction where the differences between ancestral/descendant shapes (summed across all the branches of the tree) sum up to the minimum possible. Note that the criterion of choosing ancestral shapes so that the ancestral/descendant differences sum up to a minimum properly reflects the idea that the common node of two identical taxa must necessarily be assigned the same condition; any shape different from C and D placed at their common node will imply summed differences greater than the minimum possible. This also illustrates why a shape like the one in Figure 9.3b could never have worked—two identical terminals were connected via a different shape. Of course, in other cases when there are more differences among the terminal taxa in the tree and no two terminals are identical, intermediate shapes (i.e. identical to none of the observed taxa) may well be the ones that minimize the sum of differences. Still, the shapes to select for ancestral assignment should by any possible shape, with the target criterion being only minimization of summed landmark displacements.

Note that shapes are reconstructed from the positions of individual landmarks; the sum of point displacements for each individual landmark along the branches of the tree measures the extent to which similarity in the positions of the landmark can be accounted for by common ancestry. That is, parsimony. The tree (A(B(CD))) is compatible with the minimum possible overall displacement; no tree can do better than that, but other trees can do worse. An example is in Figure 9.3d, the tree (B(C(AD))), where the similar shapes are separated. The reconstruction shown in Figure 9.3d is the best that can be achieved given that tree, but (if sizes are rescaled as before) the differences along branches sum up to 2.000, not 1.000 (i.e. twice the difference between the shapes A=B and C=D). The tree in Figure 9.3d allows explaining by common ancestry fewer of the observed similarities, and this is correctly reflected in the criterion of choosing ancestral shapes so as to minimize ancestor/descendant differences.

9.2.4 Shape Optimization in More Detail

The preceding section shows that there is no need to convert the data into a discrete form, or into continuous, unidimensional characters. In phylogenetic morphometrics, the data are the shapes themselves, as 2D or 3D coordinates, and the ancestors are to be reconstructed as optimal shapes. Given that the shape depends on the position of all landmarks, and that the difference between two shapes results from the linear sum of distances between each landmark, it is possible to reconstruct shapes by separately calculating optimal ancestral positions one landmark at a time. Therefore, in more formal terms, the score S for a given set of ancestral landmark positions can be calculated as,

$$s = \sum_L^{\#lands} \sum_i^{\#nodes} dist_{(i,anc_i)}$$

where $dist_{(i,anc_i)}$ is the distance between the point in the node i and the point in its ancestor, for the landmark in question. That distance is simply calculated as the actual, physical distance in Euclidean space, by the Pythagorean Theorem:

$$dist_{(i,anc_i)} = \sqrt{(x_i - x_{anc_i})^2 + (y_i - y_{anc_i})^2 + (z_i - z_{anc_i})^2} \qquad \text{[Formula 9.1]}$$

The optimal ancestral shapes are those that minimize S, and are a byproduct of the optimization of the individual landmarks.

But even if the landmarks can be optimized one by one, their positions are determined by the x, y, z coordinates. This is in contrast with discrete characters—which have different qualitative conditions—and with continuous characters—which can take values in only one dimension. Catalano et al. (2010: 547, their fig. 6) noted that S cannot be calculated by optimizing the coordinates separately; the three values must be considered together for every point. This requires the use of special methods; there is no way to adapt algorithms for one-dimensional variables for this purpose. Goloboff and Catalano (2011, 2016) have developed effective methods to find optimal ancestral positions, for either diagnosis of existing trees or tree searches. The details of the algorithms are beyond the scope of this book; only a brief general discussion is provided here. Of interest is that this approach to evaluate landmarks is a natural extension of Farris optimization into a 2D or 3D space. When the points are collinear, the results of applying this approach are identical to those obtained with Farris optimization, both in terms of the positions of the ancestral points, and the scores.

9.2.4.1   Fermat points and iterative refinement of point positions

Given a single landmark, with a position already determined in the ancestor and the two descendants of a given node, the point that needs to be assigned to the node in question (Figure 9.4a) is the one that minimizes the sum of the lengths of the segments r, s, and t. This is known as the *geometric median*; in the particular case of three points, this point is known as the *Fermat point* (a well-known point in triangle geometry, named after Pierre de Fermat 1601–1665), and can be calculated analytically. For more than two descendants, the geometric median can be found only with heuristics.

One of the methods used by TNT to calculate optimal positions is then an iterative refinement of points. The initial points could be obtained by any means (even randomly, but best results are obtained if initial positions are closer to optimal; see next Section). Each of the ancestral nodes is selected in turn, and the geometric median is calculated given the ancestor and descendants. If the point changes position, then it is necessary to recalculate positions for all the nodes connected to it (i.e. its ancestor and descendants). This process continues iteratively until either no points change their position, or a maximum number of iterations is completed.

9.2.4.2   Using grid templates for better point estimates

The main limitation of the iterative refinement just described is that (just as in discrete characters; Goloboff 1997) the ancestral positions of a landmark on a given tree can easily form local optima (see fig. 3 of Goloboff and Catalano 2011 for a landmark example). Some method, ideally based on tree-traversing is then needed to approximate as much as possible the optimally global positions of the points, before applying the iterative refinement.

The best assignment for a given node in a down-pass traversing of the tree may well result from considering suboptimal points in the descendants. A suboptimal point in one of the descendants may have some extra cost (locally) that is less than the extra distance needed to connect with the point(s) in the other descendant (see

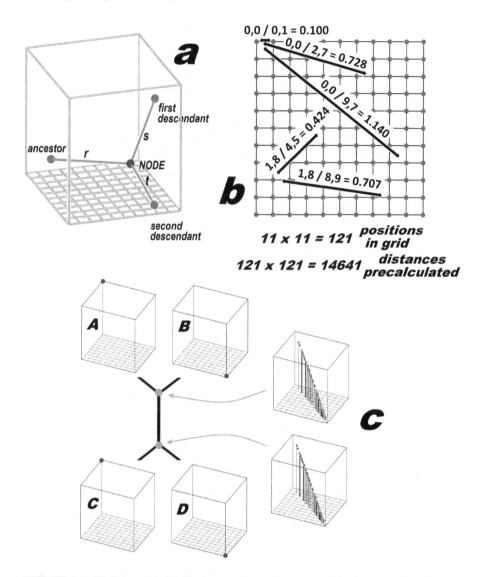

FIGURE 9.4 (a) For a node leading to two descendants, the optimal ancestral assignment (given assignments in ancestor and descendants) is the one that minimizes the sum of segments *r*, *s*, and *t*. This is known as the Fermat point, and can be calculated analytically. Iterating through the branches of the tree allows approximating globally optimal points, by sequentially solving the Fermat point for each node; this iterative process does not guarantee finding globally optimal assignments. (b) Better initial estimates for the iterative process can be obtained by dividing the space with a grid and step-matrix calculations. The distances between points available in the grid are precalculated. Using smaller cells estimates initial point positions with more precision, but needs more time to complete calculations. (c) Just as in the acctran/deltran case, there can be ambiguity in the optimization of landmark positions. By randomly choosing points that appear equally optimal in the initial grid estimation, it is possible to obtain a representation of the ambiguity in the ancestral location of a given landmark. In the example, calculations were repeated 100 times, with different random seeds.

Goloboff and Catalano 2011, their fig. 2). For considering the costs of locally sub-optimal assignments in a down-pass, the method of step-matrix optimization (see Chapter 3) is perfectly suited. The space of possible assignments for each ancestral node is then divided, forming a 2D or 3D *optimization grid*; the assignments will be more or less coarse, depending on the number of cells in the grid. The grid limits are the minimum and maximum values along each coordinate, so as to span the entire range of variation. The distances between all the positions in the grid can be precalculated (as in Figure 9.4b). In a two-pass optimization, a step-matrix algorithm is then used to find the best possible node positions (given the constraints imposed by the grid). As in a standard step-matrix optimization, the cost of placing the point at every position in the grid is calculated for every internal node during the down-pass, and the final positions are calculated in an up-pass.

The points observed in the terminals will usually not agree with any of the points in the grid; they can be added to the grid, and this improves the initial estimation. During the up-pass, if a point can be optimally located at more than one grid position, one of those positions is chosen at random (but see Section 9.2.7.1). Iterative refining (as in Section 9.2.4.1) from the point positions found by the step-matrix optimization produces substantially better results than starting iterations from, for example, randomly placed points.

The more cells that are used for the grid, the better the starting points for iterative refinement. However, the grids are necessarily limited to a relatively low number of cells. Remember that the time needed for step-matrix optimization increases with the square of the number of states (see Section 3.5), and the number of states in the grid will increase with the square of the number of cells in 2D, or the cube in 3D. Because of this, the use of grids for the initial point estimates is much more time-consuming than the subsequent iterative refinement. For 3D landmarks and a grid divided in five parts along each axis, there are $6 \times 6 \times 6 = 216$ "states" (i.e. possible positions); adding the terminal positions would make that even worse. A $6 \times 6 \times 6$ grid still produces substantially better results than arbitrary starting points for the refinement, but it is certainly coarse. A method to improve the initial estimation without using grids with more cells is by *shaking the grid* (Goloboff and Catalano 2011); this consists of re-optimizing a given number of times, randomly shifting the grid position by small amounts. A better way in which the initial estimation can be globally improved is by using *nested grids* (Goloboff and Catalano 2011). With nested grids, for each node to be improved, a new grid of the same number of cells but smaller dimensions is created around (one of) the optimal point(s) found by the initial grid, with a user-specified window (e.g. spanning one of the initial cells). Calculations similar to the ones used in the larger grid then re-position the points with more precision, with just one additional double pass over the tree. Successive nestings can continue improving point positions, but in most datasets the best tradeoff between time and results is obtained with just one or two levels of nesting.

9.2.4.3 Missing entries and inapplicable characters

For the usual treatment using *GLS*, a landmark missing in some specimens is problematic. In geometric morphometrics, those landmarks which can be observed in some specimens but are missing (e.g. destroyed, distorted) in others are often excluded from calculations. In a phylogenetic context, where few and poorly preserved specimens may be all that is available, that is highly problematic. Luckily, the framework of phylogenetic morphometrics as outlined by Catalano et al. (2010), based on linear differences, allows easily accommodating missing entries. Missing entries simply add nothing to the scores; landmark displacements in terminal taxa are calculated only for the observed landmarks.

Character inapplicability is a different question, and it does affect landmark characters. The problem is exactly the same as for discrete characters. For example, a landmark in a part that has been lost in some taxa cannot be properly coded as missing; it is not just that the landmark has not been observed—it will never ever be. For evaluating trees, ancestors are always assigned some position for the landmark, which is exactly what causes problems in discrete characters—this may easily lead to illogical reconstructions, where an ancestor is reconstructed as having a specific position for a landmark placed in a part that does not exist. The framework for inapplicable characters outlined by De Laet (2005) and Goloboff et al. (2021a), based on seeking reconstructions that maximize homology over all the components of the character hierarchy (see Chapter 3, Section 3.11), is not easily translated to landmark characters. The translation is hampered because "homoplasy" (easily defined in discrete character by reference to pre-established categories) can be defined in the case of landmark characters only by reference to score differences. Inapplicable landmarks or configurations, therefore, can be scored as missing entries, but with the understanding that this expedient solution is less than ideal.

9.2.5 Landmark Dependencies, Scaling

The landmark coordinates for a configuration are usually given in arbitrary scales, which can be further modified when aligning (i.e. resizing) the shapes. If the configuration is just to be mapped on a given tree, that is not a problem. The problem arises when the scores representing the fit of the configuration to different trees are to be compared with the scores coming from other characters (or other landmark configurations), and this is to be used for selecting trees. The scores for a landmark configuration need to be placed, somehow, on a common scale with the rest of the characters. Otherwise, the configuration may be either completely uninfluential of the results, or completely overcome all the other characters.

As implemented in TNT, every configuration is treated as "a character". Such a character is complex, determined by several landmarks, and every landmark is in turn determined by their 2D or 3D coordinates. The idea is to make the scores in that character more or less equivalent—in terms of their ability to discriminate among possible trees—to a single discrete character. The need to establish some equivalence is suggested also by the fact that the landmarks used to represent a shape are not wholly independent; their scores cannot each be counted separately as if they were different characters. Furthermore, sometimes the same shape can be represented with different numbers of (semi) landmarks, depending on the desired precision of the representation. If many landmarks are used to represent a given part, then each landmark should be allowed a milder influence on the analysis, as those landmarks are not independent. This obviously means that those landmarks that *are* independent will be weighted less than they should; a possible approach is to treat the landmarks where dependence is suspected and those where it is not as separate configurations.

A reasoning similar to that used for rescaling continuous characters (Section 9.1.4) can be applied here, taking into account the maximum possible differences among observed shapes. For simplicity, TNT calculates the maximum distance between two taxa for each landmark i ($maxd_i$), and sums up these distances over the whole configuration. Rather than changing the actual coordinates, a factor ($1 / \Sigma maxd_i$) is used to multiply the score for the whole configuration so that the maximum possible summed differences cost unity. Note that the maximum difference for a landmark may occur between taxa A and B, and the maximum difference for another between taxa A and C; in such case, the sum of maximum landmark differences may not be achievable for any comparison between observed configurations. This is the default setting in TNT.

With the rescaling just described, the same factor $1 / \Sigma maxd_i$ is used to multiply the score contribution from each landmark. But some landmarks may vary more than others, and a small difference in one landmark that is highly constrained (e.g. that can move only a few mm) may be more important than a larger difference in another landmark with more variability (e.g. one that can move several cm). Another way to rescale the contributions, then, is to take this into account, and make the maximum possible point distance between each of the n landmarks contribute the same toward the overall score of the configuration. This is done by using a separate factor for each landmark i, $1 / (n \times maxd_i)$; this insures that the (sum of) maximum possible differences for the configuration are weighted to unity, but now making the maximum possible displacement within each landmark also cost the same. In TNT, this alternative way to rescale the landmark contribution to the overall score is enabled with the `inscale` option.

In some cases, it may be preferable to decompose a configuration in *two*, and let TNT weight each independently of the other (as recommended, e.g. by Palci and Lee 2019). An example is when representing shape with many semilandmarks: each will receive a lower "weight". This approach to weighting landmarks decreases the problem of the arbitrariness in the number of semilandmarks used to represent a shape, but the weights depend on the total number of landmarks in the configuration (whether or not they are "semi"). If a shape is represented by means of some landmarks (i.e. placed on actual anatomical features) and some semilandmarks, increasing the number of semilandmarks will also reduce the "weight" of the landmarks. Therefore, mixtures of landmarks and semilandmarks are good candidates for being considered as separate configurations (=characters).

Alternatively, instead of the factors used to uniformize the landmark contributions that TNT determines from the observed variation in the landmarks, the user can define any factor. It is also possible to rescale the coordinates of the landmarks, changing (resizing) the values effectively stored in memory, so that a unit factor for all landmarks still produces the desired equivalences with other characters (see Section 9.4.2.5). This can be useful for analyses under implied weighting (which depend on the actual scale, just as discussed under Section 9.1.4 for continuous characters).

9.2.6 IMPLIED WEIGHTING AND MINIMUM POSSIBLE SCORES

The evaluation of the different landmarks and configurations can be made under implied weights, so as to take into account the homoplasy, but a few considerations are in order. The first is how to effect the calculations of homoplasy. Operationally, homoplasy can be defined as the number of steps, S (i.e. landmark distances), beyond the minimum possible that any tree can require, S_{min}. Then, homoplasy is $h = S - S_{min}$, and the familiar formula (Formula 7.2) for implied weighting $f = k / (k + h)$ is applied. The minimum S_{min} for each landmark could in principle be calculated separately; this problem is identical to the Euclidean Steiner problem. It is important that S_{min} is not overestimated, because then h would appear to be negative in a tree with the actual S_{min} and Formula 7.2 loses meaning.

To make sure that the minimum is correctly calculated, a separate search can be done for each landmark with TNT (i.e. deactivating all the other characters, and all the other landmarks in the configuration); each search would find the tree with the minimum possible amount of displacement for the landmark, and these values can be stored for subsequent homoplasy calculations. This process could be time-consuming in large datasets, and therefore the default in TNT is to calculate the minimum by choosing the triplet of most different taxa. No tree can have a total landmark displacement smaller than the one in that triplet, and so the requirement that h is never negative is fulfilled. Note, however, that the minimum possible displacement on any tree with numerous taxa will usually be larger than that in the triplet of most distinct taxa; therefore, the use of triplets will generally overestimate the homoplasy of landmark characters.

9.2.6.1 Weighting landmarks or configurations

A landmark configuration is treated as a character, but the character is complex and composed of several landmarks. This leaves two possible courses of action for weighting based on homoplasy. One possibility is to calculate the homoplasy of each individual landmark, and weight each separately according to its own homoplasy. This would take into account that some of the landmarks in a configuration may be more reliably correlated with groupings than others. The (weighted) score of the configuration is then the sum of weighted landmark scores.

The other possibility is to calculate the overall homoplasy of the configuration, $h_{conf} = \Sigma h_i$ (where h_i is the homoplasy of landmark i), then scoring the character with $h_{conf} / (k + h_{conf})$. In this case, all the landmarks in the configuration will receive the same weight. The decision to apply this approach or the separate weighting of landmarks may well depend on the way in which the particular configuration changes. In TNT, it is possible to use one approach for some configurations and the other for others.

9.2.6.2 The minimum ($\Sigma Smin$) may not be achievable on any tree

As the configuration is a character formed by various parts, those parts can be in conflict with each other. In other words, the tree that requires the minimum displacement for landmark A may not be the same tree that requires the minimum displacement for landmark B. Therefore, the value of ΣS_{min} may not be achievable on any one tree. As an analogy in the realm of discrete characters, Goloboff et al. (2021a) also noted that the result from merging several inapplicable characters into a single step-matrix character may lead to a character that, in a sense, contains homoplasy *within* itself.

That no tree can achieve ΣS_{min} is unproblematic when the reliability is being separately calculated for each landmark; some trees will imply a higher reliability of some landmarks, other trees for others, and that is perfectly fine. But when the implied weight is to be calculated for the entire configuration, things are different. The interpretation depends on how the configuration is viewed. The default in TNT is to use the sum of minima, and in this case no possible tree will have $h_{conf} = 0$ for the configuration. But one could also prefer to have $h_{conf} = 0$ when the minimum possible displacements *for the configuration taken globally* have been achieved. The minima for each landmark can be defined for the user to any value, and are not rechecked when weighting configurations globally (if weighting landmarks individually, when finding a tree of S smaller than the S_{min} value calculated or defined by the user, S_{min} is made equal to S). The minimum then could be set for each landmark i to the value of S_i in a tree where the *configuration* has the minimum unweighted score (e.g. searching under prior weights for that configuration alone, and making sure that the inscale option is off). For some landmarks, this will be larger than the minimum achievable on some possible trees, but the overall number of steps ΣS_{conf} for the configuration will be the minimum possible—thus preventing cases of negative "homoplasy" when calculating $h_{conf} = \Sigma S_i - \Sigma S_{conf}$.

9.2.7 AMBIGUITY IN LANDMARK POSITIONS

In this case, it is necessary to distinguish between ambiguity in the observed positions of the landmark in observed terminals, and ambiguity in the positions reconstructed for the landmark in the ancestors. With the present methods, unfortunately, no satisfactory treatment of either kind is easy, especially as a consequence of the high computational cost of calculating ancestral landmark positions—the problem is practical more than theoretical.

The ambiguity observed in the terminals may result from either intrinsic variability (e.g. within the population) or uncertainty in the measurements. As in the case of continuous characters (see Section 9.1.3), the ideal approach would be to represent terminals by defining a radius around each point (e.g. where the point has 90% probability of being located). The computing machinery in TNT could handle this possibility (e.g. by just calculating distance between the two points with Formula 9.1 and then subtracting the radius around the point representing an observed terminal[4]); this treatment of ambiguity requires no conceptual modifications of the approach outlined here, but has not been implemented in practice. The most common approach to deal with intra-taxon variability is using the average shape of the several available specimens to represent the taxon—this represents each landmark with a single point (the 2D or 3D equivalent of scoring continuous characters with a single value). Another alternative to represent taxon variability that can be implemented with current tools is representing the variable taxon as two terminals with the extremes of variation in the matrix, and forcing monophyly of the terminals used to represent variability.

The second problem is the ambiguity in ancestral landmark positions. That ambiguity is usually much less than for discrete characters, as the positions are determined from the geometric medians (i.e. real valued coordinates, with ties less likely). When an internal node leads to two descendants, the ancestral point will be placed at the Fermat point of the triangle formed by the positions of its descendants and its own ancestor. Fermat points of triangles with all angles within 120° are placed inside the triangle, with the lines connecting the point to the vertices of the triangle forming 120° with each other; Fermat points of triangles with one angle above 120° are placed on that same vertex. The ambiguity in placement of ancestral points will result most commonly from nodes leading to a terminal with missing entries and a terminal where the landmark is observed. In that case, TNT chooses by convention the middle point along the line connecting the point in the ancestor and the observed terminal (but any point along the line would do).

As a grid is used for the step-matrix initial placement of points, more than a single grid cell may appear as optimal in that initial phase. In that case, the implementation in TNT randomly chooses one of the equally optimal cells, by default. Because of this, changing the random seed in complex datasets produces slightly different score values (the random seed also determines the sequence of grid shakes). Even in those cases where a single landmark point at a given internal node is truly optimal, the fact that reconstructions with a slightly suboptimal score (found by alternative random seeds) display the point in the near vicinity also speaks to the ambiguity inherent to the reconstruction. Thus, repeatedly reconstructing the landmark with different random seeds and plotting all the positions obtained for each node (as done in Figure 9.4c, with 100 different random seeds) can be used in TNT to gain a sense of the ambiguity in point placements. The case shown in Figure 9.4c is extremely simple (A is identical to C, and B is identical to D, or a typical acctran/deltran case in 3D), so that the point in the ancestors could be placed anywhere along the straight line connecting the two different point positions; repeating the reconstruction with different random seeds does not find a perfect straight line, but is reasonably close. Due to the difficulty of overlaying many points for each of the landmarks in a configuration, this can be done only for one landmark at a time.

9.2.7.1 Coherence in reconstructions of different landmarks

The choice for the equally optimal points in the step-matrix initial estimation is done randomly by default. Although that is rarely a problem, it can produce unusual reconstruction in case of multiple landmarks that are expected to form a relatively continuous line, for example in landmarks placed along an edge, or semilandmarks. In the case of ambiguity in the reconstruction of some ancestor, it may happen that every individual landmark for the ancestor is placed at a different position, as shown in Figure 9.5. An option implemented in TNT to produce better reconstructions is to choose among the equally optimal points produced by the step-matrix estimation so as to maximize (or minimize) the value of a given coordinate (see Figure 9.5).

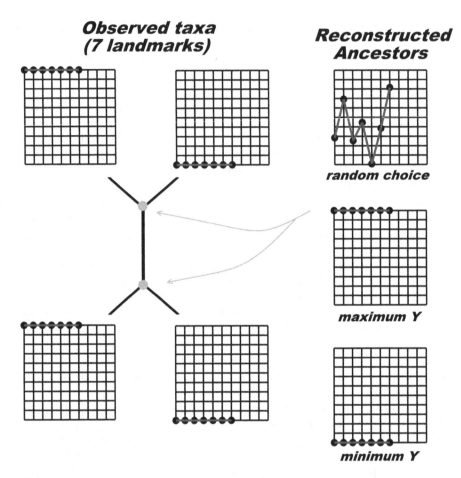

FIGURE 9.5 A configuration with seven landmarks placed along a keel, and a tree with ambiguous ancestral reconstructions. If one of the equally optimal positions in the initial step-matrix estimation is chosen randomly by TNT, the ancestral keel is reconstructed as a zig-zag line (for this dataset, this happens only when using a single thread for analysis; when using multiple threads, the same random seed is used for each landmark, aligning all the ancestral points at the same randomly chosen height). As every landmark is optimized (=reconstructed) independently, the optimality of each of the three ancestral reconstructions is identical, but choosing the optimal position that maximizes (or minimizes) the Y coordinate aligns the landmarks and produces a more appealing reconstruction.

9.2.8 DYNAMIC ALIGNMENT OF LANDMARKS

The process of superimposing shapes is normally carried out before the phylogenetic analysis *per se*. This is done by aligning all shapes against a single reference (e.g. one of the shapes). However, the changes in a configuration are to be understood in the context of a tree, and the comparison—the superimposition—should be done with its close relatives, not with a shape that is far away in the tree (or an abstract average shape). Catalano and Goloboff (2012) showed that in some cases the best scores are achievable only by simultaneously mapping and superimposing. Consider the case of Figure 9.6a, representing the relative positions of three bars, where the configuration has been superimposed by comparing it to taxon B (without sizing). Shapewise, F and A are identical; they will always be placed at the exact same coordinates by any method of prior superimposition. On the tree in question, A and F are not placed next to each other. If every segment shown in Figure 9.6a has a length of 1.00, then the score (total landmark displacements, the gray slant lines) of the tree in Figure 9.6a is 8.00. But, on that tree, taxon F differs from taxon E in the position of two bars, when only one could do if F was properly re-positioned. In Figure 9.6b, the terminal shapes themselves have been translated and rotated as needed to minimize differences along all the branches of the tree. Note that taxa A and F, even if having identical "shapes", are placed on different positions of the coordinate system; the result is that a single bar (instead of the two of Figure 9.6a) is shifted in taxon F, and the total score is 6.00. The shape of F, and the changes that lead to it, are appropriately interpreted in light of its close relatives, not of the distant taxon B.

Catalano and Goloboff (2012) described algorithms for superimposing configurations on a given tree. The superimposition produces a tree score, and this score should be used to select from among possible trees. Different trees will be scored by reference to different superimpositions, each tree being evaluated on the basis of its most favorable superimposition. Note that the size of the shapes cannot be adjusted during this process. In the case of pairwise comparisons, one shape is always taken as reference and used to normalize the size. When dynamically aligning shapes on a tree, there is no reference shape; every branch involves a comparison between a different pair of shapes. If the size was allowed to change freely, smaller shapes would always have smaller distances between landmarks—shrinking all shapes to a point would always produce the best fit. Thus, the shapes must first be aligned and sized against a fixed reference shape (e.g. the outgroup taxon), and only then aligned on a tree.

This criterion of a dynamic alignment is not in conflict with the approach to phylogenetic morphometrics as outlined—instead taking it to its logical consequences. The algorithms described by Catalano and Goloboff (2012), however, are computationally intensive. They can be reasonably applied to evaluating one or a few trees, but they are not natively implemented for actual tree searches in TNT. Tree searches with pre-aligned landmarks in TNT are much faster because they do not optimize anew every rearrangement tried, instead deriving scores from methods akin to the indirect calculation of tree lengths described in Section 5.3.2 (see Goloboff and Catalano 2016 for details); such methods for speeding up searches are not available for dynamic superimpositions on the tree. Analyses with prior superimpositions are then the only practical alternative, to be considered as an approximation to the more appropriate tree superimposition. At the most, one could realign the shapes after obtaining an initial hypothesis of relationships based on a static superimposition and then repeat the phylogenetic analysis. If doing this, keep in mind that the automatic factors to take into account differences in landmark scaling (described in Section 9.2.5) are recalculated after a new superimposition. Therefore, the score for a given configuration before and after the superimposition cannot be properly

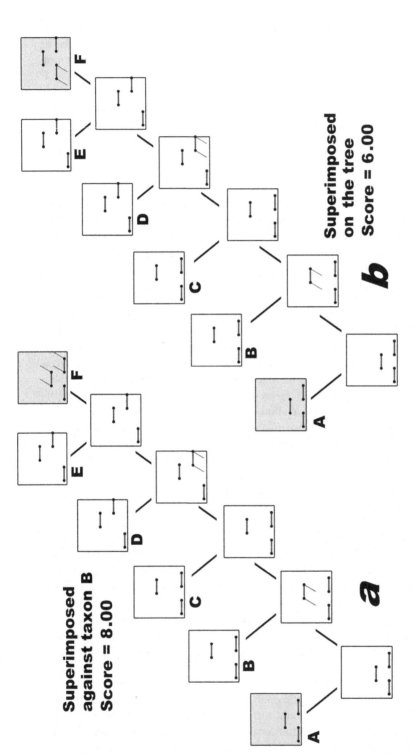

FIGURE 9.6 **(a)** If the shapes are aligned against a fixed reference, taxa with the same observed shape (e.g. A and F) will necessarily be aligned at the exact same position and angle in the coordinate system. On the tree shown, where A and F are widely separated, mapping the landmark as aligned by a prior superimposition produces a cost of 8.00 (the slant lines; assume each segment has a length of 1.00). **(b)** If F is aligned against its neighbor shapes, instead of the distant B, a better score (6.00) can be obtained, by placing F in a different position from A. This implies that the changes in shape between F and its ancestor is best interpreted in the light of local differences, rather than a fixed point of reference. Such alignments can be obtained only on given trees, and they can differ in different trees (e.g. a tree where A and F are sister groups would obviously place them at the same position in the coordinate

compared to each other if using automatic factors (only using fixed unit factors allows appropriate comparisons).

9.2.9   Other Criteria for Aligning or Inferring Ancestral Positions

In addition to the method of reconstructing ancestral shapes by minimizing linear distances between landmark points, a couple other approaches exist to reconstruct ancestral landmark positions. Both Mesquite and MorphoJ (Klingenberg 2011) use squared changes parsimony of the coordinates, with each individual coordinate optimized separately. Cophymaru (Parins-Fukuchi 2018) also treats each coordinate separately, with a Brownian motion model analyzed with a Bayesian Markov chain Monte Carlo approach. These approaches are used often to reconstruct character evolution on given trees, but rarely to infer trees (Mesquite can do primitive tree searches; Ascarrunz et al. 2019 designed their own tree search algorithm under squared changes in an R script).

In a study comparing the results for empirical datasets, Ascarrunz et al. (2019) used a linear minimization (=parsimony, as in Goloboff et al. 2006) of the separate landmark coordinates, each coordinate acting as a continuous character. In another study, this one comparing simulated datasets, Varón-González et al. (2020) also examined the results of parsimony analysis of the coordinates treated as separate continuous characters (which they call "Wagner parsimony"), as well as least squares and what they call "Euclidean parsimony" (e.g. Cavalli-Sforza and Edwards' 1967 "minimum evolution"). The use of parsimony for the individual coordinates treated as continuous characters is highly problematic and very difficult to justify on theoretical grounds. For example, and unlike least squares or phylogenetic morphometrics, it is sensitive to rotation of the coordinate system. The method of "Euclidean parsimony" treats each coordinate as a dimension in a Steiner tree problem, and there is nothing parsimonious about it other than minimizing some quantity. Varón-González et al. (2020) do not discuss the physical interpretation of finding the tree that minimizes the distances in such a hyperspace, and there seems to be none—if the "dimensions" are characters, then some of the resulting tree "lengths" cannot possibly be realized, and "shorter" trees can be less parsimonious on the actual characters.[5]

9.2.9.1   Least squares or linear changes

Separately finding the ancestral values that minimize the sum of squared changes in each coordinate amounts (by the Pythagorean Theorem) to minimizing the sum of the squared point displacements for each landmark. The minimization of squared differences is prone to preferring several small changes over a single large one, resulting in spreading of changes over several nodes of the tree (just like *GLS* does for several landmarks, and squared changes parsimony does for continuous characters). The method for finding optimal positions aside, a Brownian motion model is likely to produce similar results (as there are tight links between Brownian motion and least squares).

An example of separate optimization of each coordinate with least squares is in Figure 9.7, where a "nose" is proportionately larger in three identical taxa (F–H) that form a monophyletic group, and shorter in all successive sister groups (taxa A–E).

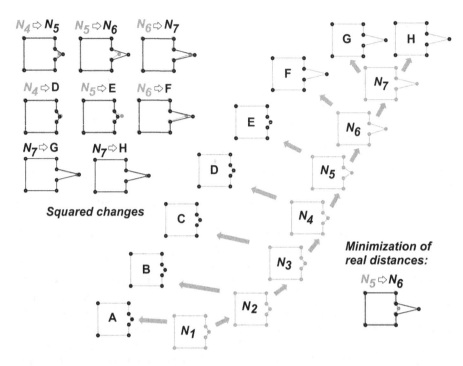

FIGURE 9.7 Separately minimizing the squared differences for the *x*, *y* landmark coordinates amounts (by Pythagoras) to minimizing the squared differences between landmark positions (thus reconstructing ancestral shapes optimal under least squares). This avoids the need for the computationally expensive reconstruction of ancestral position that minimize linear differences, but spreads change among all the branches of the tree, just as in continuous characters.

Note that the shapes are superimposed minimizing linear distances, so that only the nose landmark differs in position in the terminals (*GLS* superimposition would treat this somewhat differently). The minimization of squared differences in landmark positions reconstructs ancestors so that there is some change along most of the branches of the tree. The common ancestors of identical taxa are different. The common ancestor of the identical F and G (N_6) is smaller than both. Ancestral nodes N_4 and N_5 directly connect the identical terminals D and E, yet the reconstruction from squared changes implies that the nose of N_4 decreases in size along the branch leading to D, while increasing in size along the branch leading to N_5. The branch leading to N_5 had an increase, but from N_5 to E it decreases again. Since terminals D and E are identical, there is no need to postulate any change along the branches connecting them.

Although the spreading effect of least squares on phylogenetic conclusions is widely known, the resulting reconstructions have never been explicitly defended on the grounds of either realism or explanatory power; the only attempt of justification ever offered is by reviling the potential inconsistency of parsimony and appealing to an evolutionary model (Brownian motion) as justification. Those reconstructions can be defended only if one believes that evolution of landmarks indeed follows a model of Brownian motion, but seem indefensible otherwise. In terms of super-imposition, there have been attempts to justify a universal preference for squared changes over linear differences, and those attempts—if correct—could bear on the problem of phylogeny as well. Some of the arguments (offered, e.g. by Rohlf 2001; Klingenberg and Gidaszewski 2010) are clearly irrelevant, such as the invariance of squared changes under rotation (which applies equally well to minimization of linear distances as outlined by Catalano et al. 2010), or the fact that squared changes can be derived directly from a Brownian motion model (when it is the model itself that is subject to question).

Most recently, Klingenberg (2020, 2021) defended the exclusive use of least squares superimpositions, focusing on *GM* but tangentially touching on the subject of phylo-genetic methods. Finding a common ground for discussion of those papers is ham-pered by the fact that Klingenberg's main expertise is not in phylogenetic methods; it goes without saying that my expertise is not in *GM*. To justify the behavior of least squares in the Pinocchio effect, Klingenberg's first argument is that only compari-sons where distance is measured by squared changes form a "shape space" (Kendall 1984) that is well understood. He argues that only quantifying shape differences with squared changes produces spherical spaces for three landmarks, and that measuring shape differences with linear distances could create shape spaces with other forms (e.g. elliptical); for more than three landmarks, shape spaces are "difficult to grasp". He also points out that the statistical properties of the resulting distances are well understood only in the case of least squares, not linear distances. While all that is true in itself, if a measure of distance other than squared differences is of interest on other grounds (e.g. explanatory power), then nothing prevents studying and charac-terizing the form of the resulting shape space, or their statistical properties. This is in fact what some mathematicians have started doing, using close analogs of the linear distance proposed here (e.g. Torcida et al. 2014; González et al. 2019). Klingenberg conveniently avoids citing that work, as well as the computer program implement-ing Torcida et al.'s (2014) methods (RPS, Ferraggine and Torcida 2016, available at https://sites.google.com/site/resistantprocrustes/). Ironically, Klingenberg (2021) is not bothered by the distortion (which he describes himself, Klingenberg 2020: 339, their fig. 5; see also Rohlf 1998: 148, their fig. 1) that inevitably arises from doing approximate least squares calculations not on the spherical shape itself, but instead on a flat plane tangent to the sphere.

Perhaps Klingenberg's (2021) most curious argument to not question the Pinocchio effect of least squares is the idea that "shape changes cannot be ascribed to particular landmarks" (p. 118). He claims that (in a case like shown in Figure 9.7) one cannot know whether the nose landmark moved out, or the remaining landmarks moved back. In his view, preferring to posit movement in few instead of many landmarks is a sort of "parsimony" (nothing to do with phylogenetic parsimony, of course, which has a more specific sense than just "minimization"). He claims that the very notion of a landmark changing position is misleading, and that it is more appropriate to consider

shape changes to happen between the landmarks, rather than at particular landmarks. In a biological context, this means that shape changes result from changes in the tissues between the landmarks, not at the landmarks per se. Shape changes originate because the tissues surrounding the landmarks expand, contract, warp or distort so that they push or pull the landmarks in different directions.

(Klingenberg 2021: 119)

Landmarks are used just to track changes, true, but Klingenberg's distinction still offers no argument for preferring to spread changes "in the tissues" surrounding many landmarks instead of concentrating them on a few places that seem to change. Klingenberg's play on words is even less convincing when one remembers that (in his view) tracking individual landmarks is fine when done in evolutionary studies under a least squares paradigm—tracking individual landmarks becomes meaningless only when used as an argument against the Pinocchio effect. One cannot have it both ways.

Finally, Klingenberg (2021: 118) himself admits that different criteria for super-imposition may be appropriate for different circumstances, when he notes that the preference for minimizing linear differences can perhaps be justified in a phyloge-netic context. While least squares is generally well justified when the differences can be assimilated to errors (e.g. in measurements), that justification is certainly lacking in the case of phylogeny—where the goal is to compare substantially different forms, to seek a tree that helps explain by reference to common ancestry those differences (in form, in landmark positions), and to identify branches where evolutionary nov-elties arise. I therefore conclude that—in the phylogenetic context—no argument from first principles precludes comparing shapes by reference to the linear distances between landmark points.

9.3   CHOICE OF METHOD AND CORRECTNESS OF RESULTS

In a phylogenetic context, only minimization of linear differences—and not least squares—necessarily leads to considering the common ancestor of identical taxa (or landmarks in identical positions) as identical to their descendants. The criterion to defend that idea is parsimony: trees should be compared on the basis of how well they allow attributing shared similarities to common ancestry (Farris 1983). Parsimony is not premised on revealing truth or providing infallible results; scientists of course care about reality, but it is only through an appropriate rationalization of the evidence that we can expect to attain truth. One can sympathize with Klingenberg (2020, 2021), who finds (in connection with the Pinocchio effect) that "judgements about which changes are correct and which ones are mistaken or artefactual" (p. 118) are problematic, that no clear or self-evident criterion to know the true changes is avail-able, and that formal properties must be considered in deciding the appropriateness of different methods. The only problem with Klingenberg's (2021) position is that he is mistaken in thinking that parsimony (=minimization of linear distances) is generally defended on the grounds that it yields historically correct results—it is instead based on measuring the degree to which similarities can be traced to common ancestry, and this is correctly reflected only by linear distances. This purely formal argument to prefer linear distances is never considered by Klingenberg (2021); he only offers a caricature of the objections to least squares in a phylogenetic setting. The problem with the ancestral assignments resulting from least squares (whether in continuous characters, landmarks, or superimposition of shapes) is not that they are historically wrong, it is instead that they do not maximize the degree to which similarities are attributed to common ancestry.

Other discussions of methods (both for continuous or shape characters) focus almost solely on which method is more likely to produce historically accurate results. Brocklehurst and Haridy (2021), for example, are concerned with considering continuous characters as additive, because they think that changes skipping intermediate states may be common in evolution:

> Discussion of whether multistate characters should be ordered stems from two conflicting points of view. On the one hand, ideally the treatment of the morphological characters in the phylogenetic analysis should represent as closely as possible the evolution of the trait in question; on the other hand, researchers want to guard against making potentially invalid a priori assumptions about the evolution of that trait.

(p. 707)

Brocklehurst and Haridy (2021) entirely miss the point of considering continuous characters as additive: to reflect the relative degrees of similarity between the different conditions. That evolution may produce changes between two very different states in just one step is entirely irrelevant; it does not detract from the fact that those two states are very different, and the higher cost of transforming between more different conditions does not forbid such changes in a cladogram—only penalizes them more strongly (see also the discussion in Section 2.10.1). The "two conflicting points of view" of Brocklehurst and Haridy present an insoluble dilemma—thinking about methods in terms of their ability to make sense of observations and their internal coherence is liberating and much more fruitful than expecting magic methods that always get it right.

Several papers have made empirical or simulated comparisons between different methods for analyzing morphometric data. But even when two methods produce similar results in some particular cases (Klingenberg and Gidaszewski 2010; Ascarrunz et al. 2019; Varón-González et al. 2020), choice of method still makes a difference if one of the methods relies on more appropriate logic. An internally consistent method that rests on sound logic provides a better validation of conclusions. Similarly, several papers find that morphometric characters produce relatively poor results when analyzed by themselves, sometimes recovering only 30 to 50% of the expected groupings (Hendrixson and Bond 2009; Perrard et al. 2016; Catalano et al. 2015; Catalano and Torres 2017). Even if it were proven beyond doubt that morphometric characters are poorer indicators of phylogenetic relationships than discrete characters, this still does not mean that they must be completely ignored. A thorough analysis must consider all relevant evidence and find the tree that, collectively, best explains all of the evidence—good and bad. A proper analytical method should be able to identify bad characters and minimize their influence. This is in fact what is often observed when adding morphometric characters to a matrix of discrete characters: they have a minor effect on the results. But the only way to show that they have a minor effect is to add them to the matrix and run them together with the rest of the characters, under a valid inferential criterion. The low influence of morphometric characters should be a consequence of the analysis, rather than a premise.

To conclude, nothing mandates including morphometric characters in a matrix, but mandatory exclusion is even less warranted. Morphometric characters can be measured more easily than discrete characters, and this makes them appealing to many researchers. In some extreme cases, not much evidence is available beyond them, or they represent features of particular interest that cannot be easily discretized or analyzed by other means. They can be included in the dataset, if they are available, and they need to be analyzed with appropriate criteria. This section showed that the notion of parsimony, and the idea of selecting trees based on the degree to which they allow attributing observed similarities to common ancestry, can perfectly well accommodate morphometric characters.

9.4 IMPLEMENTATION IN TNT

Analyses with morphometric characters are mostly transparent to the user. Once the data are read, the usual components of a phylogenetic analysis (e.g. tree-scoring, mapping, tree searches, evaluations of support, etc.) can be carried out with the commands described in other sections. The main differences are in how to read the data into TNT, and a few commands that determine special behaviors or options for the treatment of morphometric characters.

9.4.1 CONTINUOUS (AND MERISTIC) CHARACTERS

Continuous characters can be read into TNT by setting the data format to continuous, with nstates cont, prior to reading the data with xread. The values for each terminal are given (as real values) in the usual form of a matrix. Ranges can be indicated with values separated by a dash, or by giving the mean plus or minus a certain value, using the symbol ± (ASCII 241; some text-editors have problems handling 241, and then you can also use ASCII 177 or a forward slash as a replacement). The data can include continuous and discrete characters, but in that case the two data-types must be in different blocks (each preceded by & and the format specification; see Section 1.11.8; the format specification for a block of continuous characters is cont). If you are going to mix continuous and other types of characters, the continuous characters cannot be preceded by any character of a different format, and you need to specify either nstates cont or nstates 32 prior to the xread. Thus,

```
nstates 32 ;
xread 'Mixed types' 15 4
& [ cont ]
fish   1.5±0.50   2.3-2.5
toad   1.0±0.33   3.4-3.7
frog   2.1±0.10   4.5-5.0
newt       ?      5.6-7.0
& [ num ]
fish   0000010100001
toad   0000200011011
frog   1111112111000
newt   111121112110
; proc/;
```

This dataset defines two continuous characters; in the fish, the first character has a range from 1.0 to 2.0 (the ± convention is useful if you have calculated averages and standard deviations for your terminal taxa). The newt is missing for the first character (this amounts to a range from the lowest to the highest possible values).

You can use any nonnegative value for continuous characters, but TNT internally rescales them between 0 and 65.000, using three decimals (i.e. to save memory, TNT uses only 16 bits to represent interval limits). Of course, transforming from 0 to 65 will cost as much as 65 steps in nonadditive characters. You can further rescale (=standardize) the values with `nstates stand N`, where N is the difference between the smallest and largest values; rescaling to unity is usually advisable. On display (e.g. mappings, lists of ancestral values, branch lengths), TNT by default uses the rescaled values stored internally. You can set TNT to display the original values, with `nstates[` (using `nstates]` instead resets the default, displaying the values in the internal scale of TNT). The conversion between scales is not exact (TNT just stores the necessary factors for each character, rather than the individual values for each cell), so there may be a small loss of precision when interconverting.

You can name continuous characters, but not their "states"; continuous characters are always additive (they cannot be made step-matrix or nonadditive). Implied weights is applicable to continuous characters with the default weighting function, scoring trees with $\Sigma h / (k+h)$, where h is the steps beyond the minimum. User-weighting functions cannot be applied to continuous characters (step costs can be defined only for integer numbers of steps; see Section 7.11), but extended implied weights can.

9.4.2 Phylogenetic Morphometrics

To avoid interference with other commands, most of the options for landmark configurations are handled with only three commands: `lmark`, `lmrealign`, and `lmbox`. These commands are used to change settings for landmarks, and save or display diagrams. Only a brief summary is provided here; more details of implementation can be found in Catalano and Goloboff (2018).

9.4.2.1 Reading and exporting data

Landmark data can be read into TNT only in blocks of their own; the format must be indicated as `[landmark 2D]` or `[landmark 3D]`. As a configuration amounts to a character, pairs (in 2D) or triplets (in 3D) of comma-separated values indicate the landmark coordinates for each terminal; you can specify any number of landmarks (of course, each terminal taxon must have the same number of landmarks for each configuration). A missing landmark is indicated with a single question mark (?); avoid using ?,?,?. If the entire configuration is missing for a taxon, just excluding the taxon from the block will leave all of its landmarks as missing entries. The configuration ends with the block (i.e. placing each landmark configuration in a block of its own) or with the pipe symbol (|). Figure 9.8 is an example; to show how the shapes are displayed, the shapes are not aligned in this example (as all points are collinear, they all completely overlap if properly aligned). Note that the `xread` command specifies that a single character will be read; another landmark character could have been included in the same block, separated from the one shown with the pipe |; that

```
xread 1 4
& [ landmark 3d ]
fish 0,0,0  0,1,0  0,2,0  0,3,0
frog 3,0,0  2,1,0  1,2,0  0,3,0
toad 0,0,0  0,0,1  0,0,2  0,0,3
newt 0,0,0  1,1,1  2,2,2  3,3,3
;
```

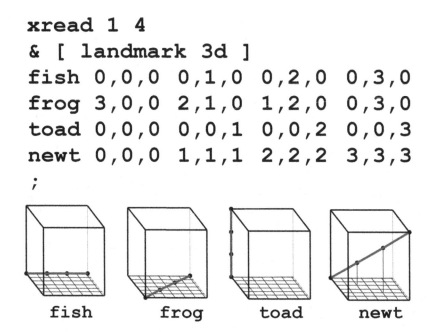

fish   frog   toad   newt

FIGURE 9.8 Example of the format for landmark input in TNT. Note that (to facilitate visualization) these "shapes" have not been aligned (if properly aligned, all taxa become identical).

would have required that xread specifies that the data consist of two characters. By default, when displaying the points, TNT will connect them with lines, to help visualize "form" (see Section 9.4.2.3 for alternatives). Landmark characters can be named with the cnames command, and the "states" for each configuration correspond to names of individual landmarks (these names can then be used, e.g. in lists of landmarks to show or reweight).

A widely used format for *GM* programs is the TPS format. TPS files can be imported into TNT, with the dmerge| option:

```
tnt*>dmerge| outfile.tnt = input1.tps input2.tps input3.tps;
```

TPS files define a single configuration at a time, but the dmerge command allows easily combining several configurations into one dataset (placing every configuration in a block of its own). The resulting file (in this example, outfile.tnt) should be TNT-readable without modifications.

The data can be saved in TNT format with the xread* or xread! options (the former saves all terminals, the latter accepts specification of taxa and characters, and allows saving to a dataset the coordinates for the ancestors; see help xread for details). The data can also be saved in TPS format, with export | filename.tps C T (where C is the character number, and T is a tree number; if T not specified, saves only terminals, if T included, saves the ancestors reconstructed with current settings on tree T).

9.4.2.2 Alignment

The data read into TNT do not need to be aligned, as you can align configurations from TNT itself. The command to realign is lmrealign, or the *Data > Edit data > Landmark alignment* menu option (in Windows). The lmrealign command is followed by the type of alignment: twopoint, rftra, pairlin, or tree. The twopoint option is applicable only to 2D configurations; the two landmarks are specified in parentheses (remember that the first landmark of a configuration is number 0), and then the character to align. For example character 4 aligned on landmarks 0 and 15:

```
tnt*>lmrealign twopoint (0 15) / 4 ;
```

The options rftra and pairlin (i.e. heuristic minimization of linear distances between landmark points) take the reference taxon within square brackets, and then the list of characters to align (by default, all configurations). Thus,

```
tnt*>lmrealign rftra [0] ;
```
or
```
tnt*>lmrealign pairlin [0] ;
```

If pairlin is preceded by !, then TNT also adjusts size heuristically (i.e. by stepwise contractions or expansions of the aligned shape; the size of the reference taxon itself is unmodified).

Finally, the lmrealign tree option aligns character C (e.g. 13) on tree T (e.g. 3):

```
tnt*>lmrealign tree 3 cycles 2 level 2 / 13 ;
```

cycles is the number of complete passes to effect, and level is the strength of each pass (0–4).

9.4.2.3 Scoring trees, displaying, and saving mapped configurations

The lmark command is used to score trees and display landmark mappings. The default option for lmark is to reconstruct ancestors; in Windows versions, it displays a diagram similar to those shown in Figures 9.3–9.8, while in character-mode versions it only displays the scores. The coordinates for the reconstructed ancestors can also be shown, with lmark showhtus (use lmark noshowhtus to display just scores). The scores for whole configurations can also be displayed with the length, score or cscore commands. The lmark command also allows displaying the scores for individual landmarks in a configuration, with the lmark lscores option; lscores is followed by specification of tree(s), character(s), and landmark(s), each separated by a slash.

In the Windows versions, the graphic boxes displaying the configuration for each node of the tree can be saved as a metafile (by pressing "M" while on the pre-view screen). The display of several parts of the diagram can be toggled by pressing different letters: f frame; g grid; c connectors in 3D diagrams; l lines connecting dots to outline shape (=wireframes); s user-defined wireframes; p apomorphies (i.e. the trajectory from ancestral point to point in the node); o overlapping ancestors with their descendants; n no longer displays the boxes for any nodes (left-clicking on a node then displays the box for that node); a displays all ancestral boxes; t displays all terminal boxes. The size of the boxes is increased or decreased pressing + or − (the size of the entire tree is changed with F3–F4, and F5–F6, as in standard graphic trees). In the case of 3D configurations, the keys F1–F2 decrease and increase the tilt of the box, and F11–F12 change the rotation.

   In all versions, it is also possible to save the boxes for the ancestral configurations in SVG format, with the lmbox command. The syntax is:

```
tnt*>lmbox filename.svg T C =N;
```

where T is the tree number (mandatory), C is the character number (mandatory), and =N lists the nodes to save (default=all). The equal symbol can be preceded by options, such as the size of the box, toggles for the elements of the box, viewing angles in 3D, or color type. The elements are toggled with line+cfglops, where letters are as for the Windows diagrams (use line-cfglops to toggle off). Size is changed with size S (default is 120 pixels); the point size and line width are always the same, so a bigger size makes them proportionally smaller. The tilt and rotation (3D diagrams only) can be indicated with tilt B-E/N, and the rotation with rot B-E/N, where B is the beginning angle, E is the ending angle, and N is the number of boxes to display (i.e. different views, each differing in (E-B)/N degrees) for each node. The diagram is by default in color, but can be set to black-and-white with bw or nocolor.

   The wireframes by default sequentially connect all the landmarks in a configuration. Sometimes, the shape may be better outlined by a different set of connectors; these can be defined by the user (their display is toggled with line+s). To define connections you use lmark connect; first the character number is specified, and then the list of connections (i.e. landmark numbers separated by a dash; you can specify multiple landmarks and/or loops, such as 0-3-5-0). Note that the connectors must be defined together (separated by a dash for each new configuration); every time you use lmark connect, previous connectors are discarded and a new set is defined.

9.4.2.4   Settings for estimating coordinates of landmark points

These are the most complex options; while they do not affect the display, they can have a strong influence on the results. Remember that the coordinates (and thus scores) for landmark configurations are calculated by heuristic methods. Therefore, making the heuristics stronger will produce better results (i.e. better approximating optimality), while taking a longer time. The default settings in TNT provide a good tradeoff between time and results in most cases, but you may want to experiment with some alternative parameters for the heuristics. The parameters are changed with options for lmark (summarized in Table 9.1) or (in Windows versions) from *Settings > Landmarks*.

TABLE 9.1

List of TNT Commands and Options for Morphometric Characters

Command & options	Minimum truncation	Action(s)
nstates		
cont	cont	Set datatype to continuous
stand	s	Standardize continuous characters
[		Display values of quantitative characters in the original scales
]		Display values of quantitative characters in the internal scale
lmark	lm	Handle settings and perform landmark optimization
cells	c	Define the number of cells for grid optimization
confsample	conf	For resampling and relative support, consider the whole configuration as the unit
connect	co	Define wireframes
factors	f	Set factors for each landmark/configuration
inscale	i	Equate the contribution of each landmark in a configuration
iter	it	Perform node by node iterative improvement of ancestral states
lscores	ls	Calculate the score for configurations and individual landmark
map	m	Show ancestral landmark configurations (GUI version only)
nest	ne	Define the level of nesting in nested grid optimization
rescale	res	Scale the configurations using the factors as multipliers and fix
termpoints	te	Choose the algorithm to optimize landmark data
threads	t	Number of threads for parallel calculations (used in searches only)
usmin	u	Define minimum score for each landmark (useful for implied weighting)
wts	w	Set weights for each landmark/configuration (use =* to set to unity, e.g. for use in implied weights)
lmbox	lmb	Save landmark box(es) to the specified SVG file
lmreal	lmr	Realign landmark configurations
tree	tr	Using a tree as guide
pairlin	p	Against a reference terminal, minimizing linear distances
rftra	r	Against a reference terminal, using repeated medians

9.4.2.5 Weights, factors, minima

As discussed in Section 9.2.5, a way to take into account the partial dependences between the landmarks in a configuration is to multiply the resulting score by a factor such that the two most distant forms would differ by as much as a step in a discrete character.[6] This factor can be changed by the user to any value, with lmark factor =F C/L, where F is the factor (use F=* to return to the automatic default factors), C is the list of characters, and L is the list of landmarks; making F unity will use as score the physical distances in whatever units were used for input. By default, the factors are defined for the individual landmarks in a configuration, and so that the maximum distance in landmarks where the farthest taxa are more distant contributes more to the score of the configuration. This can be changed with `lmark inscale`, which makes the maximum distance in a landmark where taxa are spread widely to contribute the same as the maximum distance in a landmark where taxa are closer (see Section 9.2.5), thus evening the contributions of landmarks within a configuration (the default is reset with `lmark noinscale`). It is also possible to rescale all the shapes in a configuration by any factor F, with `lmark rescale =F C/L`; using =* uses the current factors, and redefines all factors for the configuration as unity (useful, e.g. for implied weighting).

The minima, needed for calculating homoplasy under implied weights, are determined by default with a triple distance (the summed differences between the three most distant taxa for the landmark). This is usually an underestimation of the minimum (and thus an overestimation of the homoplasy for the landmark). This minimum can be changed with `lmark usmin =M C/L`, where M is the user-defined minimum. Using a value of M larger than the actual minimum may cause a "negative" homoplasy. The best value of minimum is determined by doing a tree search for each landmark, with `lmark usmin =!`. The values of user minima so determined can be saved to a file (they are displayed with `lmark usmin` and no further arguments), for subsequent use (so that a search does not have to be repeated). Keep in mind that the minima change with alignment, so if defined as different from the default triple distance, they must be recalculated if the data are realigned.

When using implied weights, the default is to separately weight each landmark in the configuration according to its own homoplasy. To weight the configuration as a whole (based on the summed homoplasy of all its constituent landmarks), you need to use extended implied weights, defining a weighting set for each configuration to be weighted as a whole (see Section 7.11.2). Remember that to use extended implied weights, you need to have implied weighting enabled at the moment of reading the dataset.

9.4.2.6 Group supports

For resampling and relative Bremer supports, it is also possible to consider either the configurations as wholes, or the individual landmarks. With `lmark confsample`, the resampling deletes (or upweights) whole configurations; in the case of relative Bremer supports, the gain or loss in fit for a tree compared to another (see Section 8.2.1) is calculated for the entire configuration. Obviously, this optional resampling of whole configurations only makes sense when several configurations are included in the dataset. The default is to consider gains and decreases in fit for the individual landmarks (i.e. conflict between landmarks, which can occur within a configuration), or to resample individual landmarks; this is reset with `lmark noconfsample`.

9.4.2.7 Ambiguity

When mapping landmark configurations, and the initial grid estimate produces tied positions for an internal node of the tree, the choice of maximum (or minimum) values along a given axis can be determined with `lmark ambig`, followed by X, Y or Z, in upper or lower case, and in different sequences. An upper case indicates choice of maximum, a lower case minimum; the sequence indicates preference along given axes. Thus, `lmark ambig zXy` will first resolve ambiguity by choosing the smallest value along the z-axis; if ambiguity remains, the largest along the x-axis; and if ambiguity remains, the smallest along the y-axis (any of X, Y or Z can be omitted).

The ambiguity in the determination of the ancestral positions for a landmark can be examined with lmark multimap. This can be done only one landmark at a time. The syntax is lmark multimap R T C L, where R is the number of replicates, T the tree to map, C the character number, and L the landmark. For this, the number of threads must be set to 1 (i.e. lmark threads 1). Every replicate uses a different (random) choice among equally optimal points in the initial grid estimate (and different minor displacements of the grid, if grid shaking set). In Windows versions, this displays boxes of the point positions (e.g. as done in Figure 9.4c); in character-mode versions, it displays the coordinates for each replication.

NOTES

1 Evidently, binary recoding or bitwise representations of ranges as described in Chapter 3 produce only discrete ancestral state assignments. Finding real valued assignments requires a proper implementation, but the algorithms themselves are the same as Farris' (1970) or Goloboff's (1993d).

2 Throughout this chapter, "population" is used in the statistical sense.

3 Remember that the distribution of a continuous variable is expected to reflect different ways to express the same genotype, as far as the variable is concerned. This might make normal distributions more common (e.g. the final product depends on the interaction of many factors), although of course this need not be the rule (e.g. the environment may determine expression as a bimodal distribution). The approach provided here covers only the simpler cases.

4 The distance could become "negative" if the points are too close or the radius is big; it would have to be reset to 0 in that case.

5 An example is a dataset with four taxa (A–D) and three additive characters (indicating dimensions of leg, head, and wings); A 1,3,1; B 2,4,2; C 2,1,3; and D 1,3,2. Only the first character is informative, and determines under parsimony the (unrooted) tree AD|BC, on which the characters cannot evolve with fewer than 6 steps. If every character is considered to be a "dimension" in a 3D space, AD|BC has a "length" of 4.831. In that 3D space, the tree AB|CD is shorter, with score 4.784 and ancestors placed with "coordinates" (character values) 1.003,2.999,1.999 and 1.168,3.166,1.742. An AB|CD tree with internal nodes at 1.003 and 1.168 for the leg character requires a length of 2.165 for the character, rather than the minimum achievable on that tree (2 steps). In such "Euclidean parsimony" the coordinates for each character in the 3D ancestors of AB|CD mutually influence each other; that is, the optimal ancestral assignment for the leg character also depends on the values of the head and wings. The characters (counted individually) cannot have evolved on AB|CD with fewer than 7 steps; the length of 4.784 is an abstraction that cannot be interpreted as measuring minimum possible amounts of evolution on the separate characters. The number 4.784 is instead something with no possible concrete interpretation; it results from counting as "dimensions" three separate characters—apples, pears, and oranges. Ironically, Varón-González et al. (2020) dismiss phylogenetic morphometrics on the grounds that it is a hybrid approach, when the qualifier really applies to their "Euclidean parsimony".

6 Remember that this is done by summing largest distances between landmarks, not configurations; the summed landmark distances may not be realizable for the configuration by any two terminals (as noted in Section 9.2.5 for the maxima and Section 9.2.6.2 for the minima).

10 Scripting
The next level of
TNT mastery

Phylogenetic analysis comprises such a vast and diverse array of possibilities that no program can offer routines to do everything—and if it did, the set of instructions would be huge and unmanageable. Thus, TNT implements built-in commands and options (many of which are described in Chapters 1–9) only to carry out the most common and general operations in a phylogenetic analysis. But in addition to the standard, built-in functions, TNT can be programmed. Programming TNT requires integrating the use of many regular commands, as well as a specific language and syntax. It is not the simplest use of the program, but it is not extremely complicated, either. As one becomes more proficient in using TNT, and explores different aspects of phylogenetic analysis, one is bound to need more sophisticated analyses and routines. With a few rules and conventions (and good doses of patience and imagination), a programming language enables tackling problems that were never considered when designing TNT. Learning how to program TNT is therefore highly advisable (and much easier if one already has even a basic knowledge of some other computing language). One of the annoying problems for external programs (written in another language) that communicate with TNT is the problem of interconnections (e.g. done via output files from TNT, read and understood by the external program). Programming TNT itself avoids that problem, allowing for simpler code contained in just one file, and avoiding issues with communication. In addition, the TNT language is designed to facilitate working with trees as much as possible; in most programming languages, the user himself needs to write routines to extract information from trees—that is unnecessary when using the TNT language.

Just like one learns how to walk by little baby steps, one learns how to program in the same way. Your first scripts can be simple tasks, perhaps only the repetition of some instructions. You do not need to know in detail either the language or every possible regular command of TNT to get started; even basic tools such as loops or conditionals may give you much more flexibility. As you gradually realize the potential and utilities of programming, and start enjoying the ability to control much more tightly what TNT does, you will gradually move onto more complicated routines.

Unsurprisingly, other phylogenetic programs in addition to TNT have provided ways to automate tasks or decisions, the earliest example probably being Physys (Farris and Mickevich 1982; see Carpenter 1988). The current version of PAUP* includes a Python interpreter (although it is unclear how the interpreter connects with PAUP*; other than the existence of the python command itself, the documentation gives no information). Mesquite (Maddison and Maddison 2008) is highly

DOI: 10.1201/9780367823412-10

programmable. PSODA (Carroll et al. 2007) was virtually never used in any published phylogenetic analysis, but it included an elaborate scripting language (Carroll et al. 2009). Nona (Goloboff 1993b) also included its own scripting language, from which TNT's language evolved.

TNT can interpret scripts written in either a TNT-only language, or in C style (Kernighan and Ritchie 1988). C is one of the most widely used programming languages and most programmers know at least its rudiments; it is easy to find documentation, help, and online courses on C programming. TNT's own language is simpler in some regards, with a weaker set of rules; many tasks can be accomplished with less typing than with C; therefore, this chapter focuses on the TNT language. All the scripts used in this chapter, as well as some of the scripts used to produce the figures in this book, can be found at www.lillo.org.ar/phylogeny/eduscripts/scripts _for_TNT.zip.

10.1   BASIC DESCRIPTION OF TNT LANGUAGE

The primary design of TNT is for parsing strings that constitute commands, expected to be alphabetical sequences; anything you type when TNT is expecting a command is compared to the list of commands, and if not found, triggers an error message. The scripting language needs to interact with the normal operation of the program, and therefore it also works by means of commands, using sigils to identify parts of the input with special meaning for TNT's regular commands. A *sigil* is a special symbol not normally found in the repertoire of regular commands and options, used to indicate to TNT that what follows needs to be interpreted in a certain way (e.g. the percentage symbol % is used to indicate arguments, the numeral symbol # is used to indicate values of loops, a dollar sign $ indicates that what follows is a text string, etc.). The *scripting commands* serve to make decisions and control flow (e.g. if, loop, goto), and declare (var) and set (set) the value of user variables. Those commands are recognized only if scripting has been enabled, with the macro command (macro= enables, macro- disables). A *user variable* is a place in memory where you can store the value of some calculation for subsequent use; to facilitate remembering their meaning, user variables can be named (the act of defining and naming a variable is the *declaration*). To assign values to user variables with the set command, you use *expressions*, where you can calculate values using *operators*, such as sums, multiplications, and so on:

```
var: myvariable ;
set myvariable ntax * 10 ;
```

The valid operators and expressions are summarized in Section 10.2.2. Expressions are combination of operators, variables, constants, and commands that the programming language interprets to produce some value, for example ntax * 10. Expressions can always be enclosed in parentheses. When one of the scripting commands finds an opening parenthesis in any context where a number would be expected, the expression within parenthesis is evaluated and substituted.

    User variables provide the bridge between the scripting language and the regular
TNT commands; they can be interpreted as plain numbers, lists, or strings. Once
their value is set, when the name (or number) of a user variable is encountered within
single quotes (ASCII 39),[1] it is interpreted as the number that has been stored in that
variable—exactly as if you had typed it there. For example,

```
var: myrepls ;
set myrepls 10 ;
mult = repl 'myrepls';
```

is the same as if you had typed `mult = repl 10`; the `set` command can be used
to calculate the number of replications to do from other values, and the result—
`myrepls`—is then used as one of the options for `mult`. Through the use of single
quotes, values of user variables are recognized in any context (e.g. as options for
regular TNT command). The only remaining component of the language is *internal
variables*, which correspond to internal states of the program (e.g. number of taxa,
characters, trees). Internal variables are recognized only by scripting commands (i.e.
`if`, `loop`, `set`, etc.), not by regular commands. In any context in which a scripting
command expects a number, encountering the name of an internal variable amounts
to typing the corresponding number. Internal variables themselves are interpreted by
scripting commands as numbers (or series of numbers); thus, their values are not to
be enclosed in single quotes. As a very simple example, suppose you want a script
to run an exact search (`ienum`) if you have few taxa (`ntax` is number of taxa minus
1), or a `mult` command otherwise. Placing the following instructions in a file will
do the trick:

```
macro=;
if (ntax < 15) ienum ;
else mult ; end
proc/;
```

`if`, `else`, and `end` are scripting commands; `ienum` and `mult` are regular commands; `ntax` is an internal variable. Note that the `proc/;` at the end is mandatory
when scripting is enabled (otherwise, finding the end of the file triggers an error).
    Another example of the possibilities allowed by scripting is the use of `loop`.
Remember that the `quote` command (Chapter 1) copies from input to output. As one
output is worth a thousand words, just try the following in the TNT console:

```
tnt*>macro=; loop 1 10 quote ITER #1; stop
```

(the `macro=` command is needed only once; subsequent examples assume scripting is already enabled). As you will discover, `loop` executes repeated instructions,
changing the `loop` value every time. Any TNT command which encounters the #
symbol followed by the number of `loop` (1 in this case; the number indicates the
level of nesting when using loops within loops), will replace the value for the current
`loop` iteration in that position—just as if you had typed it there. The special symbol

(#) indicates to TNT the need to convert into the loop value, so that *any* TNT command can use the value of the loop. For an example more useful than the one just outlined,

```
tnt*>loop 5 15 piwe=#1; mult; nelsen; stop
```

will produce the strict consensus tree (granted: you probably need to search better than with just mult; this is an example only!) for concavities from 5 to 15.

The quotation marks around user variables are also a sigil, used to indicate to the TNT parser that what follows is a special kind of thing—a number, from a user variable. For example, imagine you want to do ten times as many mult replicates as taxa (minus 1) in the matrix, or ntax*10; you cannot type

```
tnt*>mult = replic ntax*10 ;                    [example of bad syntax!]
```

because ntax is an internal variable of TNT, recognized only by scripting commands. But you can save to user variable number zero the value of ntax, and then use that value within single quotes as one of the options for mult:

```
tnt*>set 0 ntax*10 ;
tnt*>mult = replic '0';
```

Scripts—being routines of some complexity—are generally placed in files. Most scripting commands can be executed from the console itself exactly as from within a file. Only if behaves differently if run from the console: whenever the condition for an if typed at the console is not fulfilled (and the if itself is not within a loop), TNT displays a message and interrupts execution rather than expecting the user to type the end matching the if. Scripting commands differ from regular TNT commands in that they cannot be truncated; some keywords (continue, end, stop) and internal variables are case sensitive (recognized only in lower case).

A useful feature of scripts is that they allow the use of *comments*. These serve to clarify the operations (with time, you will no longer remember the details of old scripts). Comments begin with /* and end with */. No nested comments are allowed; comments can contain anything except the terminating */.

BOX 10.1   RUNNING SCRIPTS

Scripts contained in files can be run with the run command, which also takes arguments; if you name the file with the extension *.run (e.g. myscript. run), then just typing the name of the file without the extension will open the file, if present in the current directory, and read arguments (i.e. myscript arg1 arg2; the list of arguments always ends with a semicolon). In this way, scripts effectively function as new TNT commands.

10.2   THE ELEMENTS OF TNT LANGUAGE IN DEPTH

10.2.1   GETTING HELP

With the help+ option and no more arguments, TNT provides basic help for scripting. This lists all the scripting commands (i.e. recognized whenever TNT expects a command), and the full list of internal variables (recognized only by scripting commands). Help+xxx provides help for internal variable xxx. Help on any scripting command is obtained in the same way as for regular commands, with help command or with command ?.

10.2.2   EXPRESSIONS AND OPERATORS

Any expression amounts to a number, and can be enclosed within parentheses. In addition to the obvious arithmetic operators (+−*/), TNT allows numerical and logical comparisons. Valid comparisons (where a and b can themselves be expressions) are:

a == b	is a equal to b?
a != b	is a different from b?
a > b	is a larger than b?
a >= b	is a larger than or equal to b?
a < b	is a smaller than b?
a <= b	is a smaller than or equal to b?
a && b	value is 1 if a and b are true (i.e. not zero), 0 otherwise
a || b	value is 1 if a or b are true (i.e. not zero), 0 otherwise
!a	value is 1 if a is false (i.e. zero), 0 otherwise (! is read as "not")

All comparisons produce either value 0 (if false) or 1 (if true). Bit operations are a & b (bit anding, a bit is 1 if 1 in both a and b), a | b (bit oring, a bit is 1 if 1 in either a or b, or both), and a ^ b (exclusive bit oring or x-oring, a bit is 1 if 1 in a or b, but not in both). When a number is expected by a scripting command, curly braces enclosing one (or more) numbers are equivalent to the set of bits; for example {1 3 7} is the number 138 (i.e. $2^1 + 2^3 + 2^7$). The ability to handle bit sets is important because most parsimonious state sets in discrete characters are bit sets (i.e. a single "number" actually defines a "set" of numbers: 138 is the set of 1, 3 and 7).

Compound expressions are possible, with precedence always evaluated from left to right (the parser simply evaluates as it goes, including logical comparisons). To delimit subexpressions, you must use parentheses. For example,

```
if ( ('i' < 10) && ('j' > 20) )
```

will execute the subsequent code only when the number stored in variable named i is less than 10 *and* the number stored in variable named j is more than 20. It is meaningful code. If you had written instead

```
if ('i' < 10 && 'j' > 20)
```
                                                      [example of bad semantics!]

the whole expression will *always* evaluate as false: the part `'i' < 10 && 'j'` can take value 1 if the number stored in `i` is less than 10 and the number stored in `j` is more than 0, or 0 otherwise, because the logical operator `&&` compares all that is to its left with the following number. Either result (0 or 1) is then compared against 20—and always smaller. Syntactically correct, but useless. Be careful with how you delimit subexpressions; use parentheses liberally.

10.2.3 FLOW CONTROL

10.2.3.1 Decisions

Decisions are done with `if`. Every `if` must be matched by an `end` or by an `else`, every `else` must be matched by an `end`. The `if` is followed by an expression, always in parentheses. The two possible forms are,

```
if (condition)
    ... code to execute when condition true ...
    end
```

or alternatively,

```
if (condition)
    ... code to execute when condition true ...
    else
    ... code to execute when condition false ...
    end
```

keep in mind that in this context, true simply means "different from 0". The code to execute is any valid TNT command. Any number of nested `if`'s (or `if`/`else`'s) can be used, each matched by its own `end`. The condition does not need to be a comparison; it can be simply a number, and if different from 0, the subordinate code is executed—otherwise, it is just ignored. For example, you may have in your script several checks that appropriately change the value of a variable you have named as `correct`, and eventually reach code like:

```
if ('correct')
    ... code to execute when everything is right ...
    else
    errmsg Something went wrong!;
    end
```

This will work as long as variable `correct` is assigned 0 when failing a check, or any number different from 0 otherwise. The `errmsg` command displays the message (all the way to the semicolon, as in `quote`), closes all input files, and interrupts execution.

10.2.3.2 Loops

A `loop` executes repetitive actions. To execute a `loop`, you specify the initial and final values, the increment, and the name of the `loop`. The full syntax is,

```
loop =name from+increment to
   ... actions or decisions to repeat ...
   stop
```

If name is defined, the value of the loop is accessed with #name. The =name can be omitted, and then the loop is unnamed, and can be accessed only by its nesting level, as #1, #2, etc. The values from and to are mandatory. If increment is undefined, it is automatically set to +1 (when to > from) or –1 (when to < from, and then loop is decreasing). If increment is defined, it sets the increase of the loop in every iteration; when the increment is a positive number, nothing is executed if to < from, and when the increment is a negative number, nothing is executed if to > from. Within the loop, the following commands are recognized: endloop (terminates), continue (moves onto the next cycle), and setloop N (resets loop to value N and continues running). Nested loop's are possible, and each must be matched by its own stop. Try:

```
tnt*>loop 1 5 loop 10 20 quote #1-#2; stop stop
```

Nesting level of loops is specific for an input file (TNT allows reading nested input files; numbering for loops starts at 1 for every file). If you need a file yyy.run to access a value of a loop defined in another file xxx.run (from where yyy.run was called), then the loop in xxx.run must be named—otherwise, it is inaccessible from yyy.run. By default, up to 10 levels of loop nesting can be used—this can be increased with the macro*L V option, where L is the nesting level, and V is the maximum number of user variables (this must be done prior to enabling scripts).

10.2.4 ARGUMENTS

An argument is any string or symbol passed to the script when invoking it with the run command, or typing the name of the script if it has the *.run extension. From inside the script, the argument can be accessed with the % symbol, followed by the number of argument; first one is %1. The name of the script itself is %0 (depending on how you handle this in your script, it may require that the full name of the file, including path, has no spaces, for otherwise the converted %0 would amount to two or more strings instead of one). As a very simple example, place the following code in a file and call it setknsearch.run (i.e. "set k and search"):

```
─── setknsearch.run ───
macro=;
proc %1;
piwe = %2;
xmult;
coll tbr; nelsen;
proc/;
```

In this way, you can use the script to analyze different datasets (first argument) with different concavities (second argument), without having to type all the commands again:

```
tnt*>setknsearch example.tnt 8;
```
or
```
tnt*>setknsearch zilla.tnt 20;
```

Alternatively, you can use the longer syntax:

```
tnt*>run setknsearch.run example.tnt 8;
```

note that in this case you need to give the full name of the script (i.e. with the *.run extension).

Attempting to convert arguments not given amounts to a blank space; for example %5 is a blank if only four arguments were actually given to the script. The number of arguments within a given script is the internal variable argnumber. For example, if your script needs at least one argument to operate,

```
if (argnumber == 0) errmsg Give arguments please!; end
```

Or, more economically,

```
if (!argnumber) errmsg Give arguments please!; end
```

The conversion after the % symbol needs a number; up to 32 arguments can be retrieved. The conversion is literal; %24 would amount to the second argument concatenated with the digit 4. To read arguments beyond 9, you enclose the number in parentheses. Try placing in a file (say, testargs.run) and running a small script like the one shown in Figure 10.1a, with different arguments and observe the behavior when typing testargs one two three; or testargs nifty trick; at the TNT console.

Because of the way in which the expression parser works, passing a user variable as argument is somewhat counterintuitive. Assume raycharles and rayliotta are two arrays (see 10.2.6.1 for how to declare arrays), and you invoke a script (arrayasarg.run) where you want to display the first five values from one or the other array (first argument). An example like shown in Figure 10.1b will plot the proper values of raycharles, rayliotta, or any other array whose name is passed as argument. Note that 'i' functions as a sort of pointer in the C language; the parentheses are the number of variable that is to be indexed with the subsequent square brackets; the outer quotes are needed to convert the array value.

10.2.5 INTERNAL VARIABLES

The internal variables allow accessing the state of the program: number of taxa, characters, trees, length of trees, homoplasy, number of nodes in a tree, and so on. The full list of internal variables is displayed with help+; if followed by the name of

```
                  arrayasarg.run
var: i ; set i %1 ;
loop 0 4
    quote Value #1 = '( 'i' )[ #1 ]';
    stop
proc/;                                    b
```

```
                  testargs.run
macro=;
var: i;
loop 1 argnumber
    set i #1; quote ARG #1 - %('i');
    stop
proc/;                              a
```

```
                  dolengths.run
macro=;
var: treelengths[ (ntrees+1) ] ;
loop 0 ntrees
    set treelengths[ #1 ] length[ #1 ];
    stop
macfloat 0;
quote TREE LENGTHS:;
quote 'treelengths[ 0-ntrees ]';
proc/;                                    e
```

```
                  lengthfreqs.run
macro=;
var: lenfreq[ 100 ];
set * lenfreq 0;
loop 0 ntrees
    set lenfreq [ length [ #1 ] ]++;
    stop
set * lenfreq /= ( ntrees + 1 );
loop 0 99
    quote #1 steps: 'lenfreq[ #1 ]';
    stop
proc/;                              c
```

```
                  taxlists.run
var: numelems dalist[ (ntax + 1) ] ;
set numelems 0 ;
loop 0 ntax
    if ( isactax [ #1 ] ) continue ; end
    set dalist [ 'numelems' ] #1 ;
    set numelems ++ ;
    stop
macfloat 0 ;
quote 'numelems' TAXA ARE INACTIVE:;
quote 'dalist [ 0 - ('numelems'-1) ]';
            /* we dont know ahead of time
               how many numbers to print! */
proc/;                                    f
```

```
                  multistrings.run
macro=; var : txt[4 2 ] jumble[ 4 ];
set 'txt[ 0 ]' $Mickey;
set 'txt[ 1 ]' $Mouse;
set 'txt[ 2 ]' $eats;
set 'txt[ 3 ]' $cheese;
macseed 0; set jumble randomlist[ 4 ];
quote Original order: ;
loop 0 3
    quote #1- $('txt[ #1 ]');
    stop
quote Random order: ;
loop 0 3
    quote #1- $('txt[ 'jumble[ #1 ]' ]');
    stop
proc/;                              d
```

```
                  singlescript.run
macro=; goto = %0;
goto findcons;
goto nodecount;
proc/;

label findcons
mult; nelsen *; proc/;

label nodecount
tchoose/; tnodes; proc/; g
```

FIGURE 10.1 Example scripts showing how to **(a)** parse any number of arguments; **(b)** pass arrays as argument; **(c)** using internal variables as index for a user variable to be increased; **(d)** store strings in an array of strings; **(e)** printing out a list instead of a single number; **(f)** create lists with a number of elements that is unknown ahead of time; **(g)** write more structured code using goto.

an internal variable, it displays the help for the variable, if followed by a semicolon, it displays the list of all variables.

Table 10.1 shows the main variables that allow accessing trees and nodes. In TNT, all tree nodes are numbered, such that any two identical trees always have their nodes numbered identically; the same node number always corresponds to the same group of taxa in that case. In two different trees, however, the same node number may correspond to different groups of taxa, sometimes making it necessary to establish node equivalencies (shown in Table 10.2). The root of the tree is always the number of taxa; the internal variable root is thus both the root of any tree and the number of taxa (remember that ntax is the number of taxa minus 1).

Several internal variables (shown in Table 10.3) indicate character settings and states. Many of the internal variables (shown in Table 10.4) access general settings

TABLE 10.1

Internal Variables (Expressions) that Define Trees and Their Nodes.

Find out nodes in trees		Lists and numbers of nodes		Comparisons and evaluations	
`anc[T N]`	ancestor of tree T, node N	`deslist[T N]`	list of immediate descendants of node N, tree T	`eqtrees[T₁ T₂]`	are trees T_1 and T_2 equal?
`comnod[T N₁ N₂ ... Nₙ]`	common node of $N_1....N_n$ in tree T	`downlist[T N]`	list of internal nodes of tree T, in a down-pass order	`fit[T C]`	fit of tree T (optional, character C), under implied weights
`firstdes[T N]`	first descendant of node N, tree T	`listsize`	number of values written to array when saving lists	`homo[T C]`	homoplasy of tree T (optional, character C)
`distnode[T N₁ N₂]`	distance (number of branches) between nodes N_1 and N_2 in tree T	`nnodes[T]`	largest node of tree T, minus 1	`ismono[T N₁ N₂ ... Nₙ]`	if $N_1, N_2...N_n$ monophyletic on tree T, common node (0 otherwise)
`getarep[T N]`	a terminal belonging to node N of tree T	`nodfork[T N]`	number of immediate descendants of node N, tree T	`length[T C]`	number of steps of tree T (optional, character C)
`isanc[T N₁ N₂]`	is node N_1 and ancestor of N_2 in tree T?	`numdes[T N]`	number of terminals belonging to node N, tree T	`mono[T]`	does tree T obey constraints for monophyly?
`isintree[T N]`	does node N exist in tree T?	`numclads[N]`	number of possible cladograms for N taxa	`rfdist opts`	RF distances, options as in `tcomp<` command
`sister[T N]`	the sister of node N in tree T. If tree polytomous and no more sisters, −1	`tnodes[T]`	number of groups (=internal nodes) in tree T	`score[T C]`	score of tree T (optional, character C)
		`tsize[T]`	number of taxa included in tree T	`sprdiff[T₁ T₂ N]`	number of SPR moves from tree T_1 to T_2 (N=number of replicates)
		`uplist[T N]`	list of internal nodes above node N in tree T, in up-pass order (if no N provided, from `root`)	`symcoeff opts`	symmetric distortion coefficient, options as in `tcomp==`

TABLE 10.2
Internal Variables (Expressions) that Allow Comparing and Identifying Groups in Trees

eqgroup[T_1 N_1 T_2]	the group of tree T_2 equivalent to group N_1 of T_1 (0=none; -1=contradictory)
gfreq[T N]	frequency of group N of tree T, in all trees in memory
grouplist[T N]	list (array) of terminal taxa that belong to group N of tree T (listsize = # of values)
grptogrp opts	list (array) of group equivalencies, as in tcomp& command
gcomp[T_1 N_1 T_2 N_2]	compare nodes N_1 of T_1 and N_2 of T_2. 0=incompatible; 1=N_1 includes N_2; 2=N_2 includes N_1; 3=equal; 4=disjunct
simgroup[T_1 N_1 T_2]	the group of tree T_2 most similar to group N_1 of T_1 (as in rfreq command)

TABLE 10.3
Internal Variables (Expressions) to Access Settings for Characters

isact[C]	is C active?
isadd[C]	is C additive?
issank[C]	is C a step-matrix character?
iscont[C]	is C a continuous character?
isinfo[C]	is C an informative character?
islmark[C]	is C a landmark character?
maxstate[C]	return largest state of C
maxsteps[C]	return maximum possible steps of C (no C indicated = whole matrix)
minsteps[C]	return minimum possible steps of C (no C indicated = whole matrix)
nstates[C]	number of states in C
weight[C]	weight of character C

TABLE 10.4
Internal Variables (Expressions) to Access General Settings of TNT

argnumber	number of arguments passed to script
getrandom[N_1 N_2]	a random number between N_1 and N_2
eqstring[S_1 S_2]	are string S_1 and S_2 identical?
exstatus	value set with last return command
isinstring[S_1 S_2]	if string S_2 contained within S_1, return position of S_1 where S_2 ends (0 otherwise)
maxtrees	current value of maxtrees
missing	current value of missing (depends on nstates)
nchar	number of characters minus 1
ntax	number of taxa minus 1
ntrees	number of trees minus 1
numbits[N]	number of ON bits in number N
outgroup	current outgroup taxon
root	number of taxa; root of trees
states[C N T]	*MP*-sets for character C, node N (of tree T, T required only if N not a terminal)
stringend[V]	user variable where string starting at variable V ends
stringsim[S_1 S_2]	string similarity between S_1 and S_2 (1 − Edit Cost/length)
time	time since program startup (or since last resettime command)
vtrees	number of trees in vault minus 1

TABLE 10.5
Internal Variables that Return or Operate on Arrays

bremlist	copy to array values of Bremer supports (options as in bsupport command)
freqlist	copy to array group frequencies of (options as in majority command)
freqdlist	copy to array group *GC* values (options as in freqdifs command)
impwtlist	copy to array cost of adding a step to every step of homoplasy, for current *k* value or weighting function
maxval[R N]	return the maximum value of N values in array R (no N, all values in R)
meanval[R N]	return the mean value of N values in array R (no N, all values in R)
minval[R N]	return the minimum value of N values in array R (no N, all values in R)
randomlist[N F]	write N random values to array (F is optional, and indicates first value)
sortlist[R N]	write to array the sorted indices of values in R (N is optional, and indicates number of values to sort)
stddev[R N]	return the maximum value of N values in array R (no N, all values in R)

and options. Others (shown in Table 10.5) return several values or operate with multiple values in arrays. The former must be copied onto arrays (instead of single valued-variables), such as bremlist, randomlist, or impwtlist; the internal variable listsize tells you the actual number of values written to the array in the last call to a multiple-state internal variable. The latter return a single number, calculating averages (or other statistics) using arrays as input, such as maxval, minval, meanval, or stddev.

10.2.6 USER VARIABLES

The var command serves to declare user variables, as well as listing the variables declared and their values. Var without arguments just lists all declared variables. The easiest way to inspect values in a variable (say, icella, for "integer cell A") is by typing var icella; (the semicolon at the end is required). Using instead var/ icella; displays the bit sets of the variable (or every bit set, if an array). To present output in a cleaner way you can use formatted output (see Section 10.3.4).

10.2.6.1 Declaration

User variables are declared with the var: option. The declaration ends with a semicolon; several variable names can be declared together. Variables for single values are declared just by name, and arrays are declared by indicating the dimensions within square brackets. Obviously, the dimensions can be specified with other user or internal variables. For example:

```
var: i j k n
    raycharles[10] rayliotta[(ntax+1)]
    twodimray[5 6];
```

This declares single numbers i, j, k and n, and arrays raycharles and ray-liotta (the first with 10 cells, the second with as many cells as taxa in the matrix). The variable twodimray has 5 rows of 6 cells each. Indexing of arrays is zero-based.

The maximum number of values that the scripting language can use is by default 1,000; this can be changed with macro*L V (where L, loop nesting, and V, number of values), prior to enabling scripting with macro=. Some of the variables declared in the previous example occupy just one cell (i, j, k); raycharles occupies 10. The array twodimray occupies 35 (five values to specify location of each array in the second dimension, and five times six for each of the five rows or arrays). The number of cells occupied by rayliotta depends on the dataset.

Variables used within a script (if no longer needed) can be undeclared with var -; undeclaring them frees the memory space. With var -; all variables declared within the file are undeclared; with var - varname, only the variable varname and any subsequently declared variables are undeclared. As a script is closed, all the variables that had been declared within the script are automatically undeclared by default, so you do not normally need to worry about undeclaring. The variables declared within a script may be left active after closing (e.g. so that they are accessible from the calling file), with the return= option (needed only in very specific cases and if you know what you are doing; using this may require that you explicitly free those variables later in the script).

A script can access values of variables declared in its calling files, accessing them by name. A caveat here is that the identification of the variables is done by checking the names, from those defined most recently to those defined earliest; using the same name (say, n) for different variables in files xxx.run and yyy.run (where yyy is called by xxx) is possible, but then n in file yyy.run will only refer to the n declared there. You have no way to access the n in xxx.run from yyy.run—this would require that the variable has a different name in xxx.run, a name not used in yyy.run.

10.2.6.2 Assignment

The value of user variables can be changed with the set command; set must be followed by the name of the variable to set (and index, if an array), and the value (it can be an expression). The name (and index, if given) of the variable can be followed by an equal sign, but this is not mandatory (and seldom used). Valid examples of assignments are,

```
set n 25 / 8;
set raycharles[3] length[3];
set twodimray[2 1] getrandom[1 10];
```

this will give variable n the value 3.125, the fourth cell of raycharles the length of the fourth tree in memory, and the second cell of third row in twodimray a random number between 1 and 10 (getrandom is an internal variable, see Section 10.3.11). Note that, in the case of arrays of two or more dimensions, only the individual cells can be set. Attempting to set an incomplete specification of dimensions, like

```
set twodimray [3]   100;                        [example of bad syntax!]
```

would "disassemble" the array and triggers an error message. In `twodimray[3]` there is a value contained:[2] the memory location of the array for the fourth row; if you change this, the value becomes inaccessible, unless you properly "re-assemble" the array yourself (in the unlikely case you *really* want to do this, the error message can be deactivated to let you modify anything you want, not advised unless you are an expert). In the case of one-dimensional arrays, like `raycharles`, setting the array with no index is, although bad practice, allowed (and actually sets the 0th value).

A common error is trying to set the variable using single quotes. For example, if the value 5 is stored in variable n, and you want to change the value of n to 10, code like

```
set 'n' 10;                              [example of possibly bad syntax!]
```

will actually give variable number 5 the value of 10, and leave n itself unmodified. The parser works by matching names of variables with their number (if using the correct syntax, the cell that corresponds to the name n). If you just give the parser an actual number (i.e. `'n'`, as in the previous example) the number is used. You can use this in some cases to write simpler code, but you need to make sure you use it properly.

Instead of setting the variable to a new number, it is possible to change the variable by a specified value, given with an equal sign preceded by an operator. The syntax for this is

```
set n +=        increment n by the value
set n -=        decrement n by the value
set n *=        multiply n by the value
set n /=        divide n by the value
set n &=        bit anding of n and the value
set n |=        bit oring of n and the value
set n ^=        bit x-oring of n and the value
```

Two special cases (`set n++` and `set n--`) require no value specification (increasing or decreasing by one, respectively).

Of course, you can use any valid integer number to index an array; an index can be a # symbol followed by the `loop` nesting level, another user variable, or an internal variable. An example is a script that calculates the length (and frequencies) of all trees in RAM (`lengthfreqs.run`), shown in Figure 10.1c. In the `loop`, the index used for user variable `lenfreq` is the `length` (an internal variable) of every tree, thus increasing the corresponding frequency counter automatically.

The `set *` option allows operating between arrays. With `set * a b + c`, every cell in array a is set to the sum of cells in arrays b and c. Multidimensional arrays can be used (if dimensions are different, only matching dimensions are set). The possible operators between arrays b and c are + − * / & | ^. Operators < and > select the smallest or largest of b and c, respectively. Logical comparisons save value 1 (if true) or 0 (if false) to array a; possible comparisons are ==, !=, ?> (is value in b larger than in c?), ?< (the opposite), ?>= and ?<=. If b or c are numbers (instead of variable names), that constant value is used (to get from a user variable, enclose value within parentheses). If no array c specified, then just copies b to a. An alternative syntax is set * a += b which sums values of cells in b to those in a (other operators that can be used this way are −=, *=, /=, &=, |=, and ^=). This can be used to easily initialize or operate between arrays without using a loop (as done at the beginning of example in Figure 10.1c).

Another useful option of the set command is the sorting of arrays; set! a b sorts array b, writing into array a the indices of b values in increasing order. For example if b = {12, 8, 6, 7}, then a becomes {2, 3, 1, 0}. Both arrays must have the same size, and be one-dimensional. An alternative for sorting is the internal variable sortlist, which allows sorting only the first N values of an array.

When set is used for internal variables that return a series of values (e.g. those in Table 10.5), the user variable provided *must* be an array of the appropriate size. For example, bremlist saves the Bremer support values to the array, with options given exactly as in the bsupport command. Therefore, you can assign:

```
tnt*>set raycharles bremlist ;
```

this is proper syntax, at least as long as raycharles is large enough to contain values for all internal nodes. But you cannot assign

```
tnt*>set n bremlist;                                [example of bad syntax!]
```

because (as declared previously) n is not an array. The same applies to all internal variables that return arrays (e.g. freqlist, freqdlist, impwtlist, randomlist, etc.). In the cases where the number of values written into the array may vary (e.g. freqlist may use different cutoff values, thus producing trees with fewer nodes), the actual number of values written into the array is saved to internal variable listsize (note that any subsequent call to an internal function returning arrays will modify the value of listsize; you may need to save it to a user variable for subsequent use).

The setarray command reads series of values. The syntax specifies the dimensions ($d_1, d_2 \ldots d_n$, separated by commas), then the name of the variable, and $d_1 \times d_2 \times \ldots \times d_n$ values, ending with a semicolon. This can be used to easily enter data into a script, in the form of a matrix. The var command can save to the log file the values of an array of name name in setarray format, with var name *. Finally, besides the set and setarray commands, it is possible to save values to arrays from the tables where TNT reports values of any type of calculation—see Section 10.3.5.

Aside from writing numerical values in user variables, it is possible to write strings. An individual value of a user variable (of type double) can hold several letters or symbols (of type char). To specifically set a variable as a string, you use the set command, followed by the name of the variable, and the dollar sign, $. Everything between the dollar sign $ and a terminating semicolon is copied onto the string (to copy a semicolon onto the variable itself, use a dot followed by a comma, or the ASCII conversion option; see Section 10.3.4). The values in the variable are accessed as a string by preceding the name of the variable with the dollar sign (see Sections 10.2.6.3 and 10.3.4.1 for further details on how to access). Thus,

```
tnt*>var: txt[4]; set txt $Mickey Mouse eats cheese;
tnt*>quote The string is: $txt;
```

produces as output:

```
The string is: Mickey Mouse eats cheese
```

As both strings and numbers are merely series of bits, a value can be output in either format—of course, outputting what has been set as a number as if it were a string will not produce any meaningful output, and vice versa.

Keep in mind that the null character ('\0') must always be written to the string, internally, which will use one more space; up to eight symbols can be held in a single cell (blanks count, of course). For strings longer than eight spaces you need to declare the variable as an array. Unlike numerical assignments, TNT does not check whether string assignments exceed the bounds of the declared variables; be careful and make sure you declare arrays large enough to contain the string (otherwise, when writing the string you may be changing values of other variables).

BOX 10.2   ARRAYS OF STRINGS

While unidimensional arrays are usually sufficient for storing strings, sometimes you may need to use arrays of strings. In that case, you can declare the array as two-dimensional, making sure that the last dimension will have enough space to hold the entire string. An example is shown in the `multistrings.run` script (Figure 10.1d). The script just copies and stores each of the words `Mickey`, `Mouse`, `eats`, and `cheese` at a different entry of an array.

The assignment is done as shown in Figure 10.1d. Note that `'txt[0]'` is the correct way to assign a string in an array of strings (and often necessary to access arrays in multidimensional variables), because `txt[0]` refers to the first internal cell of variable `txt`, and the single quotes refer to the value stored at that cell—the cell where the array actually begins.

The strings are then output in the original order, followed by a randomized order (with the `randomlist` internal variable; as time is used to initialize seed—`macseed 0`—every time you run you get a different order of the four words). Referencing uses the same logic as assignment; the interpreter expects a number after the $ symbol; the use of parentheses around `'txt[0]'` is mandatory for accessing strings (it is optional for assignment), to avoid conflict between the $ and ' symbols.

10.2.6.3 Access

The values contained in user variables are accessed by enclosing the name (or number) of the variable within single quotes (ASCII 39). In arrays, the indices are enclosed within square brackets; the closing quote must be placed after the closing square bracket. The parser of TNT simply works by reading characters, one byte at a time, from input. The use of the single-quote sigil is what tells TNT that what follows must be converted into a number—this works, internally, by simply copying a text representation of the number into the input stream. Once converted, the "number" is simply a text sequence to TNT. Because the single quotes indicate to TNT that a conversion must be done, a user variable can be used (within quotes) in any context where a number is expected. As in the case of assignments, you can use any valid number to index an array (e.g. a # symbol, or user or internal variable). Therefore,

```
tnt*>var: n rayconiff[10] ;
tnt*>set n 1 / 3 ; set rayconiff [5] 4321 ;
tnt*>quote Var n='n', ray[5]='rayconiff[5]';
```

produces as output:

```
Var n=0.333333, ray[5]=4321
```

The conversion of variables can be done using different numbers of significant digits; the default number of digits to use when converting is six—for changing the precision, use the macfloat N option (where N = number of significant digits). See Section 10.3.4 for a tighter control of the conversion.

In addition to converting the values kept in user variables into numbers, they can be converted into strings. For this, you need to precede the name (or number) of the variable to convert with a dollar sign. This can also be applied to some special cases (the meaning of which, I hope, will be obvious to the reader): $dataset, $taxon N, $character N, and $state C S; $ttag N displays the string for branch label of node N. State sets can be converted into strings with $bitset, $dnabitset, or $protbitset. Other fixed strings are listed with help set. These fixed strings can only be accessed, not written into (or, not written into except by TNT itself). For options to control string output format, see Section 10.3.4.

BOX 10.3 CONVERTING USER VARIABLES INTO LISTS OF VALUES

In the case of arrays, if the indexing is a range (two values separated by a dash), instead of a single number, the conversion lists all the values in the range. In the case of multidimensional arrays, only the last dimension can be handled in this way. In this way, the conversion effectively works as a *list* of values, rather than a single value. If the second digit is followed by &N, where N is a number, the list will use ASCII character N as separator of the values (this can be used,

e.g. to easily calculate sums of values of an array, using as separator &43, the
+ symbol). This list can be used in any context, like a `quote` command or a
list of elements. An example is in Figure 10.1e. The `macfloat 0` option is used
because otherwise the range conversion of treelengths would use all the significant digits (six, by default) for every value, not needed in this case because tree
lengths are (usually) measured in integer steps. The conversion of individual
values (i.e. a single number as index for the array) uses significant digits only if
needed. The conversion into a range can be used for creating lists (of taxa, characters, trees, etc.) with a number of elements that is unknown prior to running
the script. A simple example is in the script `taxlists.run`, in Figure 10.1f.

10.2.7 Efficiency and Memory Management

Some tips may help improve the efficiency of TNT scripts. As in any piece of code,
avoiding redundant calculations will usually save time. But the particular implementation of the scripting language in TNT also means that other aspects, perhaps not
immediately apparent, can have an effect on the speed. The names for loops and user
variables are stored in internal tables, which TNT scans from last to first defined
when a name is encountered. Naming loops or user variables with short or very
dissimilar names (or not naming loops at all) makes it easier for TNT to fetch the
corresponding value. If user variables that will be assigned or accessed often are
defined last, this will also facilitate finding the name in the table. There is, of course,
a well-defined limit for the speed that can be achieved with scripts. Scripts are *interpreted* (that is, analyzed for meaning by TNT, then executed), unlike actual programs
which are *compiled* and converted into machine instructions (you can easily find
information on the difference between interpreted and compiled code on the inter-
net). The main goal of TNT scripts is ease and simplicity, not efficiency.

Managing memory may be more important than managing speed—improper
memory management may make scripts unusable, not just slow. The maximum
number of values that can be stored as user variables is by default 1,000, but many
routines (e.g. large datasets, many trees) may require storing more than just 1,000
numbers. This can be changed with `macro*L V`, where L is the maximum nesting
for loops, and V is the maximum number of user variables that will be allocated when
enabling scripts. The number V can itself be an expression, using internal variables,
so that you decide how many cells to use based on the dimensions of the dataset. By
default, the macros will use no more than 102 Kb of memory (this is the memory
used to expand loops into instructions, store values and names of user variables, and
a few other tasks). Only 102 Kb may be insufficient, and this can be changed with the
`macro[S` option, where S is the size in Kb. Both `macro*` and `macro[` have to be
done prior to enabling scripting with `macro=`. As an example,

```
macro*10 (ntax + 1 * 10) ;
macro [ (ntax + 1 * 8 * 10 / 100) ;
macro=;
```

this code at the beginning of a script will make sure that as many cells as number of taxa times 10 are available for user variables (`macro*`), that the memory is allocated as needed for those cells (with `macro[`; every cell is 8 bytes), and only then connect the scripting (`macro=`).

Another limit that may need to be changed for some routines is the maximum number of nested input files allowed. You may have a script that calls another script (or a different part of the same script; see Section 10.3.1), which in turn calls another. Also treated by TNT as nested input files are the instructions for some regular (e.g. `resample`, `sectsch`) or scripting commands (e.g. `tbrit`, `combine`). The maximum level of nested input files allowed is by default 10; this is sufficient for most purposes, but otherwise it can be increased (prior to initiating the script) with the `mxproc` command. It may also be necessary to increase the maximum length of taxon names (e.g. branch-legends use this same limit), with the `taxname+` command (see Chapter 1, Section 1.11.5).

10.3   OTHER FACILITIES OF THE TNT LANGUAGE

10.3.1   GOTO

The `goto` option can be used to write more structured scripts, placing the different routines in different files or different parts of the same file. Opening a script file with `procedure` or `run` begins reading the file right from the beginning, but the `goto` command directs parsing to a specific label. When opening with `goto`, the file is scanned for `label` followed by the string given to `goto`, and parsing of commands begins from then on. If you create two files, named as `masterscript.run` and `actor.run`, you can invoke the appropriate code segments in `actor` from `masterscript`:

```
—— masterscript.run ——
macro=;
goto = %0;
goto actor.run findcons;
goto actor.run nodecount;
proc/;
```

```
—————— actor.run ——————
label findcons
mult; nelsen *; proc/;

label nodecount
tchoose/; tnodes; proc/;
```

`Master` and `actor` are placed in two different files in this example; this means you have to copy both files to the same directory for the routine to work; merging everything in a single file is usually preferable. Since you can define a default target for `goto` (so that you do not need to specify the file to parse every time), and `%0` represents the name of the script itself, you can keep everything in the same file, as shown in `singlescript.run` (Figure 10.1g). The example shown is, of course, trivial, but in a real case you can place complex routines at specific parts of the script—anything you would place in a file, can be placed under a `label`, making scripts better organized and more readable. Note that any `label` must be closed with a `proc/` or a `return N` (where N is the "exit status" of the script, subsequently retrievable with internal variable `exstatus`). A published example with extensive use of `goto` for a more structured script is Sterli et al.'s (2013) script for bootstrapping node ages.

10.3.1.1 Handling errors and interruptions

A useful option of the `goto` command is to define a default target file and `label` to go to when an error occurs (including any error messages executed by the script with `errmsg`). The syntax is `goto[filename labelname`. Once this is defined, any subsequent error or user interruption (with `<escape>`) will direct parsing to that specific location. This can be used to handle errors (including out of memory or any internal TNT errors) more gracefully (e.g. resetting general options that were changed by the script, such as collapsing or reporting options, to their original values).

10.3.2 PROGRESS REPORTS

Many scripts perform time-consuming operations. In those cases, it is possible to output a progress bar, with the `progress` command. The syntax is `progress done todo message`, where `done` is the work done, `todo` is the work to do, and `message` is the text of the progress bar (the `message` ends with a semicolon). The report of TNT must be off (with `report-`) to run `progress`. For a cleaner display, you can silence all output, with `silent = all`, prior to entering the time-consuming part of the script. The progress bar is erased with `progress/` (call it when work is finished). For example, the script `reporting.run` (in Figure 10.2a) will find the (first) single terminal which, if removed from consensus, increases consensus resolution the most. The script works by brute force and may take some time for many taxa or many trees. As the script tries different prunings, it uses the `progress` command.

When both `todo` and `done` are 0, the output is a tilting bar (no indication of percentage of work done). The `progress` indicator is updated every second (or more, if called infrequently), and also checks for user-interrupts (i.e. the user pressing `<escape>`). Therefore, a routine that displays a `progress` indicator can be interrupted (and the interruption can be handled gracefully with `goto[`).

10.3.3 HANDLING INPUT FILES

The input files described so far are expected to have commands that TNT will interpret. It is also possible to parse input files to get information from specific parts of the file and handling it via scripting—for example, to retrieve output produced by other programs called by TNT via the `system` command. The command to open these input files is `hifile`. Once you have opened the file, you can skip N bytes in the files with `hifile skip N`, N lines with `hifile skipline`, or N strings with `hifile skips` (i.e. blank-delimited words). In addition to skipping, you can also seek for a specific character (symbol) with `hifile seek C N` (where C is the symbol, also specifiable as an ASCII character number with `&`, and N, number of times to find it), or a specific string with `hifile seeks S N` (where S is the string to seek; if you need a whitespace within the "string", use `&32`; N is the number of times to find it). For both skipping and seeking, if no number of times (N) specified, it skips or seeks one time only.

By appropriately skipping and seeking you can take input from specific parts of the file. If a handled input file is open, then several internal variables return

```
┌──────────── reporting.run ────────────┐
│ silent = all ;                         │
│ report - ;                             │
│ var: mxnod initre bestis ;             │
│ set initre ntrees ;                    │
│ nelsen*;                               │
│ set mxnod tnodes[ntrees]; /*initialize */│
│ set bestis (-1);                       │
│ keep 'initre' ;                        │
│ loop 0 ntax                            │
│    progress #1 ntax Cutting ;          │
│    nelsen * / #1;   /*try a pruning */ │
│    if ( 'mxnod' < tnodes[ntrees] )     │
│       set mxnod tnodes[ntrees];        │
│       set bestis #1 ; /*update value */│
│    end                                 │
│    keep 'initre'; /*discard tried tree */│
│    stop                                │
│ progress/;                             │
│ silent - all ;                         │
│ if ( 'bestis' >= 0 )                   │
│    quote BEST PRUNE: $taxon 'bestis';  │
│    quote Pruning it gives 'mxnod' nodes.;│
│ else                                   │
│    quote No improvement found ;        │
│    end                                 │
│ proc/;                            a    │
└────────────────────────────────────────┘
```

```
┌──────────── readnprocess.run ──────────┐
│ macro=;                                 │
│ hifile open filelist.txt;               │
│ var: i txt[10]; set i 0; set -;         │
│ loop 1 1                                │
│    set txt $$hifstring;                 │
│    if ( hifeof ) endloop ; end          │
│    quote FILE 'i': $txt ;               │
│    p $(txt).tnt ; mult; nelsen;         │
│    set i ++ ; setloop 1 ;               │
│    stop                                 │
│ hifile close ;                          │
│ proc/;                            b     │
└─────────────────────────────────────────┘
```

file filelist.txt must contain a list of data files in TNT format

```
┌──────────── pectinator.run ────────────┐
│ var: to bot;                            │
│ set bot root;                           │
│ loop 2 ntax                             │
│    set to #1 - 1;                       │
│    edit] 0 'bot' #1 'to' ;              │
│    stop                                 │
│ proc/;                            c     │
└─────────────────────────────────────────┘
```

```
┌──────────── easyswaps.run ─────────────┐
│ var: besL this t; set besL length[ 0 ];│
│ report - ; sil = all ; resettime ;     │
│ tbrit 0                                 │
│    progress percswap 100 At 'besL' steps;│
│    set this length [ 0 ] ;             │
│    if ( 'this' >= 'besL' ) continue ; end│
│    set besL 'this' ;                    │
│    resetswap ;                          │
│    stop                                 │
│ progress/; sil - all ; set t time ;    │
│ quote BEST LENGTH: 'besL' ('/.1t' sec.);│
│ proc/;                            d     │
└─────────────────────────────────────────┘
```

```
┌──────────── reconstructions.run ───────┐
│ var: listat[(nnodes[0]+1)] numrecs ;   │
│ set numrecs 0 ; ttag * 0 ; report - ;  │
│ iterrecs 0 0 listat                    │
│    ttag<.; sil-all;                     │
│    loop 0 nnodes[0] ttag+#1 'listat[#1]'; stop│
│    ttag > RECONS. 'numrecs'; ttag ; sil = all;│
│    set numrecs ++ ;                     │
│    endrecs ;                      e     │
│ quote TOTAL RECONSTRUCTIONS: 'numrecs'; proc/;│
└─────────────────────────────────────────┘
```

FIGURE 10.2 Example scripts showing how to **(a)** use a `progress` bar to indicate to the user that calculations advance; **(b)** open and handle information from input files; **(c)** edit the first tree in memory to create a pectinate tree; **(d)** perform *TBR* branch-swapping using `tbrit`, **(e)** handle individual most parsimonious reconstructions with `iterrecs`.

information from the file: `hifchar` (next byte in file), `hifline` (number of lines parsed), `hifnumber` (next number in file), `hifeof` (whether end-of-file was reached), `hifspy` (tells you what the next byte in file is, but without reading it, so that next call to `hifchar` retrieves it), `hiftext` (a series of bytes, copied onto an array with `set`; default number of bytes to copy is the size of the array, but you can specify a different number, and a specific terminating byte; see `help+hiftext` for details), and `$hifstring` (the next string in the input file; it can be formatted with formatting options, see Section 10.3.4). A simple script, `readnprocess.run` (Figure 10.2b), shows the use of `hifile`. The file is opened first; note that the `loop` will continue until the end-of-file is found (it is reset every time with `setloop`). The string is copied onto a user variable (every call to `$hifstring` reports the next string; to reuse, you must copy it onto some permanent location). The script takes (any number of) names of TNT input files from a file `filelist.txt`, and reads and processes every file (the strings retrieved from the file, of course, could be used as anything you need—lists of taxa, characters, etc.).

You can have multiple handled files open simultaneously; you do this by just repeatedly calling `hifile open`. The active file is the last one opened, or the one set with `hifile active`. The skipping and seeking options by default operate on active file, but the file to operate on can be specified, giving the name of the file within square brackets. The maximum number of files that can be open simultaneously is five, but you can change the limit to be up to N files with `hifile = N` (this must be done prior to enabling `macros`, and there is still a limit—although rather high—on the number of files that can be simultaneously open, imposed by your operating system). When the input file is no longer needed, it can be closed with `hifile close` (closing the active file by default; if `close` followed by a filename within square brackets, it closes the specified file).

Another useful option to handle input files is the `forfiles` command. The syntax is

```
forfiles fname
    ... command(s) ...
    stop
```

For every file matching `fname` (wildcards can be used) this will invoke command(s). Within `forfiles`, the `killfiles` command interrupts subsequent iterations, and `proc/` terminates current iteration [every invocation of the command(s) is treated as a new level of input file]. The current file is retrieved with `$curfile`; the variable `curisdir` indicates whether current file is a directory (same `fname` string may be equally matched by directories or actual files). If you need to indicate spaces within `fname`, use the ^ symbol (just like file names in other contexts). File names containing symbols used as sigils for scripting (symbols ', ", $, or #) are skipped. Within the instructions, all output is silenced by default; to produce visible output, you need a `silent – all` command. As an example, the `forfiles` option can be used to scan directories for specific files:

```
tnt*>forfiles *.tnt sil-all;p $curfile;mult;nelsen;stop
```

This will read every TNT file in current directory, do a `mult` search, and plot the consensus.

10.3.4 FORMATTED OUTPUT

The output of the program is produced mostly via the `quote` command, and managed with the commands `silent`, `warn` and `report`. `Silent` silences (or not) the output to `console`, text `buffer`, or `file`; `warn` handles TNT warnings (which you do not want to see in scripts!); `report` turns off the progress report of time-consuming operations. The `quote` command can be used to produce specific output formats. `Quote` by default places a carriage return at the end; this can be changed with `lquote=` (to make quotes "literal"; to reset the default, use `lquote-`). The `lquote` command also controls whether ASCII conversion can be performed by the `quote` command; with the `lquote[` option, &N as input for quote is converted on output into ASCII character N (the same is true of many other options: setting a user variable as a string with set, the errmsg command, etc.). This can be used to print out characters that would not be printable otherwise, such as single quotes (&39), double quotes (&34), dollar (&36) or percentage signs (&37), etc. You get a complete list of ASCII single-byte characters with

```
tnt*>lquote [; loop 33 255 quote Nr. #1 is 1 stop
```

Note characters 0–32 are skipped in this example because they represent tabs, sounds or spaces. In non-Windows versions only, character 0 (&0) is interpreted as erasing the last line written to stderr, and character 1 (&1) is interpreted as backspacing and erasing one character in stderr. To avoid ASCII conversion (e.g. to have & read as &), use lquote]. If the output is silenced (e.g. with silent = all), you need to de-silence it (silent - all) for quote to produce visible output; the same effect is achieved if the text to quote is preceded by a colon (i.e. quote: text de-silences output only temporarily, for execution of the quote command only).

User variables converted into numbers normally use the number of digits specified with macfloat. A conversion on the fly can be done if the first single quote is followed by a slash. The complete format specification for converting a variable named result into output is '/+W.Dresult', where W and D are numbers. The + symbol indicates that the sign of the value is always output (even if positive; if no + indicated, sign shown only for negative values); W is the width that the converted number will occupy (if W is negative, left flushed); .D is the number of significant digits. Any of +, W, or .D can be excluded, in which case the default is used for the corresponding option.

10.3.4.1 Handling strings

It is possible to format strings, or access only part of them. Also, when converting a string, by default TNT adds a blank at the end (this is done with the set + option); to output strings literally (i.e. without adding a blank at the end), use set –. These format options apply to either user-defined or pre-defined strings (e.g. taxon, character, or state names). Usage is (for all examples, assume string contains "hypothetical"):

- $string:N displays the first N characters, adding blanks if needed to complete a width of N. Examples: $string:5 displays "hypot" and $string:20 displays "hypothetical".
- $string:-N displays the last N characters. Examples: $string:-5 displays "tical" and $string:-20 displays "hypothetical".
- $string:+N displays the characters following the first N. Examples: $string:+5 displays "hetical", and $string:+20 displays "" (nothing),
- $string<X displays characters preceding X (case sensitive). Examples: $string<t displays "hypo", $string<c displays "hypotheti", and $string<K displays "hypothetical".
- $string>X displays characters following first occurrence of X. Examples: $string>t displays "hetical", $string>c displays "al", and $string>K displays "hypothetical".

In the case of pre-defined strings that need specification of a number, the format specification goes right after the string type, before the number. Thus, if taxon number N is Xus_yus and you want to extract the generic and specific names only, correct usage is:

```
$taxon<_N      generic name ("Xus")
$taxon>_N      specific name ("yus")
```

The same applies to character or state names, or any fixed string that requires specification of a number. The branch labels (stored with the ttag command) can be treated as strings in the same manner; they are also interpreted on tree plots to indicate branch lengths, widths, or colors (see Section 10.4.1). Branch labels (e.g. containing support values) can be copied onto user variables as numbers, but in this case they need to be within parentheses: set n ($ttag 31) will store in variable n the support for node 31 (skipping the parentheses, with set n $ttag 31, will copy onto variable n the text "ttag" followed by the text "31", not the number).

10.3.5 ARRAYS INTO AND FROM TABLES

The values of any (one-dimensional) array (say, raycharles) can also be displayed in the form of a table, with maketable raycharles (using either of the two available table formats, set with table= and table-). The name of the array can be followed by the title of the table (terminated with a semicolon). The number of significant digits used in the table is changed with macfloat. Only positive numbers are printed as such; negative numbers indicate a dash (−1), a Y ("yes", −2), an N ("no", −3), a blank (−4), or an X (inapplicable, −5). If the name of the array is followed by =N, the table displays only the first N values. If scripting was enabled as integer-only (with macfloat− prior to macro=), and the name of the array is preceded by *, then maketable expects the name of two arrays, and a double table is displayed (i.e. two dash-separated values per cell, as in the minmax* regular command). The use of maketable allows showing results produced by scripts in a format that looks exactly like the normal output of TNT.

Just like arrays can be converted into tables, tables can be copied onto arrays. With maketable + raycharles, the values from any regular TNT command that would display a table are copied onto array raycharles (triggering an error if raycharles cannot hold enough values). The table is not displayed at all, just copied onto the array. Negative numbers have the same meaning as when converting in the opposite direction. If several tables are displayed, the values in raycharles correspond to the last table displayed; the internal variable listsize gives the number of values effectively copied onto raycharles. This can be used to store in arrays the results of any TNT command that displays a table. For arrays with more than one dimension (e.g. the twodimray[5 6] used as an example earlier), only the last one can be used to receive the values from a table, for example maketable + twodimray[2] is a valid assignment, but maketable + twodimray is not (the same applies for the conversion in the opposite direction, from array to table). In case the reader is wondering, `maketable + raycharles` followed by `maketable rayliotta` copies the values from `rayliotta` onto `raycharles`, and followed by `maketable raycharles` does nothing (copying `raycharles` onto itself).

10.3.6 AUTOMATIC INPUT REDIRECTION

The `goto` command (Section 10.3.1) allows taking instructions from specific parts of a file, but `goto` itself is a command; it can only be recognized by TNT when expecting a command, not when expecting options *within* regular TNT commands. Assume you want to deactivate taxa (with `taxcode-`), and the list of taxa to deactivate is in a file called `taxlist.txt`; you cannot just do something like

```
tnt*>taxcode - goto taxlist.txt ;                      [example of bad syntax!]
```

For such cases, TNT offers the possibility of *automatic input redirection*, with `@@`. The `@@` option requires that you specify the name of the file, and the name of the `label` to go to. The file itself must have the string `label tag`, followed by any contents you want to be parsed by TNT; the end of the contents is indicated, again, with `@@`. That is,

```
┌─ taxlist.txt ─┐
│ label mylist  │
│ fish fly bee  │
│ @@            │
└───────────────┘
```

you would then "stick" the contents of `taxlist.txt` into the options for `tax-code` with

```
tnt*>taxcode - @@ taxlist.txt mylist ;;
```

Note the double semicolon; that is because the automatic input redirection can also take arguments (interpretable with `%` within the automatically parsed file), the list of which ends with a semicolon; another semicolon is needed to end execution of the `taxcode` command itself. This was exemplified in Chapter 1, Section 1.11.10, to read datasets contained in multiple files—the automatic input redirection was in that case being made in the middle of the execution of an `xread` command.

If a default target file has been defined with `goto=`, then any subsequent invocation of `@@` automatically goes to that target file. Only the `label` needs to be specified for `@@` in that case. You can use this if you will be repeatedly reading from the same file (i.e. taking input from different `label`'s from the same file).

10.3.7 DIALOGS

In Windows versions only, it is possible to define custom dialogs, with the `opendlg` command. The `opendlg` command uses a simple syntax to define the dialog, control its behavior, and allow user variables to record the results of choices made by the user. See help opendlg for details. A simple dialog is distributed with the TNT package (dialog.run, in tnt_scripts.zip); more involved dialogs are in http://www.lillo.org.ar/phylogeny/tnt/scripts/delcor.run (published by Giannini and Goloboff 2010) and https://github.com/atorresgalvis/TNT-scripts-for-phylogenomics (published by Torres et al. 2021).

10.3.8 Editing Trees and Branch Labels

The scripts can modify the trees held in memory with the edit command (with edit], changes are done silently, without showing the resulting tree). Edit (or edit]) must be followed by the number of tree to edit, the clade to which both source and target node belong (use the root of the tree to contain any two nodes), and then two numbers, source and target node. If source is an ancestor of target, then edit reroots the clade defined by source so that target is now the sister group of all the members of clade source (if source equals root, then the entire tree is rerooted). Keep in mind that equivalent groups may have different node numbers after changing the tree; you need to identify groups properly for every move. Edit can be used in combination with tread to define trees. Remember that tread can use groups (defined with agroup) or lists of taxa. In tread, a triple dot (e.g. Genus...) amounts to listing all the taxa matching the part before the dots and not yet included in the tree (... by itself is all the taxa not yet included), and lists of taxa can be given with :substring (i.e. all the taxa containing substring anywhere in the name; see Chapter 1, Section 1.11.4).

For an example, tree 0 is made fully pectinate with a script like pectinator.run (in Figure 10.2c), using the edit command. Making symmetric trees (or as symmetric as possible; only trees with certain numbers of taxa can be fully symmetric) is more involved (the supplementary material of Goloboff and Wilkinson 2018, www.lillo.org.ar/phylogeny/published/Goloboff_Wilkinson_Supplementary_Material.zip, contains a script mksymtree.run that divides the tree in groups as even as possible).

The branch labels or tags can also be modified, using the ttag command; this allows erasing tags (ttag<N, where N is the node or list of nodes to erase; use ttag << to erase tags for all terminal branches, and ttag <> to erase tags for all internal nodes), or adding text to the tag for node N (ttag +N). Tags can be interpreted to provide specific branch lengths, widths, or colors, and text colors. The scripts can therefore be used to produce formatted trees (see Section 10.4.1), by writing to branch labels suitable instructions.

10.3.9 Tree Searching and Traversals

As in the loop command, the scripting offers the possibility of executing repeated instructions, except that every time the instructions are called, different trees can be visited (for user-designed tree searches), or nodes of a tree in a specific sequence (for handling special calculations on a tree).

The commands that facilitate designing special searches are sprit, tbrit, and bandb. Sprit and tbrit perform *SPR* and *TBR* rearrangements; their syntax is very simple:

```
tbrit 0
    ...code to execute for every
      TBR rearrangement of tree 0...
    stop
```

within sprit or tbrit, the commands continue (skip remaining instructions and move onto next rearrangement), resetswap (substitute tree being swapped and begin swapping on that tree), and endswap (terminate) are recognized. The internal variable percswap reports the proportion of moves that have been done. This allows designing search routines using special criteria; an example for selecting a tree on the basis of its length is in Figure 10.2d (script easyswaps.run). The bandb command works similarly, except that it executes the instructions for every possible subtree and tree (see help bandb).

In addition to tree searches, it is possible to traverse the nodes of a tree, with the travtree command, executing a set of instructions for every node visited. The syntax for traversing tree T from node N is

```
travtree type T N varname ... commands to execute ... endtrav
```

For each iteration, the number of node being visited is written into variable varname; the type of traversal can be up (up-pass or preorder traversal, above node N; if up followed by terms it also does terminals, otherwise it only does internal nodes), down (down-pass or postorder traversal, toward node N, terminals never included), below (travels from node N toward root), path (travels from node N_1 to N_2; two nodes instead of one must be specified for path), and des (visits all immediate descendants of node N). The variable varname itself can be preceded by a minus sign (–), in which case node N_1 itself is not included in the traverse. Within travtree, the commands killtrav (ends current traveling) and skipdes (only for type up, excludes all the descendants of current node) are recognized. Instructions within travtree are treated as a nested input file; thus, proc/; skips the rest of instructions and moves onto the next node to be visited. Note that the order in which nodes are traveled in scripting commands may not agree with the sequence with which nodes are plotted in tree diagrams. If you want to make sure that scripts use the same sequence as in plotting, use travtree+ (off with travtree-). Tree-traversing can be nested (with up to 5 levels; the maximum level can be set to N with travtree = N).

A special kind of case for tracking correspondence between nodes of trees is the sectsch command, which internally creates reduced datasets and trees, and reanalyzes them with the commands given by the user. Using track as one of the options for sectsch, and a user-defined search (i.e. commands to execute given within square brackets), it is possible to map the nodes of the big (original) tree onto the reduced tree, and vice versa, with the expressions biginsect, sectonod, and nodtosect (use help+ to see details). Within sectsch, ntax corresponds to the number of taxa in the reduced dataset; the number of taxa in the whole dataset is given by bigntax. This option is the one used (in the script secboot.run) to evaluate supports with a quicker analysis for the reduced datasets, mapping the values onto the corresponding node of the original tree (see Chapter 8, Section 8.4.4.2).

10.3.10   MOST PARSIMONIOUS RECONSTRUCTIONS (*MPRs*)

TNT has regular commands that produce most parsimonious optimizations, but these simply show the results as state sets. The *MP*-sets (i.e. the union of states that occur in any *MPR*) for discrete characters can be retrieved with the internal variable states (see Table 10.4); the state sets are retrieved as bit sets. If the character number is specified as. (a dot), then states returns an array, with the values for the specified node for every character; if the node number N is specified as. then states returns an array with the values for all nodes for the specified character (if both character and node number are dots, then a two-dimensional array is required to store the values). That is useful, but it still does not produce individual reconstructions (only certain combinations of the states returned will produce *MPRs*).

The recons regular TNT command generates all possible most parsimonious reconstructions for the chosen tree(s) and character(s), but it does not allow manipulating them or using them for special calculations. For that purpose, the iterrecs command generates all possible most parsimonious reconstructions, and makes every one of them accessible to the scripting language. The syntax is

```
iterrecs T C rayname... commands to execute ... endrecs
```

This generates all *MPRs* for character C on tree T, saving the state assigned to each node in every reconstruction to the array rayname (rayname must be large enough to contain states for all nodes). The states are saved as a single number (as they are unique in individual *MPRs*, there is no need to treat them as bitsets). The iterrecs command allows choosing best possible reconstructions having (or not having) a certain state at any given node, selecting a random sample of reconstructions, and intervals by which to increase values in continuous characters; see help iterrecs in TNT for details. Within iterrecs, the string killrecs interrupts execution, and recsteps returns the cost of current reconstruction (if forcing states, this may be larger than minimum length). The command iterrecs can be nested (maximum level of nesting is set with iterrecs = N; default is 5). Nesting iterrecs may require that mxproc is reset as well. A minimalistic example of using iterrecs to display and count all possible *MPRs* for character 0 on tree 0 is in Figure 10.2e (reconstructions.run). Published examples of the use of iterrecs are the scripts of Catalano et al. (2009) and Giannini and Goloboff (2010).

10.3.11   RANDOM NUMBERS AND LISTS, COMBINATIONS, PERMUTATIONS

The internal variable getrandom[i j] returns a random number in the interval i–j. Every invocation of getrandom returns the next number of the series; series are initialized by the random seed for scripts, with the `macseed` command (use `macseed 0` to initialize with time). The expression

```
set raycharles randomlist [N] ;
```

writes a random ordering of N elements into array `raycharles`. For example, the following commands (note change of prompt when reading `loop` input):

```
tnt*>macro=; macfloat 0;
tnt*>macseed 54321; var:raycharles[6];
tnt*>loop 1 5
loop> set raycharles randomlist [6] ;
loop> quote 'raycharles[0-5]';
loop>stop
```

produces as output

```
5 2 4 1 3 0
4 5 2 1 3 0
4 5 3 1 0 2
3 0 2 4 1 5
4 0 3 5 2 1
```

Other random seeds will of course produce different orderings. Note that the first *m* elements in `raycharles` are also a random selection of *m* out of the six elements.

Some analyses may require enumerating *all* combinations or permutations of elements. That task is made easier with the `combine` command, the syntax of which is

```
combine X min/max varname ... commands ... endcomb
```

This enumerates all combinations of `min` out of X elements, `min+1` out of X, ... `max` out of X. If `/max` is omitted, then `max` = `min`. For every combination, the commands are executed, with list of elements written into variable `varname` (must be an array); `listsize` equals number of elements minus 1. The set of instructions to execute for every combination is treated as a new level if input file; you can move onto the next combination with `proc/`; nested combinations may require changing the `mxproc` limit; maximum possible recursion of `combine` is the same as for `travtree` and `iterrecs`. As an example of the use of `combine`, try,

```
tnt*>macro=; macfloat 0 ; var: ray[5];
tnt*>report-; combine 5 2/3 ray
combine>sil-all; quote 'ray[0-listsize]'; endcomb
```

(note change of prompt when reading input for `combine`). Instead of combinations, it is also possible to generate all permutations of the X elements, using an asterisk (*) instead of `min/max`. Another, more useful example of `combine` is in generating completely undecisive matrices (Goloboff 1991a). As those matrices comprise all combinations of 0's and 1' (but no characters with fewer than two 1's or two 0's needed), the lists of taxa having state 1 can be generated with combine, using xread= to modify an all-0 matrix (created before with xread/), as shown in the doundec.run example (in Figure 10.3a).

10.4   GRAPHICS AND CORRELATION

10.4.1   PLOTTING GRAPHIC TREES

Obviously, any phylogenetic program with some pride must be able to plot decent trees. Windows versions of TNT can save graphic trees as metafiles, but the design of the tree can be controlled more tightly with the ttag command, and the use of appropriate branch labels. The ttag& option saves the tree-tags as an SVG file. By default, it saves a black-and-white tree, just printing the branch labels literally. With the color option, it converts codes from the branch labels into specific branch colors, thicknesses and lengths; the color can be indicated individually for the text (which can be in multiple lines, with a different color for each line). With the xysave option, the SVG file is left open and the coordinates for every tree node in the diagram are saved to an array, so that you can add text, lines, squares, or any SVG instructions to the tree diagram.

The simplest use of the ttag& option is just saving a black-and-white tree to a file: ttag& myfile.svg. Remember that ttag& saves just the previously defined tag-tree. If you need to save multiple trees (only one per SVG file can be saved!), you then must disconnect the tags (ttag-), store the new tree as tag-tree (either ttag*N, or ttag=;tplot N;), and then call ttag& again. With ttag&, you can also control the font size (fontsize N, or bfontsize for branches and tfontsize for terminals), label position (legup N, raises legends N points—if N negative, lowers them), branch lengths (blength N; if using color and length of a branch indicated with :L, then L is relative to N), and vertical separation between branches (bheight N).

For color diagrams, you simply add the color option. With the color option, the branch labels (in the tagged tree, ttag) are interpreted to indicate width, color, and type of branch, as well as text color, as shown in Figure 10.3b. By suitably changing the tags with your script (see Section 10.3.8 for how to edit the tags), you can have a lot of control on how the tree looks, on a branch-per-branch basis. In the tags, first code expected is color; it can be a single-digit code (0–9) for pre-defined colors, or an RGB code (values of Red, Green, Blue, separated by commas) in parentheses (see Figure 10.3b); the RGB code can be followed by branch thickness and length. If the first color in the RGB code is negative, then it plots a dashed line. If RGB code is white (i.e. 255,255,255), it then plots a "hollow" branch (i.e. white with a frame). The color code is separated from the branch legend by a forward slash; text color is by default the same as the branch color (can be changed with an RGB code within curly braces; RGB code can be followed by text stroke). New lines are indicated with a backslash.

With the xysave varname option of the ttag& command you save to user variable named varname (a 2-dimensional array) the coordinates of every node.

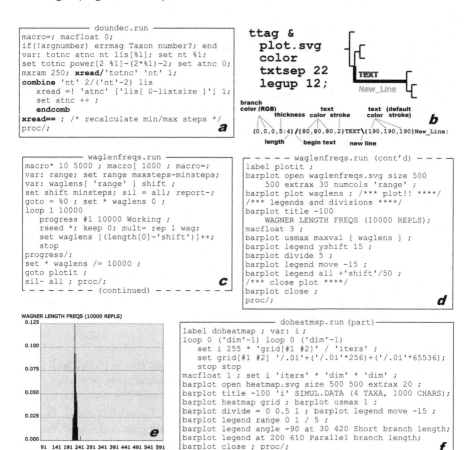

FIGURE 10.3 (a) Example script showing how to use combinations generated by `combine` to generate an undecisive dataset; (b) The interpretation of branch-legends when printing tree-tags in color (with the `ttag&` option); (c, d) A script to calculate frequency distributions of *RAS* Wagner trees with different random seeds, and plot those values with the `barplot` command; (e) The plot (unedited) produced by TNT for the script in (c,d); (f) Example script (part) showing how to use the `barplot` command to produce a heatmap.

This can be used to subsequently add text, rectangles (with continuous or dashed contour, with any degree of opacity), and lines to the tree diagram. This can be used for example to graphically indicate groups in the tree. Check `help ttag` for details on the use of `xysave`.

10.4.2 BAR PLOTS

The `barplot` command saves bar plots, in SVG format. The command has many options (see `help barplot` for details). Basically, `barplot open file.svg` opens the file, which remains open for plotting. Subsequent calls to `barplot` can plot arrays (`barplot plot arrayname`), and handle legends (`barplot legend`) or other details of the plot (`font`, `reset`; scale is handled with `usmax` and `usmin`).

When opening the file, you specify the number of columns to plot (this can subsequently be changed with reset); in the simplest case, the number of columns should be set to the number of cells in the array to plot. Columns can be shifted (e.g. to plot several variables side-by-side, with `barplot reset shift`). Consider the script in Figure 10.3c (`waglenfreqs.run`), which saves frequencies of lengths for Wagner trees to an array called `waglens`. Those frequencies are then saved as a bar plot in the `label plotit` part of the script (Figure 10.3d). Figure 10.3e shows the diagram exactly as produced by TNT (with no modifications). In command-driven versions, if you have the `*.svg` file extension associated with a specific browser (or know where the binary for your browser is), you can also have TNT calling the browser to display the plot with the `system` command (i.e. `system waglenfreqs.svg` in this case). Another (more elaborate) example of the use of `barplot` is in Figure 7.8b (with the script `plusntru.run`), where `reset` was used to start plotting again from the left, so that proportions of correct groups appear as a fraction of the number of groups found by one method and not the other.

10.4.2.1   Heat maps

After the bar plot is open (in the usual fashion) a heatmap can also be plotted with `barplot heatmap array`. The array must be a two-dimensional array, printed as a grid, with color of every cell indicated as an RGB code. Remember that to convert an RGB code into a single integer you use (R*65536)+(256*G)+B, and that the R, G, or B values must be integers (i.e. with `macfloat 0`, or converting with `/.0`). The example in Figure 10.3f shows part of a script (full code at www.lillo.org.ar/phylogeny/eduscripts/, omitted here for brevity), `doheatmap.run`, that takes an array (called `grid`) containing the frequency with which the right tree is recovered in a 4-taxon problem with different branch lengths.[3] The option `barplot heatmap` automatically uses the dimensions of the array for the grid; when grid is a 50 × 50 array, the plot right out of TNT looks like the one shown in Figure 10.4a (it may need some edition for publishable quality, but very little). In general, using more than 75 columns or rows for the diagram is pointless (human eye stops perceiving differences when cells are so tiny, and the SVG files become very large).

10.4.3   CORRELATION

Pearson correlation coefficients for the values in two (unidimensional) arrays can be calculated with `var&` followed by the names of two arrays. The names of the two arrays can be followed by `*weightarray`, in which case the values in `weightarray` are considered as weights. Weights must be between 0 and 1 (if some weights exceed unity, all weights are rescaled; having a point of weight W is equivalent to having W points with those coordinates instead of one). If a number N follows after the arrays, only the first N values of the arrays are considered in the regression. `Var&` reports the results to the console or file, but after calculation, the internal variables

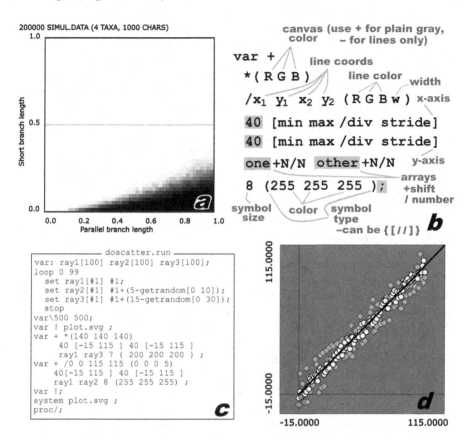

FIGURE 10.4 (a) The heatmap produced by the code in script `doheatmap.run`; (b) The different parameters that can be specified in scatter plots with the `var+` option; (c) Example of a script to plot two pairs of variables using `var+`; (d) The plot produced by the script `doscatter.run`.

`regr`, `regalfa`, and `regbeta` contain the correlation coefficient r, the intercept and the slope, respectively. In the example that follows, program output is marked in bold (exact numbers depend obviously on random seed for scripts, macseed),

```
tnt*>loop 0 99
loop> set ray1[#1] #1;
loop> set ray2[#1] #1+(5-getrandom[0 10]); stop
tnt*>var & ray1 ray2;
```
Correlation of ray1 (X) and ray2 (Y):
   Y=0.097228+0.997834*X (r=0.993574)
```
tnt*>set rvalue regr ;
tnt*>set alfa regalfa; set beta regbeta ;
tnt*>quote R='rvalue'- intercept='alfa' - slope='beta';
```
R=0.993574- intercept=0.097228 - slope=0.997834

Spearman's coefficient is useful for detecting rank correlations (e.g. those that deviate from lineal). This is available only as an internal variable, `spearman[array1 array2 N]` (where N is the number of values; if omitted, all values). If `*R` is added before the closing square bracket, it does R randomizations (permutations) of the values (using current random seed) and returns proportion of permutations with absolute value of correlation equal to or greater than observed (with > or < instead of *, returns proportion of permutations with a larger or smaller value, respectively). Thus, you can type after these instructions:

```
tnt*>set rvalue spearman [ray1 ray2] ;
tnt*>quote Spear-value 'rvalue';
Spear-value 0.993713
```

10.4.4 SCATTER PLOTS

The type of diagram most often needed for phylogenetic comparisons is a scatter plot. Scatter plots are obtained with `var+`. By default, a minimalistic plot is drawn (with ASCII extended characters) into the text buffer. In Windows version, it is possible to save the plots as metafiles. But the best control can be obtained when saving the scatter plots to SVG files. For this, you need to first open the SVG file, with `var!` `filename.svg`. The dimensions of the frame can be specified with `var \X Y` (this needs to be done before opening the SVG file, otherwise the file will be opened with the default values, 1500 × 1500). After opening the file, a title (on top of the diagram) can be specified with `var!!` `title` (ending with a semicolon).

The plot itself is invoked with `var+`. The syntax of the `var+` option lets you specify several details of the plot (see Figure 10.4b). You specify first the canvas and frame. The default is to print only a basic set of x, y axes and the dots (or squares, rubies, or lines), with no background. If an SVG file is opened, every time you call `var+` it prints again on top of the same diagram; as this is transparent, it lets you superpose plots for different variables; only the first call to `var+` needs to define a color for the canvas. With `var++` a gray background is used; with `var+*(R G B)` you can specify any canvas color. Using `var +−` (plus minus), it only prints the lines specified with / (see next); this can be used to add details to the diagram (e.g. to add standard deviations, as done for Figure 7.10).

Second, a series of lines can be specified after the canvas. Each is indicated as / x_1 y_1 x_2 y_2 `(R G B width)`; x_1, y_1, x_2, y_2 are coordinates for beginning and end points of the line (in the reference scale of the diagram). The RGB code and width are optional (default is black and 1 unit wide). Up to 15 lines can be indicated in one call to `var+`, but as you can repeat calls to `var+`, any number of lines can be added to the diagram. Note that if using `var+−` to print only the lines, then specification of scale for the x and y−axes is mandatory (otherwise, lines cannot be placed in the plot).

Third, the scale for x and y−axes follows the lines (if present). Two numbers are mandatory, width and height of the diagram in the text buffer; each can be followed (within square brackets) by specification of minimum and maximum values (if unspecified, TNT uses minimum/maximum of each array), and this can be followed optionally by indication of the interval for secondary tick marks (as `/divsize`) and the number of tick marks for a primary tick mark (axes are smooth if no interval for tick marks indicated).

Fourth are the arrays to print (one-dimensional arrays). The name of each array can be followed by a "shift" (i.e. start to print from the Nth value, instead of the first[4]), with +N; you can also specify printing only N out of the total number of values in the array with /N (N must be equal to or smaller than the number of values in the array, or the number left after shifting it). If you have not defined /N, you can also shave-off the last N values of the array with −N. Each of the two arrays can use its set of +N, /N or −N. If a single array is indicated, the values for each cell are printed—in this case, you cannot specify the size and type of symbol to be used in plotting.

Fifth, and last, is the size, type, and color of the symbol to print. The size is a single number after the second array. Color is indicated as an RGB code; in parentheses, the printer uses a circle (default). With RGB code within square brackets, the printer uses squares; within curly braces, a diamond. Using slashes (i.e. /R G B /), it prints a line (of the width specified as size) connecting every dot.

A very simple example showing how to use these options is in the script `dos-catter.run` (Figure 10.4c). The script declares three arrays with 100 values each, assigning to the first one the same number of cell. The second and third arrays are assigned the same number, plus or minus a random number between −5 and +5 (for `ray2`), or between −15 and +15 (for `ray3`). The script then plots `ray3` against `ray1` (with gray diamonds), and `ray2` against `ray1` (with white circles), producing the diagram shown in Figure 10.4d. As expected, there is a greater dispersion in `ray3` (incidentally, that is why it is plotted below `ray2`; the plot for `ray3` would obscure the plot of `ray2` if on top). The canvas is indicated as a rather dark gray, RGB=(140,140,140), to illustrate how the color is handled; a 5-unit wide black line indicating the diagonal (i.e. perfect lineal correlation) is added. Figure 10.4d shows the diagram exactly as it comes out of TNT; only minor modifications could be required for publication.

Finally, the file needs to be closed (with `var!;`); this is required because a special code needs to be placed at the end of the file for it to be a valid SVG. All the explanation given so far concerns a single diagram. The only detail remaining to be explained is that the `var` command can place several scatter plots tiled in a single file. For this, instead of specifying just the size, you also specify the separator size, number of columns, and rows, with `var \ width height/separator x cols x rows` (note `cols` and `rows` are separated by an x symbol). Again, this must be specified before opening the file with `var! filename.svg`. The current tile is by default 0–0, and you can move between tiles with `var!= C,R` (makes column C, row R the current tile), or with `var!+ C,R` (moves current tile C columns and R rows). Any call to plot a diagram, draw lines, or add a title, will be done on the current tile.

10.5 SIMULATING AND MODIFYING DATA

The commands to generate or modify datasets are part of the regular set of TNT commands—they do not require that scripting be enabled with `macro=`. However, many comparisons may require simulating or modifying data, and this can be easily

```
┌──────────── simuldat.run ────────────┐
macro=; var: niters ; set niters 400;
var: eqwt['niters'] iwt['niters'] lam;
sil =all ; report - ; coll tbr ;
loop 0 ('niters'-1)
  progress #1 'niters' Simulating ;
  rseed * ; xread /100 50 ; ra 1 ;
  set lam getrandom[ 150 250 ]/1000 ;
  xread&('lam') 0 2 .; hold 110/3; tva>.;
  piwe-; resam cut 10 [mul=hol;]; tva>0 ;
  piwe=10; resam cut 10 [mul=hol;]; tva>0;
  keep 0; tva<.;
  set eqwt[ #1 ] rrfdistp 0 / 1;
  set  iwt[ #1 ] rrfdistp 0 / 2;
  stop
progress/; goto = %0 ; goto doplots ;
proc/;

label doplots ;
var: minis maxis ;
set maxis maxval [ eqwt ] ;
if ( 'maxis' < maxval [ iwt ] )
   set maxis maxval [ iwt ] ; end
set minis minval [ eqwt ] ;
if ( 'minis' < minval [ iwt ] )
   set minis minval [ iwt ] ; end
var \ 750 750 ; var ! simulation.svg ;
var !! 'niters' CASES, X=EQWTS Y=IMPWTS;
var ++
   /'minis' 'minis' 'maxis' 'maxis'
   40['minis' 'maxis'] 40['minis''maxis']
   eqwt iwt 10 (255 255 255 ) ;
var! ;
proc/;
```

a

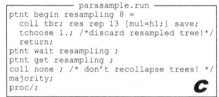

400 CASES, X=EQWTS Y=IMPWTS

b

```
┌──────────── parasample.run ────────────┐
ptnt begin resampling 8 =
   coll tbr; res rep 13 [mul=hl;] save;
   tchoose 1.; /*discard resampled tree!*/
   return;
ptnt wait resampling ;
ptnt get resampling ;
coll none ; /* don't recollapse trees! */
majority;
proc/;
```

c

FIGURE 10.5 **(a)** Script to simulate data (with exponential distribution, lambda 0.15–0.25) for 100 characters and 50 taxa, and measure the *RF* distance to model tree from trees produced by resampling under equal and implied weights (with all groups of positive *GC*, with `cut 10`); **(b)** The values calculated with script in (a); **(c)** A minimalistic script for parallel calculation of resampling.

handled in scripts. The `xread` command allows generating data anew (with the exponential model of Goloboff et al. 2017, or a Poisson model with branch lengths of model tree specified by the user), or modifying preexisting data. Using these facilities, doing simulations to test alternative methods is very easy. Examples of scripts that modify or generate new data are in Prevosti and Chemisquy (2010), Sansom (2015), or Watanabe (2015). A simpler example is in Figure 10.5a, a script that generates datasets with the exponential model (`xread&`), and resamples (with `cut 10`) every dataset with both equal and implied weights, comparing each resampled tree (*RF* distance rescaled by number of groups present in the trees, the `rrfdistp` internal variable) with the model tree (i.e. tree number 0 in the tree vault). Finally, the script plots the *RF* distances for equal and implied weighting (Figure 10.5b), showing that the tree obtained by implied weighting tends to be more similar to the model tree than the model obtained by equal weights.

10.6   A DIGRESSION: THE C INTERPRETER OF TNT

In addition to TNT language, TNT can also interpret C-style scripts (Goloboff and Morales 2020), with the `runc` command. The interpreter uses code from PicoC, an excellent C interpreter from Zik Saleeba (2019), with about 150 functions and 45 values added to a "TNT library" (and about 5,000 additional lines of code added to the code of PicoC itself). Although the syntax for the interpreter is standard C, the main interest is in the added functions, which mimic most of the facilities available in the TNT interpreter, and the direct interface to TNT. The list of functions in the TNT library can be obtained in TNT with runc = tnt (in addition to the TNT library, PicoC includes the ctype, math, stdio, stdlib, string, time, and unistd libraries; those are always included in any TNT script, with no need to #include them). When invoking a script with the runc command, TNT parses C-style instructions; any error occurred during execution of the script displays an error message and returns control to the main command loop of TNT (expecting standard commands again). In C scripts, calls to TNT commands are effected via the tnt() function, the prototype of which is

```
tnt (char * cmds)
```

where cmds is a string formatted as in fprintf. All the commands are passed to TNT internally, just as if you were typing commands at the console; data and trees are modified only by calls to commands to read data or change trees. All internal states of TNT can also be accessed as pre-defined values (full list is accessed with runc+). In general, TNT-style scripts require less typing, as the syntax is less strict than C's—the same results can be accomplished with fewer keystrokes. The advantage of C is in routines that need a high level of abstraction, structures, or bit operations; those tend to be simpler in a C context. The C interpreter keeps track of all memory allocations (up to 5,000, by default; this number can be increased), so that freeing pointers after using them is facilitated (or not needed at all: all pointers are automatically deallocated when exiting script and going back to the TNT parser). For a more in-depth discussion of how to use the C parser of TNT, see Goloboff and Morales (2020). Examples of C-style TNT scripts are in www.lillo.org.ar/phylogeny/eduscripts/. Scripts mixing both languages are also possible.

BOX 10.4   RUNNING TNT IN PARALLEL

In Linux and Mac versions, the ptnt command allows running in parallel, with the PVM system (Parallel Virtual Machine, Geist et al. 1994). This launches new instances of TNT, and intercommunicates them via a PVM daemon. A port into the MPI system (Message Passing Interface, MPI Forum 1993) is in preparation by M. Morales and myself, now close to completion; this will operate with the same set of commands. See help ptnt for a (terse) explanation of how to parallelize TNT. A very simple parallelization (with only basic communication to monitor progress with ptnt wait), parallelizing resampling into 8 processes, is shown in the parasample.run script (Figure 10.5c).

10.7   SOME GENERAL ADVICE ON HOW TO WRITE SCRIPTS

The best use of scripts is to help you solve specific problems. There is not much point in mastering the scripting language (or any computer language, for that matter) unless you use it to solve specific problems. Scripts help bypassing the need to write stand-alone programs from scratch, and facilitate writing proof-of-concept code that lets you check whether some idea or algorithm is worth pursuing. But the utility is still in testing specific problems. It is not that you will first learn every detail of the TNT language, and *then* you will be able to test new phylogenetic ideas. The ideas must always come first; once you have a project, then you can start thinking of how to program that with scripts. It is very difficult to learn programming in the void; always let interesting scientific questions be the guide. In the same vein, it is pointless to memorize all the possibilities and internal variables of the language; learn to rely on the scripting help system (help+) for full lists of scripting commands and internal variables.

As you write your script, try to think of everything that the script will need before you sit down and start typing. For complex routines, some people find it useful to do flowcharts of the calculations needed before beginning to program. Adding new calculations once the script is close to finished often requires breaking the flow of the code, and may make further development of the script harder. As your script will probably need different types of calculations, keeping them in separate modules will make the script easier to follow and modify. Several scripts used as examples in this chapter use goto to better organize the code; as your scripts become more complex, this modularity becomes more important. Scripts that are intended just as proof-of-concept can use exhaustive enumeration of possibilities, without worrying about the time needed for the calculation; always keep in mind what the script is intended for. For example, the scripts used to test Faith and Trueman's (2001) and Salisbury' (1999) methods (in Chapter 7, Section 7.7.2) are intended only to illustrate problems with the methods, and thus are usable only for very small cases—those scripts are useful for what they are intended, nothing more.

It is very important to test that every part of your script functions as intended, at every stage of development. If you write an entire 10-page script without ever testing the individual parts, an error produced at the early steps of execution will be carried over subsequent steps, making identification of the source of erroneous results very difficult. If you develop your script part by part, and carefully test every stage for correctness of results, this is minimized. Then, as you write the first ten lines of your script, start checking that those ten lines work as you want. This applies to all stages of development; you can write and test first only the code that simulates the data, without worrying for the subsequent calculations involving tree searches and tree distances; then test that the tree searches are working as you expect, and so on. If possible, double-check results with independent calculations; for example independently calculating identical results by exhaustive enumeration of possibilities in small cases will increase your confidence that the faster algorithm produces correct results. Such checks can be incorporated in early stages of your script, then eliminated to save time.

As you write your script, try to be organized; indentation is not mandatory, but very useful to indicate subordinate code. Comments explaining the most difficult or intricate parts of the script will help others understand your code, and yourself if you have to reuse it in the distant future. Use variable names that represent something, but at the same time keeping the names relatively brief. Some people like the "Camel Case" notation, where spaces between words are eliminated and first letter of every word is capitalized, for example `BestModifiedTreeLength`. Personally, I find typing whole words tiresome, and I prefer shorter names, using just the first syllables of every word, for example `besmodtreelen`. The style to use may depend on whether you intend the script just for yourself or for a more public use—if others will need to understand your script, it may be wise to use more explicit names for variables.

As you write your script, especially if you intend for other people to use it, program defensively and always prepare for the worst. Sometimes it seems that users consciously try to figure out how to trick your program or script into failing in ways that you did not predict. Check input sanity. If your script requires that a dataset has been read into memory, has it? Are the trees needed for calculation really present in the tree buffer? Such kinds of checking make your code robust, and together with relatively descriptive error messages, help users understand why your script cannot run in certain circumstances, and make the script much easier to use.

As you start working with your first scripts, be aware that the script will never work as intended in the first try. Errors in scripts are eliminated little by little, testing small and easy cases; such a process is known as *debugging*. To debug TNT scripts, you can resort to (a) printing messages (i.e. with the `quote` command) as the script proceeds, to find out exactly what intermediate values are and where the script is halting (this may produce a lot of output; once you are sure that the script is working, you remove the `quote`'s and silence output with `silent = all`); (b) using the `macreport=` option, which produces a report for every comparison, `loop` iteration, and assignment made in the script, (c) commenting out portions of code, to find out where the problematic logic is. In the case of long output, saving the output of the script to a file and inspecting it with a text editor may be much easier than looking at thousands of lines of output going by at full speed. The error messages TNT outputs when interpreting scripts may help find errors. Finally, when correcting a non-functional script seems to have become impossible, be prepared to give up. Sometimes it is wiser to abandon the script entirely and start one anew; stubbornly clinging to a script that used the wrong approach from the start may in the long run require more work than using what you have learnt in that failed attempt to start from scratch with a better design.

NOTES

1 If you are not running a parallel version of TNT, the double quote (ASCII 34) is exactly equivalent to the single quote. The two (ASCII 34 and 39) have a different meaning within instructions for slaves.

2 A technical digression on how multidimensional arrays are organized internally by the TNT interpreter may help understand this. All the values available for scripts are stored internally by TNT in a single lineal memory segment; multidimensional arrays are mapped onto that segment (that is also how computer memory works, by the way). If variable `txt[4 3]` starts at internal position 0, then the first 4 internal cells will contain values that serve to locate the beginning of each of the 4 3-cell unidimensional arrays.

That is, the values stored in internal cells are $0 \rightarrow 4$, $1 \rightarrow 7$, $2 \rightarrow 10$, $3 \rightarrow 13$. The array beginning at internal cell number 4 (that is, cells 4–6) contains the values for 0,0, for 0,1, and for 0,2. The array beginning at internal cell number 7 (that is, cells 7–9) contains the values for 1,0, for 1,1, and for 1,2. And so on. The numbers within square brackets indicate to the interpreter shifts from the starting position.

3 The black zone, incidentally, is the famous "Felsenstein zone", where parsimony is inconsistent. Note it is small and confined to the lower-right side of the diagram.

4 This can be useful to print for example support values, which are undefined for terminals, but must be saved to arrays referencing node numbers, that is, starting at the first terminal. Without "shifting", you need to copy to another temporary array with a `loop`.

References

Aberer, A., Krompas, D., and Stamatakis, A. 2013. Pruning rogue taxa improves phylogenetic accuracy: An efficient algorithm and web service. *Syst. Biol.* 62, 162–166.

Aberer, A., and Stamatakis, A. 2011. A simple and accurate method for rogue taxon identification. In: *IEEE BIBM 2011.* Atlanta, Georgia, November 2011.

Adams, D., Rohlf, F., and Slice, D. 2013. A field comes of age: Geometric morphometrics in the 21st century. *Hystrix, the Italian J. Mammal.* 24, 7–14.

Adams, E. 1972. Consensus techniques and the comparison of taxonomic trees. *Syst. Zool.* 21:390–397.

Akaike, H. 1973. Information theory as an extension of the maximum likelihood principle. In: Petrov, B.N., and Csaki, F. (eds.), *Second International Symposium on Information Theory.* Akademiai Kiado. New York: Academic Press, pp. 267–281.

Anisimova, M., and Gascuel, O. 2006. Approximate likelihood-ratio test for branches: A fast, accurate, and powerful alternative. *Syst. Biol.* 55, 539–552.

Anisimova, M., Gil, M., Dufayard, J., Dessimoz, C., and Gascuel, O. 2011. Survey of branch support methods demonstrates accuracy, power, and robustness of fast likelihood-based approximation schemes. *Syst. Biol.* 60, 685–699.

Archie, J. 1985. Methods for coding variable morphological features for numerical taxonomic analysis. *Syst. Zool.* 34, 326–345.

Archie, J. 1989. Homoplasy excess ratios: New indices for measuring levels of homoplasy in phylogenetic systematics and a critique of the consistency index. *Syst. Zool.* 38, 253–269.

Arias, J., and Miranda, D. 2004. Profile parsimony (PP): An analysis under Implied Weights (IW). *Cladistics* 20, 56–63.

Ascarrunz, E., Claude, J., and Joyce, W. 2019. Estimating the phylogeny of geoemydid turtles (Cryptodira) from landmark data: An assessment of different methods. *PeerJ* 7, e7476.

Baker, A. 2003. Quantitative parsimony and explanatory power. *Br. J. Phil. Sci.* 54, 245–259.

Baker, R., and DeSalle, R. 1997. Multiple sources of character information and the phylogeny of Hawaiian Drosophila. *Syst. Biol.* 46, 654–673.

Bansal, M., Burleigh, J.G., Eulenstein, O., and Fernández-Baca, D. 2010. Robinson—Foulds supertrees. *Algorithms Mol. Biol.* 5, 1–12.

Baroni, M., Semple, C., and Steel, M. 2006. Hybrids in real time. *Syst. Biol.* 55, 46–56.

Barrett, M., Donoghue, M., and Sober, E. 1991. Against consensus. *Syst. Biol.* 40, 486–493.

Barthélemy, J., and McMorris, F. 1986. The median procedure for n-trees. *J. Classif.* 3, 329–334.

Barthélemy, J., McMorris, F., and Powers, R. 1992. Dictatorial consensus functions on n-trees. *Math. Soc. Sci.* 25, 59–64.

Baum, B. 1992. Combining trees as a way of combining data sets for phylogenetic inference, and the desirability of combining gene trees. *Taxon* 41, 3–10.

Baum, B., and Ragan, M. 1993. A reply to A.G. Rodrigo's "A comment on Baum's method for combining phylogenetic trees". *Taxon* 42: 637–640.

Baum, D., and Larson, A. 1991. Adaptation reviewed: A phylogenetic methodology for studying character macroevolution. *Syst. Zool.* 40, 1–18.

Beiko, R.G., and Hamilton, N. 2006. Phylogenetic identification of lateral genetic transfer events. *BMC Evol Biol.* 6, 15.

Berger, S., and Stamatakis, A. 2010. Accuracy of morphology-based phylogenetic fossil placement under maximum likelihood. In: *ACS/IEEE International Conference on Computer Systems and Applications-AICCSA 2010*, Hammamet, Tunisia, pp. 1–9.

Berry, V., and Gascuel, O. 1999. On the interpretation of bootstrap trees: Appropriate threshold of clade selection and induced gain. *Mol. Biol. Evol.* 13, 999–1011.

Bininda-Emonds, O. 2004. The evolution of supertrees. *Trends. Ecol. Evol.* 19, 315–322.

Bininda-Emonds, O., and Bryant, H. 1998. Properties of matrix representation with parsimony analyses. *Syst. Biol.* 47, 497–508.

Bininda-Emonds, O., Gittleman, J., and Purvis, A. 1999. Building large trees by combining phylogenetic information: A complete phylogeny of the extant Carnivora (Mammalia). *Biol. Rev.* 74, 143–175.

Bock, W. 1977. Foundations and methods of evolutionary classification. In: Hecht, M., Goody, P., and Hecht, B. (eds.), *Major Patterns in Vertebrate Evolution*. Plenum Press, New York, pp. 851–895.

Böcker, S., Canzar, S., and Klau, G. 2013. The generalized Robinson-Foulds metric. In: Darling, A., and Stoye, J. (eds.), *Algorithms in Bioinformatics. WABI 2013*. Lecture Notes in Computer Science. Springer, Berlin, Heidelberg, vol. 8126, pp. 156–169.

Bookstein, F. 1978. *The Measurement of Biological Shape and Shape Change*. Lecture Notes in Biomathematics 24. Springer-Verlag, Berlin, 191 pp.

Bookstein, F. 1986. Size and shape spaces for landmark data in two dimensions. *Stat. Sc.* 1, 181–242.

Bookstein, F. 1989. Principal warps: Thin-plate splines and the decomposition of deformations. *IEEE Trans. Pattern Anal. Mach. Intell.* 11, 567–585.

Bookstein, F. 1991. *Morphometric Tools for Landmark Data, Geometry and Biology*. Cambridge University Press, Cambridge, 435 pp.

Bookstein, F. 1994. Can biometrical shape be a homologous character? In: Hall, B. (ed.), *Homology: The Hierarchical Basis of Comparative Biology*. Academic Press, New York, pp. 197–227.

Bordewich, M., and Semple, C. 2005. On the computational complexity of the rooted subtree prune and regraft distance. *Ann. Combinatorics* 8, 409–423.

Bremer, K. 1988. The limits of amino-acid sequence data in angiosperm phylogenetic reconstruction. *Evolution* 42, 795–803.

Bremer, K. 1990. Combinable component consensus. *Cladistics* 6, 369–372.

Bremer, K. 1994. Branch support and tree stability. *Cladistics* 10, 295–304.

Brocklehurst, N., and Haridy, Y. 2021. Do meristic characters used in phylogenetic analysis evolve in an ordered manner? *Syst. Biol.* 70, 707–718.

Bron, C., and Kerbosch, J. 1973. Algorithm 457: Finding all cliques of an undirected graph. *Commun. ACM* 16, 575–577.

Brower, A. 2006. The how and why of branch support and partitioned branch support, with a new index to assess partition incongruence. *Cladistics* 22, 378–386.

Brown, J., Parins-Fukuchi, C., Stull, G., Vargas, O., and Smith, S. 2017 Bayesian and likelihood phylogenetic reconstructions of morphological traits are not discordant when taking uncertainty into consideration: A comment on Puttick et al. *Proc. R. Soc. B* 284, 20170986.

Bruen, T., and Bryant, D. 2008. Parsimony via consensus. *Syst. Biol.* 57, 251–256.

Cain, A., and Harrison, G. 1960. Phyletic weighting. *Proc. Zool. Soc. London* 135, 1–31.

Camin, J., and Sokal, R. 1965. A method for deducing branching sequences in phylogeny. *Evolution* 19, 311–326.

Carpenter, J. 1988. Choosing among multiple equally parsimonious cladograms. *Cladistics* 4, 291–296.

Carpenter, J. 1994. Successive weighting, reliability, and evidence. *Cladistics* 10, 215–220.

Carpenter, J. 1996. Uninformative bootstrapping. *Cladistics* 12, 177–181.

Carpenter, J., Goloboff, P., and Farris, J. 1998. PTP is meaningless, T-PTP is contradictory: A reply to Trueman. *Cladistics* 14, 105–116.

Carroll, H., Clement, M., Ebbert, M., and Snell, Q. 2007. *PSODA: Better Tasting and Less Filling Than PAUP*. Brigham Young University Faculty Publications, p. 228. https://scholarsarchive.byu.edu/facpub/228.

Carroll, H., Teichert, A., Krein, J., Sundberg, K., Snell, Q., and Clement, M. 2009. An open source phylogenetic search and alignment package. *Int. J. Bioinf. Res. Appl.* 5, 349–364.

Carter, M., Hendy, M., Penny, D., Szekely, L., and Wormald, N. 1990. On the distribution of lengths of evolutionary trees. *SIAM J. Disc. Math.* 3, 38–47.

Catalano, S., Ercoli, M., and Prevosti, F. 2015. The more, the better: The use of multiple landmark configurations to solve the phylogenetic relationships in Musteloids. *Syst. Biol.* 64, 294–306.

Catalano, S., and Goloboff, P. 2012. Simultaneously mapping and superimposing landmark configurations with parsimony as optimality criterion. *Syst. Biol.* 61, 392–400.

Catalano, S., and Goloboff, P. 2018. A guide for the analysis of continuous and landmark characters in TNT (Tree Analysis using New Technologies). ResearchGate Technical Report, doi: 10.13140/RG.2.2.23797.27360.

Catalano, S., Goloboff, P., and Giannini, N. 2010. Phylogenetic morphometrics (I): The use of landmark data in a phylogenetic framework. *Cladistics* 26, 539–549.

Catalano, S., Saidman, B., and Vilardi, J. 2009. Evolution of small inversions in chloroplast genome: A case study from a recurrent inversion in angiosperms. *Cladistics* 25, 93–104.

Catalano, S., and Torres, A. 2017. Phylogenetic inference based on landmark data in 41 empirical data sets. *Zool. Scripta* 46, 1–11.

Cavalli-Sforza, L., and Edwards, A. 1967. Phylogenetic analysis: Models and estimation procedures. *Evolution.* 21:550–570.

Chappill, J.A. 1989. Quantitative characters in phylogenetic analysis. *Cladistics* 5, 217–234.

Chen, D., Diao, L., Eulenstein, O., Fernández-Baca, D., and Sanderson, M. 2003. Flipping: A supertree construction method. In: Janowitz, M., Lapointe, F., McMorris, F., Mirkin, B., and Roberts, F. (eds.), *DIMACS Series in Discrete Mathematics and Theoretical Computer Science*. Bioconsensus. American Mathematical Society, Providence, RI, vol. 61, pp. 135–160.

Coddington, J. 1988. Cladistic tests of adaptational hypotheses. *Cladistics* 4, 3–22.

Coddington, J., and Scharff, N. 1994. Problems with zero-length branches. *Cladistics* 10, 415–423.

Colless, D. 1980. Congruence between morphometric and allozyme data for *Menidia* species: A reappraisal. *Syst. Zool.* 29, 288–299.

Colless, D. 1981. Predictivity and stability in classifications: Some comments on recent studies. *Syst. Zool.* 30, 325–331.

Congreve, C., and Lamsdell, J. 2016. Implied weighting and its utility in palaeontological datasets: A study using modelled phylogenetic matrices. *Palaeontology* 59, 447–462.

Cooper, N., Thomas, G., Venditti, C., Meade, A., and Freckleton, R. 2016. A cautionary note on the use of Ornstein Uhlenbeck models in macroevolutionary studies. *Biol. J. Linn. Soc.* 118, 64–77.

Cotton, J.A., and Wilkinson, M. 2007. Majority-rule supertrees. *Syst. Biol.* 56: 445–452.

Cranston, P., and Humphries, C. 1988. Cladistics and computers: A chironomid conundrum? *Cladistics* 4, 72–92.

Csürös, M. 2008. Ancestral reconstruction by asymmetric Wagner parsimony over continuous characters and squared parsimony over distributions. In: Nelson, C., and Vialette, S. (eds.), *Comparative Genomics*, Volume 5267 of Lecture Notes in Computer Science. Springer, Berlin, pp. 72–86.

Cummins, C., and McInerney, J. 2011. A method for inferring the rate of evolution of homologous characters that can potentially improve phylogenetic inference, resolve deep divergence and correct systematic biases. *Syst. Biol.* 60, 833–844.

Darwin, C. 1859. *On the Origin of Species by Means of Natural Selection, or the Preservation of Favoured Races in the Struggle for Life.* John Murray, London.

DasGupta, B., He, X., Jiang, T., Li, M., Tromp, J., and Zhang, L. 1997. On distances between phylogenetic trees. In: *Proceedings of the 8th Annual ACM-SIAM Symposium on Discrete Algorithms (SODA).* New Orleans, pp. 427–436. www.siam.org/meetings/archives. php.

Davis, J., Nixon, K., and Little, D. 2005. The limits of conventional cladistic analysis. In: Albert, V. (ed.), *Parsimony, Phylogeny, and Genomics.* Oxford University Press, Oxford, pp. 119–147.

Day, W., McMorris, F., and Wilkinson, M. 2008. Explosions and hot spots in supertree methods. *J. Theo. Biol.* 253, 345–348.

De Laet, J. 1997. A reconsideration of three-item analysis, the use of implied weights in cladistics, and a practical application in Gentianaceae. Ph. D. Thesis, K.U.Leuven, Faculteit der Wetenschappen, 232 pp.

De Laet, J. 2005. Parsimony analysis of unaligned sequence data: Maximization of homology and minimization of homoplasy, not minimization of operationally defined total cost or minimization of equally weighted transformations. *Cladistics* 31, 550–567.

De Laet, J., Farris, J., and Goloboff, P. 2004. Treatment of multiple trees in resampling analyses. P. 590 in Grandcolas, P. Abstracts of the 23rd Annual Meeting of the Willi Hennig Society. "Phylogenetics and Evolutionary Biology". *Cladistics* 20, 583–608.

De Laet, J., and Smets, E. 1998. On the three-taxon approach to parsimony analysis. *Cladistics* 14, 363–381.

Deepak, A., Dong, J., and Fernández-Baca, D. 2012. Identifying rogue taxa through reduced consensus: NP-Hardness and exact algorithms. In: *ISBRA'12 Proceedings of the 8th International Conference on Bioinformatics Research and Applications.* Springer-Verlag, Heidelberg, pp. 87–89.

Didier, G. 2017. Time-dependent-asymmetric-linear-parsimonious ancestral state reconstruction. *Bull. Math. Biol.* 79, 2334–2355.

Didier, G., Chabrol, O., and Laurin, M. 2019. Parsimony-based test for identifying changes in evolutionary trends for quantitative characters: Implications for the origin of the amniotic egg. *Cladistics* 35, 576–599.

Dong, J., and Fernández-Baca, D. 2009. Properties of majority rule supertrees. *Syst. Biol.* 58: 360–367.

Dong, J., Fernández-Baca, D., and McMorris, F.R. 2010a. Constructing majority-rule supertrees. *Algorithms Mol. Biol.* 5, 1–16.

Dong, J., Fernández-Baca, D., McMorris, F.R., and Powers, R. 2010b. Majority-rule (+) consensus trees. *Math. Biosci.* 228, 10–15.

Donoghue, M., Olmstead, R., Smith, J., and Palmer, J. 1992. Phylogenetic relationships of Dipsacales based on rbcL sequences. *Ann. Missouri Bot. Gard.* 79, 333–345.

Efron, B. 1979. Bootstrap methods: Another look at the jackknife. *Ann. Statist.* 7, 1–26.

Efron, B., Halloran, E., and Holmes, S. 1996. Bootstrap confidence levels for phylogenetic trees. *Proc. Natl. Acad. Sc. USA* 93, 7085–7090.

Eldredge, N., and Cracraft, J. 1980. *Phylogenetic Patterns and the Evolutionary Process: Method and Theory in Comparative Biology.* Columbia University Press, New York, 349 pp.

Estabrook, G.F. 1992. Evaluating undirected positional congruence of individual taxa between two estimates of the phylogenetic tree for a group of taxa. *Syst. Biol.* 41, 172–177.

Estabrook, G.F., McMorris, F., and Meacham, C. 1985. Comparison of undirected phylogenetic trees based on subtrees of four evolutionary units. *Syst. Zool.* 34, 193–200.

Estabrook, G.F., Strauch, J., and Fiala, K. 1977. An application of compatibility analysis to the Blackith's data on orthopteroid insects. *Syst. Zool.* 26, 269–276.

Eulenstein, O., Chen, D., Burleigh, G., Fernández-Baca, D., and Sanderson, M. 2004. Performance of flip supertree construction with a heuristic algorithm. *Syst. Biol.* 53, 299–308.

Faith, D., and Cranston, P. 1991. Could a cladogram this short have arisen by chance alone? On permutation tests for cladistic structure. *Cladistics* 7, 1–28.

Faith, D., and Trueman, J. 2001. Towards an inclusive philosophy for phylogenetic inference. *Syst. Biol.* 50, 331–350.

Faivovich, J. 2002. On RASA. *Cladistics* 18, 324–333.

Faivovich, J., Haddad, C., Baêta, D., Jungfer, K.H., Álvares, G., Brandão, R., Christopher Sheil, C., Barrientos, L., Barrio-Amorós, C., Cruz, C., and Wheeler, W. 2010. The phylogenetic relationships of the charismatic poster frogs, Phyllomedusinae (Anura, Hylidae). *Cladistics* 26, 227–261.

Faivovich, J., Haddad, C., Garcia, P., Frost, D., Campbell, J., and Wheeler, W. 2005. Systematic review of the frog family Hylidae, with special reference to Hylinae: Phylogenetic analysis and taxonomic revision. *Bull. Am. Mus. Nat. Hist.* 294, 1–240.

Farris, J. 1969. A successive approximations approach to character weighting. *Syst. Zool.* 18, 374–385.

Farris, J. 1970. Methods for computing Wagner trees. *Syst. Zool.* 19, 83–92.

Farris, J. 1971. The hypothesis of nonspecificity and taxonomic congruence. *Ann. Rev. Ecol. Syst.* 2, 277–302.

Farris, J. 1973. On comparing the shapes of taxonomic trees. *Syst. Zool.* 22, 50–54.

Farris, J. 1982. Simplicity and informativeness in systematics and phylogeny. *Syst. Zool.* 31, 413–444.

Farris, J. 1983. The logical basis of phylogenetic analysis. In: Platnick, N.I., and Funk, V.A. (eds.), *Advances in Cladistics.* Columbia University Press, New York, NY, pp. 7–36.

Farris, J. 1988. *Hennig86.* Program and documentation, distributed by the author. New York.

Farris, J. 1989. The retention index and the rescaled consistency index. *Cladistics* 5:417–419.

Farris, J. 1990. Phenetics in camouflage. *Cladistics* 6, 91–100.

Farris, J. 1991. Excess homoplasy ratios. *Cladistics* 7, 81–91.

Farris, J. 1995. Conjectures and refutations. *Cladistics* 11, 105–118.

Farris, J. 2000. Corroboration versus "strongest evidence". *Cladistics* 16, 385–393.

Farris, J. 2001. Support weighting. *Cladistics* 17, 389–394.

Farris, J. 2002. RASA attributes highly significant structure to randomized data. *Cladistics* 18, 334–353.

Farris, J. 2003. Efficient calculation of completely connected sets. In: Muona, J. (ed.), Abstracts of the 21st Annual Meeting of the Willi Hennig Society. *Cladistics* 19, 148–163.

Farris, J. 2007. Coding of continuous characters, revisited. *Darwiniana* 45 (Suppl.), 9–10.

Farris, J. 2008. Parsimony and explanatory power. *Cladistics* 24, 825–847.

Farris, J., Albert, V., Källersjö, M., Lipscomb, D.L., and Kluge, A. 1996. Parsimony jackknifing outperforms neighbor-joining. *Cladistics* 12, 99–124.

Farris, J., and Goloboff, P. 2008. Is REP a measure of "objective support"? *Cladistics* 24, 1065–1069.

Farris, J., Källersjö, M., Crowe, T., Lipscomb, D., and Johansson, U. 1999. Frigatebirds, Tropicbirds, and Ciconiida: Excesses of confidence probability. *Cladistics* 15, 1–7.

Farris, J., Källersjö, M., and De Laet, J. 2001. Branch lengths do not indicate support—even in maximum likelihood. *Cladistics* 17, 298–299.

Farris, J., Källersjö, M., Kluge, A., and Bult, C. 1994. Permutations. *Cladistics* 10, 65–76.

Farris, J., and A. Kluge. 1985. Parsimony, synapomorphy, and explanatory power: A reply to Duncan. *Taxon* 34, 130–135.

Farris, J., and Kluge, A. 1986. Synapomorphy, parsimony, and evidence. *Taxon* 35, 298–304.

Farris, J., Kluge, A., and Carpenter, J. 1990. Weighting: Successive, spectral, paradoxical, and phantasmal. In: McHugh, J., and Bruneau, A., The eight Annual Meeting of the Willi Hennig Society. *Cladistics* 6, 197–204.

Farris, J., and Mickevich, M. 1982. *Physys: Phylogenetic System*. Mainframe program and documentation, privately distributed by the author, Port Jefferson, New York.

Felsenstein, J. 1978. Cases in which parsimony and compatibility methods will be positively misleading. *Syst. Zool.* 27, 401–410.

Felsenstein, J. 1981. A likelihood approach to character weighting and what it tells us about parsimony and compatibility. *Biol. J. Linn. Soc.* 16, 183–196.

Felsenstein, J. 1985a. Phylogenies and the comparative method. *Amer. Nat.* 125, 1–15.

Felsenstein, J. 1985b. Confidence limits on phylogenies: An approach using the bootstrap. *Evolution* 39, 783–791.

Felsenstein, J. 1988. Phylogenies and quantitative characters. *Ann. Rev. Ecol. Syst.* 19, 445–471.

Felsenstein, J. 1993. *PHYLIP, ver. 3.5c*. Department of Genetics, University of Washington, Seattle.

Felsenstein, J. 2004. *Inferring Phylogenies*. Sinauer Associates, Sunderland, MA, 664 pp.

Felsenstein, J. 2005. *PHYLIP (Phylogeny Inference Package) version 3.6*. Distributed by the author. Department of Genome Sciences, University of Washington, Seattle. http://evo lution.genetics.washington.edu/phylip.html.

Felsenstein, J., and Kishino, H. 1993. Is there something wrong with the bootstrap on phylogenies? A reply to Hillis and Bull. *Syst. Biol.* 42, 193–200.

Ferraggine, V., and Torcida, S. 2016. RPS (Resistant Procrustes Software): una herramienta novedosa para el análisis morfométrico resistente. 4to Congreso Nacional de Ingeniería en Informática/Sistemas de Información, Salta, Argentina.

Finden, C., and Gordon, A. 1985. Obtaining common pruned trees. *J. Classification* 2, 255–276.

Fischer, M. 2019. On the uniqueness of the maximum parsimony tree for data with up to two substitutions: An extension of the classic Buneman theorem in phylogenetics. *Mol. Phyl. Evol.* 137, 127–137.

Frost, D., Grant, T., Faivovich, J., Bain, R., Haas, A., Haddad, C., de Sá, R., Channig, A., Wilkinson, M., Donnellan, S., Raxworthy, C., Campbell, J., Blotto, B., Moler, P., Drewes, C., Nussbaum, R., Lynch, J., Green, D., and Wheeler, W. 2006. The amphibian tree of life. *Bull. Am. Mus. Nat. Hist.* 297, 1–370.

Gatesy, J., Mathee, C., De Salle, R., and Hayashi, C. 2002. Resolution of a supertree/supermatrix paradox. *Syst. Biol.* 51, 652–664.

Gatesy, J., O'Grady, P., and Baker, R. 1999. Corroboration among data sets in simultaneous analysis: Hidden support for phylogenetic relationships among higher-level artiodactyl taxa. *Cladistics* 15, 271–313.

Geist, A., Beguelin, A., Dongarra, J., Jiang, W., Manchek, R., and Sunderam, V. 1994. *PVM: Parallel Virtual Machine—A Users' Guide and Tutorial for Networked Parallel Computing*. MIT Press, Cambridge, MA, 299 pp.

Giannini, N., and Goloboff, P. 2010. Delayed-response phylogenetic correlation, an optimization-based method to test covariation of continuous characters. *Evolution* 64, 1885–1898.

Giribet, G. 2003. Stability in phylogenetic formulations, and its relationship to nodal support. *Syst. Biol.* 52, 554–564.

Giribet, G., and Wheeler, W. 2002. On bivalve phylogeny: A high-level analysis of the Bivalvia (Mollusca) based on combined morphology and DNA sequence data. *Invert. Biol.* 121, 271–324.

Giribet, G., and Wheeler, W. 2007. The case for sensitivity: A response to Grant and Kluge. *Cladistics* 23, 294–296.

Goldman, N. 1990. Maximum likelihood inference of phylogenetic trees, with special reference to a Poisson process model of DNA substitution and to parsimony analyses. *Syst. Zool.* 39, 345–361.

Goloboff, P. 1991a. Homoplasy and the choice among cladograms. *Cladistics* 7, 215–232.

Goloboff, P. 1991b. Random data, homoplasy, and information. *Cladistics* 7, 395–406.

Goloboff, P. 1993a. Estimating character weights during tree search. *Cladistics* 9, 83–91.

Goloboff, P. 1993b. *NONA: A Tree-Searching Program*. Program and documentation. www.lillo.org.ar/phylogeny/Nona-PeeWee/Pars-win.exe.

Goloboff, P. 1993c. *Pee-Wee: Parsimony and Implied Weights*. Program and documentation. www.lillo.org.ar/phylogeny/Nona-PeeWee/Pars-win.exe.

Goloboff, P. 1993d. Character optimization and calculation of tree lengths. *Cladistics* 9, 433–436.

Goloboff, P. 1995a. A revision of the South American spiders of the family Nemesiidae (Araneae, Mygalomorphae). Part I: Species from Peru, Chile, Argentina, and Uruguay. *Bull. Amer. Mus. Nat. Hist.* 224, 1–189.

Goloboff, P. 1995b. Parsimony and weighting: A reply to Turner and Zandee. *Cladistics* 10, 91–104.

Goloboff, P. 1996. Methods for faster parsimony analysis. *Cladistics* 12, 199–220.

Goloboff, P. 1997. Self-weighted optimization: Character state reconstructions and tree searches under implied transformation costs. *Cladistics* 13, 225–245.

Goloboff, P. 1999. Analyzing large data sets in reasonable times: Solutions for composite optima. *Cladistics* 15, 415–428.

Goloboff, P. 2005. Minority rule supertrees? MRP, compatibility, and minimum flip may display the *least* frequent groups. *Cladistics* 21, 282–294.

Goloboff, P. 2007. Calculating SPR distances between trees. *Cladistics* 24, 591–597.

Goloboff, P. 2014a. Extended implied weighting. *Cladistics* 30, 260–272.

Goloboff, P. 2014b. Hide and vanish: Data sets where the most parsimonious tree is known but hard to find, and their implications for tree search methods. *Mol. Phyl. Evol.* 79, 118–131.

Goloboff, P. 2014c. Oblong, a program to analyse phylogenomic data sets with millions of characters, requiring negligible amounts of RAM. *Cladistics* 30, 273–281.

Goloboff, P. 2015. Computer science and parsimony: A reappraisal, with discussion of methods for poorly structured datasets. *Cladistics* 31, 210–225.

Goloboff, P., and Arias, J. 2019. Likelihood approximations of implied weights parsimony can be selected over the Mk model by the Akaike information criterion. *Cladistics* 35, 695–716.

Goloboff, P., Carpenter, J.M., Arias, J.S., and Miranda, D. 2008a. Weighting against homoplasy improves phylogenetic analysis of morphological data sets. *Cladistics* 24, 758–773.

Goloboff, P., and Catalano, S. 2011. Phylogenetic morphometrics (II): Algorithms for landmark optimization. *Cladistics* 27, 42–51.

Goloboff, P., and Catalano, S. 2016. TNT version 1.5, including a full implementation of phylogenetic morphometrics. *Cladistics* 32, 221–238.

Goloboff, P., Catalano, S., Mirande, M., Szumik, C., Arias, J., Källersjö, M., and Farris, J. 2009a. Phylogenetic analysis of 73 060 taxa corroborates major eukaryotic groups. *Cladistics* 25, 211–230.

Goloboff, P., De Laet, J., Ríos-Tamayo, D., and Szumik, C. 2021a. A reconsideration of inapplicable characters, and an approximation with step-matrix recoding. *Cladistics* 37, 596–629.

Goloboff, P., and Farris, J. 2001. Methods for quick consensus estimation. *Cladistics* 17, S26–S34.

Goloboff, P., Farris, J., Källersjö, M., Oxelmann, B., Ramírez, M., and Szumik, C. 2003a. Improvements to resampling measures of group support. *Cladistics* 19, 324–332.

Goloboff, P., Farris, J., and Nixon, K. 2003b. TNT: Tree analysis using new technology, vers. 1.0. Program and documentation. www.lillo.org.ar/phylogeny/tnt.

Goloboff, P., Farris, J., and Nixon, K. 2008b. TNT, a free program for phylogenetic analysis. *Cladistics* 24, 774–786.

Goloboff, P., Mattoni, C., and Quinteros, S. 2006. Continuous characters analyzed as such. *Cladistics* 22, 589–601.

Goloboff, P., Mirande, M., and Arias, J. 2009b. On weighting characters differently in different parts of the cladogram. In: Szumik, C., and Goloboff, P. (eds.), A summit of cladistics: Abstracts of the 27th Annual Meeting of the Willi Hennig Society and VIII Reunión Argentina de Cladística y Biogeografía. *Cladistics* 25, 1–36.

Goloboff, P., and Morales, M. 2020. A phylogenetic C interpreter for TNT. *Bioinformatics* 36, 3988–3995

Goloboff, P., Pittman, M., Pol, D., and Xu, X. 2018a. Morphological data sets fit a common mechanism much more poorly than DNA sequences and call into question the Mkv model. *Syst. Biol.* 68, 494–504.

Goloboff, P., and Pol, D. 2002. Semi-strict supertrees. *Cladistics* 18, 514–525.

Goloboff, P., and Pol, D. 2005. Parsimony and bayesian phylogenetics. In: Albert, V. (ed.), *Parsimony, Phylogeny, and Genomics*. Oxford University Press, Oxford, pp. 148–159.

Goloboff, P., and Simmons, M. 2014. Bias in tree searches and its consequences for measuring groups supports. *Syst. Biol.* 63, 851–861.

Goloboff, P., and Szumik, C. 2015. Identifying unstable taxa: Efficient implementation of triplet-based measures of stability, and comparison with Phyutility and RogueNaRok. *Mol. Phyl. Evol.* 88, 93–104.

Goloboff, P., and Szumik, C. 2016. Problems with supertrees based on the subtree prune-and-regraft distance, with comments on majority rule supertrees. *Cladistics* 32, 82–89.

Goloboff, P., Torres, A., and Arias, J. 2017. Weighted parsimony outperforms other methods of phylogenetic inference under models appropriate for morphology. *Cladistics* 34, 407–437.

Goloboff, P., Torres, A., and Arias, J. 2018b. Parsimony and model based phylogenetic methods for morphological data: A response to O'Reilly et al. (2017). *Palaeontology* 61, 625–630.

Goloboff, P., Torres, A., and Catalano, S. 2021b. Parsimony analysis of phylogenomic datasets (II): Evaluation of PAUP*, MEGA, and MPBoot. *Cladistics* 38, 126–146.

Goloboff, P., and Wilkinson, M. 2018. On defining a unique phylogenetic tree with homoplastic characters. *Mol. Phylogenet. Evol.* 122, 95–101.

González, P., Bonfili, N., Vallejo Azar, M., Barbeito-Andres, J., Bernal, V., and Pérez, S. 2019. Description and analysis of spatial patterns in geometric morphometric data. *Evol. Biol.* 46, 260–270.

González-José, R., Escapa, I., Neves, W., Cúneo, R., and Pucciarelli, N. 2008. Cladistic analysis of continuous modularized traits provides phylogenetic signals in Homo evolution. *Nature* 453, 775–778.

Goodman, M., Olson, C., Beeber, J., and Czelusniak, J. 1982. New perspectives in the molecular biological analysis of mammalian phylogeny. *Acta Zool. Fenn.* 169, 19–35.

Gordon, A. 1980. On the assessment and comparison of classifications. In: Tomassine, R. (ed.), *Analyse de donnes et informatique*. INRIA, Le Chesnay, France, pp. 149–160.

Gordon, A. 1986. Consensus supertrees: The synthesis of rooted trees containing overlapping sets of labelled leaves. *J. Classif.* 3, 335–348.

Goremykin, V., Nikiforova, S., and Bininda-Emonds, O. 2010. Automated removal of noisy data in phylogenomic analyses. *J. Mol. Evol.* 71, 319–331.

Gower, J. 1975. Generalized Procrustes analysis. *Psychometrika* 40, 33–51.

Grandcolas, P. 2015. Adaptation. In: Heams, T., Huneman, P., Lecointre, G., and Silberstein, M. (eds.), *Handbook of Evolutionary Thinking in the Sciences*. Springer, Dordrecht, pp. 77–93.

Grant, T., and Kluge, A. 2003. Data exploration in phylogenetic inference: Scientific, heuristic, or neither. *Cladistics* 19, 379–418.

Grant, T., and Kluge, A. 2005. Stability, sensitivity, science and heurism. *Cladistics* 21, 597–604.

Grant, T., and Kluge, A. 2007. Ratio of explanatory power (REP): A new measure of group support. *Mol. Phyl. Evol.* 44, 483–487.

Grant, T., and Kluge, A. 2008a. Credit where credit is due: The Goodman—Bremer support metric. *Mol. Phyl. Evol.* 49, 405–406.

Grant, T., and Kluge, A. 2008b. Clade support measures and their adequacy. *Cladistics* 24, 1051–1064.

Grant, T., and Kluge, A. 2010. REP provides meaningful measurement of support across datasets. *Mol. Phyl. Evol.* 55, 340–342.

Guindon, S., and Gascuel, O. 2003. A simple, fast and accurate algorithm to estimate large phylogenies by maximum likelihood. *Syst. Biol.* 52, 696–704

Gutmann, W. 1977. Phylogenetic reconstruction: Theory, methodology, and application to Chordate evolution. In: Hecht, M., Goody, P., and Hecht, B. (eds.), *Major Patterns of Vertebrate Evolution*. Plenum Press, New York, pp. 645–669.

Harshman, J. 1994. The effect of irrelevant characters on bootstrap values. *Syst. Biol.* 43, 419–424.

Hecht, M., and Edwards, J. 1976. The determination of parallel or monophyletic relationships: The proteid salamanders—a test case. *Amer. Nat.* 110, 653–677.

Hecht, M., and Edwards, J. 1977. The methodology of phylogenetic inference above the species level. In: Hecht, M., Goody, P., and Hecht, B. (eds.), *Major Patterns of Vertebrate Evolution*. Plenum Press, New York, pp. 3–51.

Hedges, S. 1992. The number of replications needed for accurate estimation of the bootstrap P value in phylogenetic studies. *Mol. Biol. Evol.* 9, 366–369.

Hein, J., Jiang, T., Wang, L., and Zhang, K. 1996. On the complexity of comparing evolutionary trees. *Discrete Appl. Math.* 71, 153–169.

Hendrixson, B., and Bond, J. 2009. Evaluating the efficacy of continuous quantitative characters for reconstructing the phylogeny of a morphologically homogeneous spider taxon (Araneae, Mygalomorphae, Antrodiaetidae, Antrodiaetus). *Mol. Phyl. Evol.* 53, 300–313.

Hennig, W. 1966. *Phylogenetic Systematics*. University of Illinois Press, Urbana, 263 pp.

Hovenkamp, P. 2009. Support and stability. *Cladistics* 25, 107–108.

Hillis, D. 1991. Discriminating between phylogenetic signal and random noise in DNA sequences. In: Miyamoto, M., and Cracraft, J. (eds.), *Phylogenetic Analysis of DNA Sequences*. Oxford University Press, Oxford.

Hillis, D., and Bull, J. 1993. An empirical test of bootstrapping as a method for assessing confidence in phylogenetic analysis. *Syst. Biol.* 42, 182–192.

Hoang, D., Vinh, L., Flouri, T., Stamatakis, A., and von Haeseler, A. 2018. MPBoot: Fast phylogenetic maximum parsimony tree inference and bootstrap approximation. *BMC Evolutionary Biology* 18, 11.

Höhna, S., Landis, M., Heath, T., Boussau, B., Lartillot, N., Moore, B., Huelsenbeck, J., and Ronquist, F. 2016. RevBayes: Bayesian phylogenetic inference using graphical models and an interactive model-specification language. *Syst. Biol.* 65, 726–736.

Holder, M., Lewis, P., and Swofford, D. 2010. The Akaike information criterion will not choose the no common mechanism model. *Syst. Biol.* 59, 477–485.

Hormiga, G., Scharff, N., and Coddington, J. 2000. The phylogenetic basis of sexual size dimorphism in orb-weaving spiders (Araneae, Orbiculariae). *Syst. Biol.* 49, 435–462.

Horovitz, I. 1999. A report on "one day symposium on numerical cladistics". *Cladistics* 15, 177–182.

Huelsenbeck, J. 1991. Tree-length distribution skewness: An indicator of phylogenetic information. *Syst. Zool.* 3, 257–270.

Huelsenbeck, J., Alfaro, M., and Suchard, M. 2011. Biologically inspired phylogenetic models strongly outperform the no common mechanism model. *Syst. Biol.* 60, 225–232.

Humphries, C. 2002. Homology, characters and continuous variables. In: MacLeod, N., and Forey, P. (eds.), *Morphology, Shape and Phylogeny*. Taylor & Francis, London, pp. 8–26.

Janies, D., Studer, J., Handelman, S., and Linchangco, G. 2013. A comparison of supermatrix and supertree methods for multilocus phylogenetics using organismal datasets. *Cladistics* 29, 560–566.

Källersjö, M., Albert, V., and Farris, J. 1999. Homoplasy increases phylogenetic structure. *Cladistics* 15, 91–93.

Källersjö, M., Farris, J., Kluge, A., and Bult, C. 1992. Skewness and permutation. *Cladistics* 8, 275–287.

Karp, R. 1972. Reducibility among combinatorial problems. In: Miller, R., and Thatcher, J. (eds.), *Complexity of Computer Computations*. Plenum Press, New York, pp. 85–103.

Kearney, M., and Clark, J. 2003. Problems due to missing data in phylogenetic analyses including fossils: A critical review. *J. Vert. Paleont.* 23, 263–274.

Keating, J., Sansom, R., Sutton, M., Knight, C., and Garwood, R. 2020. Morphological phylogenetics evaluated using novel evolutionary simulations. *Syst. Biol.* 69, 897–912.

Kendall, D. 1984. Shape-manifolds, Procrustean metrics and complex projective spaces. *Bull. London Math. Soc.* 16, 81–121.

Kendrick, W. 1965. Complexity and dependence in computer taxonomy. *Taxon* 14, 141–154.

Kernighan, B., and Ritchie, D. 1988. *The C Programming Language*, 2nd ed. Prentice Hall, Englewood Cliffs.

Kishino, H., Miyata, T., and Hasegawa, M. 1990. Maximum likelihood inference of protein phylogeny and the origin of chloroplasts. *J. Mol. Evol.* 31, 151–160.

Kitching, I., Forey, P., Humphries, C., and Williams, D. 1998. *Cladistics: The Theory and Practice of Phylogenetic Systematics*. Oxford University Press, Oxford. 227 pp.

Klassen, G., Mooi, R., and Locke, A. 1990. Consistency indices and random data sets, or, how low you can get? Presented at the Fourth International Congress of Systematic and Evolutionary Biology, University of Maryland, 1–7 July 1990 (no page numbers).

Klingenberg, C. 2011. MorphoJ: An integrated software package for geometric morphometrics. *Mol. Ecol. Resources* 11, 353–357.

Klingenberg, C. 2020. Walking on Kendall's shape space: Understanding shape spaces and their coordinate systems. *Evol. Biol.* 47, 334–352.

Klingenberg, C. 2021. How exactly did the nose get that long? A critical rethinking of the Pinocchio effect and how shape changes relate to landmarks. *Evol. Biol.* 48, 115–127.

Klingenberg, C., and Gidaszewski, N. 2010. Testing and quantifying phylogenetic signals and homoplasy in morphometric data. *Syst. Biol.* 59, 245–261.

Kluge, A. 1989. A concern for evidence and a phylogenetic hypothesis of relationships among Epicrates (Boidae, Serpentes). *Syst. Zool.* 38, 7–25.

Kluge, A. 1997. Sophisticated falsification and research cycles: Consequences for differential character weighting in phylogenetic systematics. *Zool. Scr.* 26, 349–360.

Kluge, A. 2003. The repugnant and the mature in phylogenetic inference: Atemporal similarity and historical identity. *Cladistics* 19, 356–368.

Kluge, A., and Farris, J. 1969. Quantitative phyletics and the evolution of anurans. *Syst. Zool.* 18, 1–32.

Kluge, A., and Grant, T. 2006. From conviction to anti-superfluity: Old and new justifications for parsimony in phylogenetic inference. *Cladistics* 22, 276–288.

Kopuchian, C., and Ramírez, M. 2010. Behaviour of resampling methods under different weighting schemes, measures and variable resampling strengths. *Cladistics* 26, 86–97.

Kuhner, M., and Yamato, J. 2015. Practical performance of tree comparison metrics. *Syst. Biol.* 64, 205–214.

Kumar, S., Stecher, G., Li, M., Knyaz, C., and Tamura, K. 2018. MEGA X: Molecular Evolutionary Genetics Analysis across computing platforms. *Mol. Biol. Evol.* 35, 1547–1549.

Kvist, S., and Siddall, M. 2013. Phylogenomics of Annelida revisited: A cladistic approach using genome-wide expressed sequence tag data mining and examining the effects of missing data. *Cladistics* 29, 435–448.

Lanyon, S. 1985. Detecting internal inconsistencies in distance data. *Syst. Zool.* 34, 397–403.

Lemoine, F., Domelevo Entfellner, J.-B., Wilkinson, E., Correia, D., Dávila Felipe, M., De Oliveira, T., and Gascuel, O. 2018. Renewing Felsenstein's phylogenetic bootstrap in the era of big data. *Nature* 556, 452–456.

Le Quesne, W. 1969. A method of selection of characters in numerical taxonomy. *Syst. Zool.* 18, 201–205.

Lewis, P. 2001. A likelihood approach to estimating phylogeny from discrete morphological character data. *Syst. Biol.* 50, 913–925.

Li, W.H., and Zharkikh, A. 1994. What is the bootstrap technique? *Syst. Biol.* 43, 424–430.

Liu, F., Miyamoto, M., Freire, N., Ong, P., Tennant, M., Young, T., and Gugel, K. 2001. Molecular and morphological supertrees for eutherian (placental) mammals. *Science* 291, 1786–1789.

Lockwood, C., Kimbel, W., and Lynch, J. 2004. Morphometrics and hominoid phylogeny: Support for a chimpanzee—human clade and differentiation among great ape subspecies. *Proc. Natl Acad. Sci. USA* 101, 4356–4360.

Lutteropp, S., Kozlov, A., and Stamatakis, A. 2020. A fast and memory-efficient implementation of the transfer bootstrap. *Bioinformatics* 36, 2280–2281.

Lyons-Weiler, J., Hoelzer, G., and Tausch, R. 1996. Relative apparent synapomorphy analysis (RASA): The statistical measurement of phylogenetic signal. *Mol. Biol. Evol.* 13, 749–757.

MacLeod, N. 1999. Generalizing and extending the Eigenshape method of shape space visualization and analysis. *Paleobiology* 25, 107–138.

MacLeod, N. 2002. Phylogenetic signal in morphometric data. In: MacLeod, N., and Forey, P. (eds.), *Morphology, Shape and Phylogeny.* Taylor & Francis, London, pp. 100–138.

Maddison, W. 1991. Squared-change parsimony reconstructions of ancestral states for continuous-valued characters on a phylogenetic tree. *Syst. Zool.* 40, 304–314.

Maddison, W. 1997. Gene trees in species trees. *Syst. Biol.* 46, 523–536.

Maddison, W., and Maddison, D. 2008. A modular system for evolutionary analysis. www.mesquiteproject.org.

Manton, S. 1977. *The Arthropoda: Habits, Functional Morphology, and Evolution.* Oxford University Press, Oxford, 527 pp.

Margush, T., and McMorris, F. 1981. Consensus n-trees. *Bull. Math. Biol.* 43, 239–244.

Martins, E. 1994. Estimating the rate of phenotypic evolution from comparative data. *Amer. Natur.* 144, 193–209.

Mattoni, C., Ochoa, J., Ojanguren, A., and Prendini, L. 2011. Orobothriurus (Scorpiones: Bothriuridae): Phylogeny, Andean biogeography, and the relative importance of genitalic and somatic characters. *Zool. Scripta* 41, 160–176.

Mayr, E. 1969. *Principles of Systematic Zoology.* McGraw-Hill, New York, 428 pp.

McNeill, J. 1978. Purposeful phenetics. *Syst. Zool.* 28, 465–482.

Meacham, C. 1983. Theoretical and Computational Considerations of the Compatibility of Qualitative Taxonomic Characters. In: Felsenstein, J. (eds) *Numerical Taxonomy.* NATO ASI Series (Series G: Ecological Sciences). Springer, Berlin, Heidelberg, Vol. 1, pp. 304–314.

Meier, R., Kores, P., and Darwin, S. 1991. Homoplasy slope ratio: A better measurement of observed homoplasy in cladistic analyses. *Syst. Zool.* 40, 74–88.

Mickevich, M. 1978. Taxonomic congruence. *Syst. Biol.*, 27, 143–158.

Miller, J., and Hormiga, G. 2004. Clade stability and the addition of data: A case study from erigonine spiders (Araneae: Linyphiidae, Erigoninae). *Cladistics* 20, 385–442.

Minh, B., Hahn, M., and Lanfear, R. 2020. New methods to calculate concordance factors for phylogenomic datasets. *Mol. Biol. Evol.* 37, 2727–2733.

Mirande, M. 2009. Weighted parsimony phylogeny of the family Characidae (Teleostei: Characiformes). *Cladistics* 25, 574–613.

Mishler, B., Thrall, P., Hopple, J., De Luna, E., and Vilgalys, R. 1992. A molecular approach to the phylogeny of Bryophytes: Cladistic analysis of chloroplast-encoded 16S and 23S ribosomal RNA genes. *The Bryologist* 95, 172–180.

Mongiardino, N., Soto, I., and Ramírez, M. 2015a. First phylogenetic analysis of the family Neriidae (Diptera), with a study on the issue of scaling continuous characters. *Cladistics* 31, 142–165.

Mongiardino, N., Soto, I., and Ramírez, M. 2015b. Overcoming problems with the use of ratios as continuous characters for phylogenetic analyses. *Zool. Scripta* 44, 463–474.

Monteiro, L. 2000. Why morphometrics is special: The problem with using partial warps as characters for phylogenetic inference. *Syst. Biol.* 49, 796–800.

MPI Forum. 1993. MPI: A message passing interface in Supercomputing'93. Proceedings of the 1993 ACM/IEEE conference on Supercomputing, December 1993, Portlan, Oregon, pp. 878–883.

Murphy, J., Puttick, M., O'Reilly, J., Pisani, D., and Donoghue, P. 2021. Empirical distributions of homoplasy in morphological data. *Palaeontology*. https://doi.org/10.1111/pala.12535.

Nakhleh, L., Ruths, D., and Wang, L.S. 2005. RIATA-HGT: A fast and accurate heuristic for reconstructing horizontal gene transfer. In: *Proceedings of the Eleventh International Computing and Combinatorics Conference* (COCOON 05). Lecture Notes in Computer Science. LNCS no. 3595. Springer, Berlin, pp. 84–93.

Neff, N. 1986. A rational basis for a priori character weighting. *Syst. Zool.* 35, 110–123.

Nelson, G. 1979. Cladistic analysis and synthesis: Principles and definitions, with a historical note on Adanson's Familles des plantes (1763–1764). *Syst. Zool.* 28, 1–21.

Nixon, K. 2008. Paleobotany, evidence, and molecular dating: An example from the Nymphaeales. *Ann. Miss. Bot. Gard.* 95, 43–50.

Nixon, K., and Carpenter, J. 1996a. On simultaneous analysis. *Cladistics* 12, 221–241.

Nixon, K., and Carpenter, J. 1996b. On consensus, collapsibility, and clade concordance. *Cladistics* 12, 305–321.

Nylander, J., Ronquist, F., Huelsenbeck, J., and Nieves-Aldrey, J. 2004. Bayesian phylogenetic analysis of combined data. *Syst. Biol.* 53, 47–67.

O'Meara, B., Ané, C., Sanderson, M., and Wainwright, P. 2006. Testing for different rates of continuous trait evolution using likelihood. *Evolution* 60, 922–933.

O'Meara, B., and Beaulieu, J. 2014. Modelling stabilizing selection: The attraction of Ornstein—Uhlenbeck models. In: Garamszegi, L. (eds.), *Modern Phylogenetic Comparative Methods and Their Application in Evolutionary Biology*. Springer, Berlin, Heidelberg, pp. 381–393.

O'Reilly, J., Puttick, M., Parry, L., Tanner, A., Tarver, J., Fleming, J., Pisani, D., and Donoghue, P. 2016. Bayesian methods outperform parsimony but at the expense of precision in the estimation of phylogeny from discrete morphological data. *Biol. Lett.* 12, 1–5.

O'Reilly, J., Puttick, M., Pisani, D., and Donoghue, P. 2017. Probabilistic methods surpass parsimony when assessing clade support in phylogenetic analyses of discrete morphological data. *Palaeontology* 16, 105–118.

O'Reilly, J., Puttick, M., Pisani, D., and Donoghue, P. 2018. Empirical realism of simulated data is more important than the model used to generate it: A reply to Goloboff et al. *Palaeontology* 61, 631–635.

Padial, J., Grant, T., and Frost, D. 2014. Molecular systematics of terraranas (Anura: Brach-ycephaloidea) with an assessment of the effects of alignment and optimality criteria. *Zootaxa* 3825, 1–132.

Palci, A., and Lee, M. 2019. Geometric morphometrics, homology and cladistics: Review and recommendations. *Cladistics* 35, 230–242.

Parins-Fukuchi, C. 2018. Bayesian placement of fossils on phylogenies using quantitative morphometric data. *Evolution* 72, 1801–1814.

Pattengale, N., Aberer, A., Swenson, K., Stamatakis, A., and Moret, B. 2011. Uncovering hidden phylogenetic consensus in large datasets. *IEEE/ACM Trans. Comput. Biol. Bioinform.* (TCBB) 8, 902–911.

Pattengale, N., Alipour, M., Bininda-Emonds, O., Moret, B., and Stamatakis, A. 2010. How many bootstrap replicates are necessary? *J. Comp. Biol.* 17, 337–354.

Pattengale, N., Gottlieb, E., and Moret, B. 2007. Efficiently computing the Robinson-Foulds metric. *J. Comput. Biol.* 14, 724–735.

Patterson, C. 1982. Morphological characters and homology. In Joysey, K., and Friday, A. (eds.), *Problems of phylogenetic reconstruction*. Systematics Association special volume nr. 21. Systematics Association, London, pp. 21–74.

Pei, R., Pittman, M., Goloboff, P., Dececchi, T., Habib, M., Kaye, T., Larsson, H., Norell, M., Brusatte, S., and Xu, X. 2020. Potential for powered flight neared by most close avialan relatives but few crossed its thresholds. *Current Biology* 30, 1–14.

Perrard, A., Lopez-Osorio, F., and Carpenter, J. 2016. Phylogeny, landmark analysis and the use of wing venation to study the evolution of social wasps (Hymenoptera: Vespidae: Vespinae). *Cladistics* 32, 406–425.

Platnick, N. 1978. [Review of] The Arthropoda: Habits, functional morphology, and evolution. S.M. Manton. *Syst. Zool.* 27, 252–255.

Platnick, N., Griswold, C., and Coddington, J. 1991. On missing entries in cladistic analysis. *Cladistics* 7, 337–343.

Pimentel, R., and Riggins, R. 1987. The nature of cladistic data. *Cladistics* 3, 201–209.

Pisani, D., and Wilkinson, M. 2002. Matrix representation with parsimony, taxonomic congruence, and total evidence. *Syst. Biol.* 51, 151–155.

Pol., D., and Escapa, I. 2009. Unstable taxa in cladistic analysis: Identification and the assessment of relevant characters. *Cladistics* 25, 515–527.

Pol, D., and Goloboff, P. 2020. The impact of unstable taxa in Coelurosaurian phylogeny and resampling support measures for parsimony analysis. In: Pittman, M., and Xu, X. (eds.), *Pennaraptoran Theropod Dinosaurs: Past Progress and New Frontiers. Bull. Amer. Mus. Nat. History* 440, 97–115.

Prevosti, F., and Chemisquy, A. 2010. The impact of missing data on real morphological phylogenies: Influence of the number and distribution of missing entries. *Cladistics* 26, 329–339.

Purvis, A. 1995. A composite estimate of primate phylogeny. *Philos. Trans. R. Soc. Lond. B* 348, 405–421.

Puttick, M., O'Reilly, J., Oakley, D., Tanner, A., Fleming, J., Clark, J., Holloway, L., Lozano-Fernandez, J., Parry, L., Tarver, J., Pisani, D., and Donoghue, P. 2017a. Parsimony and maximum-likelihood phylogenetic analyses of morphology do not generally integrate uncertainty in inferring evolutionary history: A response to Brown et al. *Proc. R. Soc. B* 284, 20171636.

Puttick, M., O'Reilly, J., Pisani, D., and Donoghue, P. 2018. Probabilistic methods outperform parsimony in the phylogenetic analysis of data simulated without a probabilistic model. *Palaeontology* 62, 1–17.

Puttick, M., O'Reilly, J., Tanner, A., Fleming, J., Clark, J., Holloway, L., Lozano Fernandez, J., Parry, L., Tarver, J., Pisani, D., and Donoghue, P. 2017b. Uncertain-tree: Discriminating among competing approaches to the phylogenetic analysis of phenotype data. *Proc. R. Soc. B* 284, 20162290.

Ragan, M. 1992. Phylogenetic inference based on matrix representation of trees. *Mol. Phyl. Evol.* 1, 51–58.

Ramírez, M. 2003. The spider subfamily Amaurobioidinae (Araneae, Anyphaenidae): A phylogenetic revision at the generic level. *Bull. Am. Mus. Nat. Hist.* 277, 1–262.

Ramírez, M. 2005. Resampling measures of group support: A reply to Grant and Kluge. *Cladistics* 21, 83–89.

Ramírez, M., and Michalik, P. 2014. Calculating structural complexity in phylogenies using ancestral ontologies. *Cladistics* 30, 635–649.

Ranwez, V., Criscuolo, A., and Douzery, E. 2010. SuperTriplets: A triplet-based supertree approach to phylogenomics. *Bioinformatics* 26, 115–123.

Rieppel, O. 1988. *Fundamentals of Comparative Biology*. Birkhäuser Verlag, Berlin, 201 pp.

Robinson, D. 1971. Comparison of labeled trees with valency three. *J. Comb. Theory, Series B* 11, 105–119.

Robinson, D., and Foulds, L. 1981. Comparison of phylogenetic trees. *Math. Biosc.* 53, 131–147.

Rohlf, F. 1998. On applications of geometric morphometrics to studies of ontogeny and phylogeny. *Syst. Biol.* 47, 147–158.

Rohlf, F. 2001. Comparative methods for the analysis of continuous variables: Geometric interpretations. *Evolution* 55, 2143–2160.

Rohlf, F., and Bookstein, F. 1990. Proceedings of the Michigan morphometrics workshop, Ann Arbor. *The University of Michigan Museum of Zoology Special Publication* 2.

Rohlf, F., Colless, D., and Hart, G. 1983. Taxonomic congruence—a reanalysis. In: Felsenstein, J. (ed.), *Numerical Taxonomy*. Proceedings of the NATO Advanced Study Institute on Numerical Taxonomy, Springer-Verlag, Berlin, pp. 82–86.

Rohlf, F., and Slice, D. 1990. Extensions of the Procrustes method for the optimal superimposition of landmarks. *Syst. Zool.* 39, 40–59.

Rohlf, F., and Sokal, R. 1981. Comparing numerical taxonomic studies. *Syst. Zool.* 30, 459–490.

Ronquist, F., Teslenko, M., van der Mark, P., Ayres, D., Darling, A., Höhna, S., Larget, B., Liu, L., Suchard, M., and Huelsenbeck, J. 2012. MrBayes 3.2: Efficient Bayesian phylogenetic inference and model choice across a large model space. *Syst. Biol.* 61, 539–542.

Ross, H., and Rodrigo, A. 2004. An assessment of matrix representation with compatibility in supertree construction. In: Bininda-Emonds, O. (ed.), *Phylogenetic Supertrees: Combining Information to Reveal the Tree of Life*. Kluwer Academic, The Netherlands, Vol. 3, pp. 35–63.

Rzhetsky, A., and Nei, M. 1992. A simple method for estimating and testing minimum-evolution trees. *Mol. Biol. Evol.* 9, 945–967.

Saleeba, Z. 2019. A very small C interpreter. https://gitlab.com/zsaleeba/picoc.

Salisbury, B. 1999. Strongest evidence: Maximum apparent phylogenetic signal as a new cladistic optimality criterion. *Cladistics* 15, 137–149.

Salisbury, B. 2000. Strongest evidence revisited. *Cladistics* 16, 394–402.

Sanderson, M. 1989. Confidence limits on phylogenies: The bootstrap revisited. *Cladistics* 5, 113–129.

Sanderson, M. 1995. Objections to bootstrapping phylogenies: A critique. *Syst. Biol.* 44, 299–320.

Sanderson, M., and Donoghue, M. 1989 Patterns of variation in levels of homoplasy. *Evolution* 43, 1781–1795.

Sanderson, M., and Donoghue, M. 1996. The relationship between homoplasy and confidence in a phylogenetic tree. In Sanderson, M., and Hufford, L. (eds.), *Homoplasy: The Recurrence of Similarity in Evolution*. Academic Press, San Diego, pp. 67–89.

Sankoff, D., and Cedergren, R. 1983. Simultaneous comparison of three or more sequences related by a tree. In: Sankoff, D., and Kruskal, B. (eds.), *Time Warps, String Edits, and Macromolecules: The Theory and Practice of Sequence Comparison*. Addison-Wesley, Reading, MA, pp. 253–263.

Sansom, R. 2015. Bias and sensitivity in the placement of fossil taxa resulting from interpretations of missing data. *Syst. Biol.* 64, 256–266.

Sansom, R., Choate, P., Keating, J., and Randle, E. 2018. Parsimony, not Bayesian analysis, recovers more stratigraphically congruent phylogenetic trees. *Biol. Lett.* 14, 20180263. http://dx.doi.org/10.1098/rsbl.2018.0263

Schols, P., D'Hondt, C., Geuten, K., Merckx, V., Janssens, S., and Smets, E. 2004. MorphoCode: Coding quantitative data for phylogenetic analyses. *Phyloinformatics* 4, 1–4.

Schuh, R. 1978. [Review of] major patterns in vertebrate evolution.—Max K. Hecht, Peter C. Goody, and Bessie M. Hecht (eds.). *Syst. Zool.* 27, 255–260.

Schuh, R., and Brower, A. 2009. *Biological Systematics: Principles and Applications*. Cornell University Press, Ithaca, 311 pp.

Schuh, R., and Farris, J. 1981. Methods for investigating taxonomic congruence and their application to the Leptopodomorpha. *Syst. Zool.* 30, 331–351.

Schuh, R., and Polhemus, J. 1980. Analysis of taxonomic congruence among morphological, ecological, and biogeographic data sets for the Leptopodomorpha (Hemiptera). *Syst. Zool.* 29, 1–26.

Scott, E. 2005. A phylogeny of ranid frogs (Anura: Ranoidea: Ranidae), based on a simultaneous analysis of morphological and molecular data. *Cladistics* 21, 507–574.

Semple, C., and Steel, M. 2003. *Phylogenetics. Oxford Lecture Series in Mathematics and Its Applications*. Oxford University Press, Oxford, 239 pp.

Sharkey, M. 1989. A hypothesis-independent method of character weighting. *Cladistics* 5, 63–86.

Sharkey, M. 1993. Exact indices, criteria to select from minimum length trees. *Cladistics* 9, 211–222.

Siddall, M. 1995. Another monopyly index: Revisiting the jackknife. *Cladistics* 11, 33–56.

Siddall, M. 2002. Measures of support. In: DeSalle, R., Giribet, G., and Wheeler, W. (eds.), *Techniques in Molecular Systematics and Evolution*. Birkhäuser Verlag, Basel, pp. 80–101.

Siegel, A., and Benson, R. 1982. A robust comparison of biological shapes. *Biometrics* 38, 341–350.

Simmons, M., and Freudenstein, J. 2011. Spurious 99% bootstrap and jackknife support for unsupported clades. *Mol. Phyl. Evol.* 61, 177–191.

Simmons, M., and Gatesy, J. 2016. Biases of tree-independent-character-subsampling methods. *Mol. Phyl. Evol.* 100, 424–443.

Simmons, M., and Kessenich, J. 2019. Divergence and support among slightly suboptimal likelihood gene trees. *Cladistics* 36, 322–340.

Simmons, M., and Goloboff, P. 2013. An artifact caused by undersampling optimal trees in supermatrix analyses of locally sampled characters. *Mol. Phyl. Evol.* 69, 265–275.

Simon, C. 1983. A new coding procedure for morphometric data with an example from periodical cicada wing veins. In: Felsenstein, J. (ed.), *Numerical Taxonomy*. Springer-Verlag, Berlin, pp. 378–382.

Simpson, G. 1961. *Principles of Animal Taxonomy*. Columbia University Press, New York, 247 pp.

Smith, M. 2019. Bayesian and parsimony approaches reconstruct informative trees from simulated morphological datasets. *Biol. Lett.* 15, 20180632. http://dx.doi.org/10.1098/rsbl.2018.0632.

Smith, M. 2020. Information theoretic Generalized Robinson-Foulds metrics for comparing phylogenetic trees. *Bioinformatics* 36, 5007–5013.

Smith, U., and Hendricks, J. 2013. Geometric morphometric character suites as phylogenetic data: Extracting phylogenetic signal from gastropod shells. *Syst. Biol.* 62, 366–385.

Smith, M., Jonker, R., Yang, Y., and Cao, Y. 2021. R package "TreeDist". https://ms609.github.io/packages/.

Sneath, P. 1967. Trend-surface analysis of transformation grids. *J. Zool.* 151, 65–122.

Sober, E., and Steel, M. 2015. Similarities as evidence for common ancestry: A likelihood epistemology. *Brit. J. Phil. Sc.* 68, 1–22.

Sokal, R., and Rohlf, F. 1981. Taxonomic congruence in the Leptopodomorpha reexamined. *Syst. Zool.* 30, 309–325.

Sokal, R., and Sneath, P. 1963. *Principles of Numerical Taxonomy*. San Francisco, Freeman & Co., 359 pp.

Stamatakis, A. 2014. RAxML version 8: A tool for phylogenetic analysis and post-analysis of large phylogenies. *Bioinformatics* 30, 1312–1313.

Stamatakis, A., Hoover, P., and Rougemont, J. 2008. A rapid bootstrap algorithm for the RAxML web servers. *Syst. Biol.* 57, 758–771.

Steel, M. 1992. The complexity of reconstructing trees from qualitative characters and subtrees. *J. Classif.* 9, 91–116.

Steel, M., Dress, A., and Böcker, S. 2000. Simple but fundamental limitations on supertree and consensus tree methods. *Syst. Biol.* 49, 363–368.

Steel, M., and Rodrigo, A. 2008. Maximum likelihood supertrees. *Syst. Biol.* 57, 243–250.

Steel, M., and Warnow, T. 1993. Kaikoura tree theorems: Computing the maximum agreement subtree. *Information Processing Letters* 48, 77–82.

Sterli, J., Pol, D., and Laurin, M. 2013. Incorporating phylogenetic uncertainty on phylogeny-based palaeontological dating and the timing of turtle diversification. *Cladistics* 29, 233–246.

St John, K. 2017. The shape of phylogenetic treespace. *Syst. Biol.* 66, 83–94.

Stinebrickner, R. 1984. S-consensus trees and indices. *Bull. Math. Biol.* 46, 923–935.

Strimmer, K., and von Haeseler, A. 1996. Quartet-puzzling: A quartet maximum-likelihood method for reconstructing tree topologies. *Mol. Biol. Evol.* 13, 964–969.

Swofford, D. 1991. When are phylogeny estimates from molecular and morphological data incongruent? In: Miyamoto, M., and Cracraft, J. (eds.), *Phylogenetic Analysis of DNA Sequences*. Oxford University Press, Oxford, pp. 295–333.

Swofford, D. 1993. *PAUP: Phylogenetic Analysis Using Parsimony, Version 3.1*. Program and documentation, Laboratory of Molecular Systematics, Smithsonian Institution, Washington.

Swofford, D. 2001. *PAUP*: Phylogenetic Analysis using Parsimony (*and Other Methods)*. Sinauer Associates, Sunderland.

Swofford, D., and Olsen, G. 1990. Phylogeny reconstruction. In: Hillis, D., and Moritz, C. (eds.), *Molecular Systematics*. Sinauer Associates, Sunderland, pp. 411–501.

Swofford, D., Olsen, G., Waddell, P., and Hillis, D. 1996. Phylogenetic inference. In: Hillis, D., Moritz, C., and Mable, B. (eds.), *Molecular Systematics*, 2nd ed. Sinauer, Sunderland, MA, pp. 407–514.

Szalay, F. 1977. Ancestors, descendants, sister groups and testing of phylogenetic hypotheses. *Syst. Zool.* 26, 12–18.

Szumik, C., Juárez, M., Ramírez, M., Goloboff, P., and Pereyra, V. 2019. Implications of the tympanal hearing organ and ultrastructure of chaetotaxy for the higher classification of Embioptera. *Am. Mus. Novitates* 3933, 1–32.

Takezaki, N. 1998. Tie trees generated by distance methods of phylogenetic reconstruction. *Mol. Biol. Evol.* 15, 727–737.

Thiele, K. 1993. The Holy Grail of the perfect character: The cladistic treatment of morphometric data. *Cladistics* 9: 275–304.

Thompson, D. 1917. *On Growth and Form.* Cambridge University Press, Cambridge, 793 pp.

Thiele, K., and Ladiges, P. 1988. A cladistic analysis of *Angophora* Cav. (Myrtaceae). *Cladistics* 4, 23–42.

Thorley, J., and Page, R. 2000. RadCon: Phylogenetic tree comparison and consensus. *Bioinformatics* 16, 486–487.

Thorley, J., and Wilkinson, M. 1999. Testing the phylogenetic stability of early tetrapods. *J. Theoret. Biol.* 200, 343–344.

Torcida, S., Pérez, S., and Gonzalez, P. 2014. An integrated approach for landmark-based resistant shape analysis in 3D. *Evol. Biol.* 41, 351–366.

Torres, A., Goloboff, P., and Catalano, S. 2021. Parsimony analysis of phylogenomic datasets (I): Scripts and guidelines for using TNT (Tree Analysis using New Technology). *Cladistics*. doi: 10.1111/cla.12477.

Tuffley, C., and Steel, M. 1997. Links between maximum likelihood and maximum parsimony under a simple model of site substitution. *Bull. Math. Biol.* 59, 581–607.

Turner, H., and Zandee, R. 1995. The behaviour of Goloboff's tree fitness measure F. *Cladistics* 11, 57–72.

Varón, A., Sy Vinh, L., and Wheeler, W. 2010. POY version 4: Phylogenetic analysis using dynamic homologies. *Cladistics* 26, 72–85.

Varón-González, C., Whelan, S., and Klingenberg, C. 2020. Estimating phylogenies from shape and similar multidimensional data: Why it is not reliable. *Syst. Biol.* 69, 863–883.

Vitter, J. 1986. Random sampling with a reservoir. *ACM Trans. Math. Softw.* 11, 37–57.

Wägele, J. 1995. On the information content of characters in comparative morphology and molecular systematics. *J. Zool. Syst. Evol. Res.* 33, 42–47.

Watanabe, A. 2015. The impact of poor sampling of polymorphism on cladistic analysis. *Cladistics* 32, 317–334.

Waterman, M., and Smith, T. 1978. On the similarity of dendrograms. *J. Theoret. Biol.* 73, 789–800.

Wenzel, J., and Siddall, M. 1999. Noise. *Cladistics* 15, 51–64.

Wheeler, Q. 1986. Character weighting and cladistic analysis. *Syst. Zool.* 35, 102–109.

Wheeler, W. 1995. Sequence alignment, parameter sensitivity, and the phylogenetic analysis of molecular data. *Syst. Biol.* 44, 321–331.

Wheeler, W. 2012. *Systematics: A Course of Lectures.* John Wiley and Sons, Hoboken, 426 pp.

Wheeler, W. 2021a. Distance Wagner tree refinement as a heuristic approach to character-based initial tree construction. *Cladistics.*

Wheeler, W. 2021b. Phylogenetic supergraphs. *Cladistics.*

Wheeler, W., Aagesen, L., Arango, C., Faivovich, J., Grant, T., D'Haese, C., Janies, D., Leo Smith, W., Varón, A., and Giribet, G. 2006. *Dynamic Homology and Phylogenetic Systematics: A Unified Approach Using POY.* American Museum of Natural History, New York.

Whidden, C., Beiko, R., and Zeh, N. 2010. Fast FPT algorithms for computing rooted agreement forests: Theory and experiments. In: Festa, P. (ed.), *Experimental Algorithms.* Lecture Notes in Computer Science. Springer, Berlin, Heidelberg, Vol. 6049, pp. 141–153.

Whidden, C., and Zeh, N. 2009. A unifying view on approximation and FPT of agreement forests. In: *WABI 2009. LNCS.* Springer-Verlag, Berlin, Vol. 5724, pp. 390–401.

Whidden, C., Zeh, N., and Beiko, R. 2014. Supertrees based on the subtree prune-and-regraft distance. *Syst. Biol.* 63, 566–581.

Williams, P., and Fitch, W. 1990. Phylogeny determination using dynamically weighted parsimony. *Meth. Enzymol.* 183, 615–626.

Wilkinson, M. 1994a. Common cladistic information and its consensus representation: Reduced Adams and reduced cladistic consensus trees and profiles. *Syst. Biol.* 43, 343–368.

Wilkinson, M. 1994b. Weights and ranks in numerical phylogenetics. *Cladistics* 10, 321–329.

Wilkinson, M. 1995. More on reduced consensus methods. *Syst. Biol.* 44, 435–439.

Wilkinson, M. 1996. Majority rule reduced consensus and their use in bootstrapping. *Mol. Biol. Evol.* 13, 437–444.

Wilkinson, M., and Crotti, M. 2017. Comments on detecting rogue taxa using RogueNaRok. *Syst. Biodiv.* 15, 291–295.

Wilson, E. 1965. A consistency test for phylogenies based on contemporaneous species. *Syst. Zool.* 14, 214–220.

Yang, Z. 2006. *Computational Molecular Evolution.* Oxford University Press, Oxford, 357 pp.

Yang, Z., and Rannala, B. 2005. Branch-length prior influences Bayesian posterior probability of phylogeny. *Syst. Biol.* 54, 455–470.

Zelditch, M., Fink, W., and Swiderski, D. 1995. Morphometrics, homology, and phylogenetics: Quantified characters as synapomorphies. *Syst. Biol.* 44, 179–189.

Zelditch, M., Fink, W., Swiderski, D., and Lundrigan, B. 1998. Applications of geometric morphometrics to studies of ontogeny and phylogeny: A reply to Rohlf. *Syst. Biol.* 47, 159–167.

Zelditch, M., Swiderski, D., Sheets, H., and Fink, W. 2004. *Geometric Morphometrics for Biologists: A primer.* Elsevier Academic Press, London, 443 pp.

Index

A

absolute Bremer supports, 122, *124*
accuracy of support measures under simulations, 167–170, *168*
Adams consensus, 8
adaptive arguments and weighting, 58–59
additive characters and weighting, 57, 75–76
alignment, 116n4
    criteria for landmark, 215–218, *216*
    dynamic alignment of landmarks, 212–215, *214*
    landmark, 198–199, 223, **225**
ambiguity
    in individual pseudoreplicates of resampling, 159
    in landmark positions, *205*, 210–211, *212*, 226–227
    in optimization, 21–24
    in results, 120–121
anticonsensus, 32, *33*
    calculation of, **40**, 43
approximate likelihood ratio test (aLRT), 139–140
approximation for concavity ratios, 85–86, *86*
approximations for speeding up resampling, 164–170
a priori character weighting, 53–54, 58
arguments for weighting, 58–60, 98–111
asymmetric linear parsimonious reconstruction (method), 196–197
average weighting of partitions, 87, *88*, **111**, 115–116

B

bar plots (scripting), *259*, 259–260
Bayesian analysis, 98–103
    tree-distances and, 35–36
bias in tree searches, *160*, 161–164, **163–164**
binary recoding and character weighting, 75, 112
bootstrapping, 143, 144–145, *148*
    accuracy of, *168*, 169–170
    asymmetry in, 146
    and character conflict, 149
    and search bias, *160*, 161–164
branch collapsing, *18*, 21–26, **25**
    and resampling with FWR, 152
branch labels
    editing trees with, 254
Bremer supports, 122–139, *124*, **173**, 173–178
    absolute, 122
    calculation of, *124*, 134–139, **173**, 173–177

combined, 125–126
and implied weighting, 112
relative, 125–126, 177
and successive weighting, 68
Brownian motion, 194–196, 215
brute force methods for identifying wildcards, 15–16, **40**, 45–46

C

character-state transformation weighting, 87–92, *89*, 112–113
character weighting, 53–116
    implied, 68–81, *67*, 111–112, **111**
    and resampling, **142**, 146–147
    successive approximations, 60–68, *61*
C language in TNT scripts, 264–265
cliques, 49–50
    calculation in TNT, 49–50
    and implied weighting, 77, 106
    and lack of resolution, 77
    supertree methods based on, 31, 49
cluster-based consensus methods, 3–81
collapsing rules, *18*, 21–26
    and consensus topologies, 24
    and numbers of trees, 24–25, **25**
    temporary, 26, 41–45
combinable component consensus, *4*, 6–7
    applicability, 6–7
    calculation in TNT, 39, **40**
    improving resolution of, **40**, 46
    supertree equivalent of, 29–30
combined Bremer supports, 125–126
    calculation of, 177–178
comparisons between trees, 32–41, *33*, **34**, **35**, **40**
complexity and a priori weighting, 58–59
concavity of the weighting function, 72–74, *73*, **74**
    set to produce a specific weight ratio, 79
condensing, *see* collapsing rules
confidence and support, 170–172
confidence probability (measure), 121
configuration of landmarks, *195*, 197–198
conflict between characters, 7, 69–70
    and numbers of characters, 108
    and parsimony, 55–56
    and relative fit differences, 125
    and self-consistency, 62
    and supports, 121, 123, 131–132, **142**, 147–149
    and weighting, 54, *64*, 69–70, 74

Printed in the United States
by Baker & Taylor Publisher Services